ADVANCES IN
ORGAN BIOLOGY

Volume 5C • 1998

MOLECULAR AND CELLULAR BIOLOGY OF BONE

MOLECULAR AND CELLULAR BIOLOGY OF BONE

Guest Editor

MONE ZAIDI
Medical College of Pennsylvania
School of Medicine and
Veterans Affairs Medical Center
Philadelphia, Pennsylvania

Associate Guest Editors

OLUGBENGA A. ADEBANJO
Medical College of Pennsylvania
School of Medicine and
Veterans Affairs Medical Center
Philadelphia, Pennsylvania

CHRISTOPHER L. -H. HUANG
Department of Physiology
University of Cambridge
Cambridge, England

ADVANCES IN ORGAN BIOLOGY

MOLECULAR AND CELLULAR BIOLOGY OF BONE

Guest Editor: MONE ZAIDI
Veterans Affairs Medical Center

Series Editor: E. EDWARD BITTAR
Department of Physiology
University of Wisconsin-Madison

Associate Guest Editors: OLUGBENGA A. ADEBANJO
Veterans Affairs Medical Center

CHRISTOPHER L. -H. HUANG
Department of Physiology
University of Cambridge

VOLUME 5C • 1998

Stamford, Connecticut *London, England*

100 Prospect Street
Stamford, Connecticut 06901

JAI PRESS LTD.
38 Tavistock Street
Covent Garden
London WC2E 7PB
England

ISBN: 0-7623-0390-5

Printed and bound by CPI Antony Rowe, Eastbourne

CONTENTS (Volume 5C)

CONTENTS (Volume 5A)

CONTENTS (Volume 5B)

SECTION III. BONE FORMATION

LIST OF CONTRIBUTORS

E.M. Aarden — Research Scientist
Department of Cell Biology
Faculty of Medicine, Leiden University
Leiden, The Netherlands

Etsuko Abe, PhD — Research Professor of Medicine
Department of Medicine
University of Arkansas for Medical Sciences
Little Rock, Arkansas

A.B. Abou-Samra, MD — Associate Professor of Medicine
Endocrine Unit, Department of Medicine
Massachusetts General Hospital
Harvard Medical School
Boston, Massachusetts

Olugbenga A. Adebanjo, MD — Assistant Professor of Medicine
Department of Medicine
Medical College of Pennsylvania
School of Medicine and Veterans Affairs Medical Center
Philadelphia, Pennsylvania

N.E. Ajubi — Research Scientist
Department of Cell Biology
Faculty of Medicine
Leiden University
Leiden, The Netherlands

David J. Baylink, MD
Distinguished Professor of Medicine
Loma Linda University
and Associate Vice President
for Medical Affairs for Research
J.L. Pettis Veterans Affairs Medical Center
Loma Linda, California

Paolo Bianco, MD
Dipartmento di Biopatologia Umana
Universita La Sapienza
Rome, Italy

L.F. Bonewald, PhD
Associate Professor of Medicine
Department of Medicine
University of Texas Health Science Center
San Antonio, Texas

Brendan F. Boyce
Professor of Pathology
Department of Medicine
Division of Endocrinology and Metabolism
University of Texas Health Science Center
San Antonio, Texas

Alan Boyde, PhD
Professor of Mineralized Tissue Biology
Department of Anatomy and
Developmental Biology
University College London
London, England

Edward M. Brown, MD
Professor of Medicine
Endocrine-Hypertension and Renal
Divisions
Brigham and Women's Hospital
Boston, Massachusetts

Elisabeth H. Burger, PhD
Professor
Department of Oral Cell Biology
ACTA-Vrije Universiteit
Amsterdam, The Netherlands

T.J Chambers, PhD, MBBS, MRCPath
Professor and Chairman
Department of Histopathology
St. George's Hospital Medical School
London, England

Chantal Chenu, PhD	Staff Research Fellow INSERM Hôpital Edouard Herriot Lyon, France
Roberto Civitelli, MD	Associate Professor of Medicine and Orthopedic Surgery and Assistant Professor of Cell Biology and Physiology Division of Bone and Mineral Diseases Washington University School of Medicine St. Louis, Missouri
Thomas L. Clemens, PhD	Professor of Medicine Department of Molecular and Cellular Physiology and Orthopedic Surgery University of Cincinnati Medical Center Cincinnati, Ohio
Silvia Colucci, PhD	Assistant Professor of Histology Institute of Human Anatomy University of Bari Bari, Italy
Stephen C. Cowin	Department of Mechanical Engineering City University of New York New York, New York
C.G. Dacke, B.Tech, PhD, FIBiol	Reader and Head, Pharmacology Division School of Pharmacy and Biomedical Science University of Portsmouth Portsmouth, England
Sarah L. Dallas, PhD	Assistant Professor of Medicine Department of Medicine University of Texas Health Science Center San Antonio, Texas
Pietro De Togni, MD	Assistant Professor of Pathology Immunogenetics and Transplantation Laboratory University of Arkansas for Medical Sciences Little Rock, Arkansas

P.D. Delmas, MD, PhD

Professor of Medicine
INSERM
Hôpital Edouard Herriot
Lyon, France

S.J. Dixon, DDS, PhD

Associate Professor of Physiology and Oral Biology
Department of Physiology
Faculty of Dentistry
The University of Western Ontario
London, Ontario, Canada

S. Epstein, MD, FRCP

Professor of Medicine and Chief Division of Endocrinology
Medical College of Pennsylvania
Hahnemann School of Medicine
Philadelphia, Pennsylvania

R. J. Fitzsimmons, PhD

Assistant Research Professor of Medicine and Director Mineral Metabolism
Jerry L. Pettis Veterans Affairs Medical Center
Loma Linda University
Loma Linda, California

Herbert Fleisch, MD

Professor and Chairman
Department of Pathophysiology
University of Berne
Berne, Switzerland

Steven R. Goldring, MD

Associate Professor of Medicine and Chief of Rhematology
Beth Israel-Deaconess Hospital
Harvard Medical School
Boston, Massachusetts

David Goltzman, MD

Professor and Chairman
Department of Medicine
McGill University,Royal Victoria Hospital
Montréal, Québec, Canada

Grant R. Goodman, MD

Research Associate
Department of Medicine
Albert Einstein Medical Center
Philadelphia, Pennsylvania

Maria Grano, PhD
Assistant Professor of Histology
Institute of Human Anatomy
University of Bari
Bari, Italy

Ted S. Gross, PhD
Assistant Professor
Departments of Medicine and Molecular and Cellular Physiology and Orthopedic Surgery
University of Cincinnati Medical Center
Cincinnati, Ohio

Theresa A. Guise, MD
Assistant Professor of Medicine
Division of Endocrinology and Metabolism
University of Texas Health Science Center
San Antonio, Texas

Steven C. Hebert, MD
Professor of Medicine and Chief, Division of Nephrology
Vanderbelt University
Nashville, Tennessee

Janet E. Henderson, PhD
Assistant Professor of Medicine
Department of Medicine
McGill University
Montréal, Québec, Canada

M. Horton, MD, FRCP, FRCPath
Professor
Rayne Institute
Bone and Mineral Center
University of London
London, England

Osamu Ishibashi, MS
Scientist
Ciba-Geigy Japan Limited
International Research Laboratories
Takarazuka, Japan

Sheila Jones, PhD
Professor of Anatomy
Department of Anatomy and Developmental Biology
University College London
London, England

J. Klein-Nulend, PhD
Assistant Professor
Department of Oral Cell Biology
ACTA-Vrije Universiteit
Amsterdam, The Netherlands

Toshio Kokubo, PhD
Group Leader
International Research Laboratories
Ciba Geigy Japan Limited
Takarazuka, Japan

Masayoshi Kumegawa, DDS
Professor
Department of Oral Anatomy
Meikai University School of Dentistry
Saitama, Japan

Pierre J. Marie, PhD
Professor
Cell and Molecular Biology of Bone and Cartilage
Lariboisière Hospital
Paris, France

T.J. Martin, MD, DSC. FRCPA, FRACP
Professor of Medicine
St. Vincent's Institute of Medical Research
University of Melbourne
Fitzroy, Victoria, Australia

Toshio Matsumoto, MD
Professor and Chairman
First Department of Medicine
Tokushima University School of Medicine
Tokushima, Japan

Cedric Minkin, PhD
Professor
Department of Basic Sciences
University of Southern California School of Dentistry
Los Angeles, California

Ambrish Mithal, MD, DM
Professor
Department of Medical Endocrinology
Sanjay Gandhi Post Graduate Institute of Medical Sciences
Lucknow, India

Hanna Mocharla, PhD
Research Instructor
Department of Medicine
University of Arkansas for Medical Sciences
Little Rock, Arkansas

S. Mohan, PhD
Research Professor of Medicine, Biochemistry, and Physiology
J.L. Pettis Veterans Affairs Medical Center
Loma Linda University
Loma Linda, California

Baljit Moonga, PhD
Assistant Professor of Medicine
Medical College of Pennsylvania
School of Medicine and
Veterans Affairs Medical Center
Philadelphia, Pennsylvania

K.W. Ng, MBBS, MD, FRACP
Associate Professor
Department of Medicine
The University of Melbourne
St. Vincent's Hospital
Fitzroy, Victoria, Australia

Peter J. Nijweide
Professor
Department of Cell Biology
Faculty of Medicine
Leiden University
Leiden, The Netherlands

Richard O.C. Oreffo, D. Phil.
MRC Research Fellows
MRC Bone Research Laboratory
Nuffield Orthopedic Center
University of Oxford
Headington
Oxford, England

Roberto Pacifici, MD
Associate Professor of Medicine
Division of Bone and Mineral Diseases
Washington University Medical Center
St. Louis, Missouri

Michael Pazianas, MD — Associate Professor of Medicine
Division of Geriatric Medicine and Institute on Aging
University of Pennsylvania
Philadelphia, Pennsylvania

J. Wesley Pike, PhD — Professor of Medicine
Department of Molecular and Cellular Physiology
University of Cincinnati Medical Center
Cincinnati, Ohio

James T. Ryaby, PhD — Director of Research
Orthologic Corporation
Phoenix, Arizona

Ian R. Reid, MD — Associate Professor of Medicine
Department of Medicine
University of Auckland
Auckland, New Zealand

Barry Rifkin, DDS, PhD — Professor and Dean
State University of New York Dental School
Stony Brook, New York

Pamela Gehron Robey, PhD — Chief
Craniofacial and Skeletal Diseases
National Institute of Dental Research
National Institutes of Health
Bethesda, Maryland

G. David Roodman, MD — Professor of Medicine and Chief of Hematology
Audie Murphy Veterans Affairs Medical Center
University of Texas Health Science Center
San Antonio, Texas

F. Patrick Ross, PhD — Associate Professor of Pathology
Department of Pathology
Barnes-Jewish Hospital
St. Louis, Missouri

Dennis Sakai, PhD
Research Professor
Department of Basic Sciences
University of Southern California School of Dentistry
Los Angeles, California

Edna Schwab, MD
Assistant Professor of Medicine
Division of Geriatric Medicine and Institution Aging
University of Pennsylvania
Philadelphia, Pennsylvania

Geetha Shankar, PhD
Scientist
NPS Pharmaceuticals Inc.
Salt Lake City, Utah

Jay Shapiro, MD
Professor
Department of Medicine
Walter Reed Army Medical Center
Bethesda, Maryland

Stephen M. Sims, PhD
Associate Professor
Department of Physiology
Faculty of Medicine and Dentistry
The University of Western Ontario
London, Ontario, Canada

Li Sun, MD, PhD
Research Fellow
Medical College of Pennsylvania
School of Medicine and Veterans Affairs Medical Center
Philadelphia, Pennsylvania

Yasuto Taguchi, MD
Research Fellow
Department of Medicine
University of Arkansas for Medical Sciences
Little Rock, Arkansas

Yasuhiro Takeuchi, MD
Assistant Professor
Fourth Department of Internal Medicine
University of Tokyo School of Medicine
Tokyo, Japan

James T. Triffitt, PhD
Head of Department
MRC Bone Research Laboratory
Nuffield Orthopedic Center
University of Oxford
Headington
Oxford, England

A. Van der Plas
Head of Technical Staff
Department of Cell Biology
Faculty of Medicine
Leiden University
Leiden, The Netherlands

Anthony Vernillo, PhD DDS
Associate Professor
Department of Oral Medicine and Pathology
New York University College of Dentistry
New York, New York

A. Frederik Weidema, PhD
Research Associate
Laboratorium voor Fysiologie
Katholieke Universiteit Leuven
Herestraat, Leuven, Belgium

Matsuo Yamamoto, PhD
Research Fellow
Department of Medicine
University of Arkansas for Medical Sciences
Little Rock, Arkansas

Tomoo Yamate, MD, PhD
Instructor
Department of Medicine
University of Arkansas for Medical Sciences
Little Rock Arkansas

Toshiyuki Yoneda, DDS, PhD
Professor of Medicine
Department of Medicine
Division of Endocrinology and Metabolism
University of Texas Health Science Center
San Antonio, Texas

Alberta Zambonin Zallone, PhD
Professor of Histology
Institute of Human Anatomy
University of Bari
Bari, Italy

M. Zaidi, MD, PhD, FRCP, FRCPath
Professor of Medicine and Associate Dean
Medical College of Pennsylvania
School of Medicine
Associate Chief of Staff and Chief, Geriatrics and Extended Care
Veterans Affairs Medical Center
Philadelphia, Pennsylvania

FOREWORD

These volumes differ from the current conventional texts on bone cell biology. Biology itself is advancing at breakneck speed and many presentations completely fail to present the field n a truly modern context. This text does not attempt to present detailed clinical descriptions. Rather, after discussion of basic concepts, there is a concentration on recently developed findings equally relevant to basic research and a modern understanding of metabolic bone disease. The book will afford productive new insights into the intimate inter-relation of experimental findings and clinical understanding. Modern medicine is founded in the laboratory and demands of its practitioners a broad scientific understanding: these volumes are written to exemplify this approach. This book is likely to become essential reading equally for laboratory and clinical scientists.

Ian MacIntyre, FRS
Research Director
William Harvey Research Institute
London, England

DEDICATION

To Professor Iain MacIntyre,
MBChB, PhD, Hon MD, FRCP, FRCPath, DSc, FRS

In admiration of his seminal contributions to bone and mineral research that have spanned over more than four decades, and

In gratitude for introducing us into the field of bone metabolism and for his continued encouragement, assistance, and friendship over many years

PREFACE

The intention of putting this book together has been not to develop a full reference text for bone biology and bone disease, but to allow for an effective dissemination of recent knowledge within critical areas in the field. We have therefore invited experts from all over the world to contribute in a way that could result in a complete, but easily readable text. We believe that the volume should not only aid our understanding of basic concepts, but should also guide the more provocative reader toward searching recent developments in metabolic bone disease.

For easy reading and reference, we have divided the text into three subvolumes. Volume 5A contains chapters outlining basic concepts stretching from structural anatomy to molecular physiology. Section I in Volume 5B is devoted to understanding concepts of bone resorption, particularly in reference to the biology of the resorptive cell, the osteoclast. Section II in Volume 5B contains chapters relating to the formation of bone with particular emphasis on regulation. Volume 5C introduces some key concepts relating to metabolic bone disease. These latter chapters are not meant to augment clinical knowledge; nevertheless, these do emphasize the molecular and cellular pathophysiology of clinical correlates. We do hope that the three subvolumes, when read in conjunction, will provide interesting reading for those dedicated to the fast emerging field of bone biology.

We are indebted to the authors for their significant and timely contributions to the field of bone metabolism. We are also grateful to Christian Costeines (JAI Press) and Michael Pazianas (University of Pennsylvania) for their efforts in ensuring the creation of quality publication. The editors also acknowledge the support and perseverance of their families during the long hours of editing.

Mone Zaidi
Guest Editor
Olugbenga A. Adebanjo
Christopher L.-H. Huang
Associate Guest Editors

SECTION IV

MOLECULAR BASIS OF BONE DISEASE

ESTROGEN AND BONE LOSS

Roberto Pacifici

I. INTRODUCTION

Postmenopausal osteoporosis is a heterogeneous disorder characterized by a progressive loss of bone tissue which begins after natural or surgical menopause and leads to fracture within 15–20 years from the cessation of the ovarian function. Although suboptimal skeletal development ("low peak bone mass") and age related bone loss may be contributing factors, a hormone-dependent increase in bone resorption and accelerated loss of bone mass in the first five or 10 years after the menopause appears to be the

Advances in Organ Biology
Volume 5C, pages 641-659.

ISBN: 0-7623-0390-5

main pathogenetic factor (Riggs and Melton, 1986a,b) in this condition. That estrogen deficiency plays a major role in postmenopausal bone loss is strongly supported by the higher prevalence of osteoporosis in women than in men (Nilas and Christiansen, 1987), the increase in the rate of bone mineral loss detectable by bone densitometry after artificial or natural menopause (Riggs et al., 1981; Genant et al., 1982; Slemenda et al., 1987), the existence of a relationship between estrogen levels and rates of bone loss (Johnston et al., 1985; Ohta et al., 1992, 1993), and the protective effect of estrogen replacement with respect to both bone mass loss and fracture incidence (Lindsay et al., 1980; Horsman et al., 1983; Ettinger et al., 1985).

The bone-sparing effect of estrogen is mainly related to its ability to block bone resorption (Riggs et al., 1972; Heaney et al., 1978; Lindsay et al., 1980; Riggs and Melton 1986a; Slemenda et al., 1987), although stimulation of bone formation is likely to play a contributory role (Chow et al., 1992; Bain et al., 1993). Estrogen-dependent inhibition of bone resorption is, in turn, due to both decreased osteoclastogenesis and a diminished resorptive activity of mature osteoclasts.

II. CYTOKINES AND BONE REMODELING

In spite of over 20 years of investigation, the mechanism by which estrogen prevents bone loss is still controversial. Most studies conducted before the discovery of estrogen receptors in bone cells focused on the effects of estrogen on the major calciotropic hormones (Prince, 1994). However, the etiologic relevance of the changes induced by menopause and estrogen replacement on 1,25 $(OH)_2$ D (Gallagher et al., 1980), parathyroid hormone (Stock et al., 1985), and calcitonin (Tiegs et al., 1985) remains unclear. The discovery of estrogen receptors in osteoblasts (Erikson et al., 1988; Komm et al., 1988), their stromal precursors (Bellido et al., 1993) and osteoclasts (Oursler et al., 1991b, 1993) suggests that a direct effect on bone or bone marrow cells may be involved. Estrogen could interact with these cells and modulate the secretion of, or the response to, local regulatory factors or systemic hormones. Estrogen, however, could also have indirect effects on bone, perhaps by modulating the production of one or more of the bone regulating factors released by immune and hematopoietic cells of the bone marrow (Raisz, 1988). Indeed, a large number of immune and hemopoietic factors have been shown to have complex and overlapping effects on both bone formation and resorption. Among these are interleukin- (IL) 1α and β (Gowen et al., 1983, 1985; Canalis, 1986; Lorenzo et al., 1987a; Stashenko et al., 1987), IL-6 (Girasole et al., 1992; Jilka et al., 1992, Passeri et al., 1993), tumor necrosis factor (TNF) α and β (Bertolini et al., 1986; Thomson et al., 1987), M-CSF (Schneider and Relfson, 1989; Takahashi et al., 1991; Tanaka et al., 1993), and granulocyte macrophage colony-stimulating factor (GM-CSF) (Kurihara et al., 1989; Schneider and Relfson, 1989).

IL-1 and TNF are among the most powerful stimulators of bone resorption known, and well-recognized inhibitors of bone formation (Bertolini et al., 1986; Stashenko et al., 1987; Nguyen et al., 1991). These cytokines promote bone resorption *in vitro* (Gowen et al., 1983; Lorenzo et al., 1987a) and cause bone loss and hypercalcemia when infused *in vivo* (Sabatini et al., 1988; Boyce et al., 1989; Johnston et al., 1989). IL-1 and TNF activate mature osteoclasts indirectly via a primary effect on osteoblasts (Thomson et al., 1986, 1987) and inhibit osteoclast apoptosis (Hughes et al., 1995). In addition, they markedly enhance osteoclast formation by stimulating osteoclast precursor proliferation both directly (Pfeilschifter et al., 1989) and by stimulating the pro-osteoclastogenic activity of stromal cells (Suda et al., 1992; Srivastava et al., 1995). IL-1 and TNF are also powerful inducers of other cytokines which regulate the differentiation of osteoclast precursor cells into mature osteoclasts, such as IL-6 (Elias and Lentz, 1990; Girasole et al., 1992), M-CSF (Felix et al., 1989), and GM-CSF (Lorenzo et al., 1987b). Therefore, with respect to osteoclastogenesis, IL-1 and TNF should be regarded as "upstream" cytokines necessary for inducing the secretion of "downstream" factors which stimulate hematopoietic osteoclast precursors. This cascade mechanism assures that small changes in IL-1 and TNF levels results in large changes in osteoclast production.

DNA cloning has revealed two independent species of IL-1, IL-1α and IL-1β, which, despite a distant homology, exert the same biological effects (Dinarello, 1988). In human cells there is a preponderant expression of the IL-1β gene, which after antigenic stimulation can increase 200–300 times within 2–3 hours (Dinarello, 1988).

A specific competitive inhibitor of IL-1, known as IL-1 receptor antagonist (IL-Ira), has been purified from the supernatant of IgG stimulated monocytes (Hannum et al., 1990) and from the myelomonocytic cell line U937 (Carter et al., 1990). This substance exists in two forms, a 22,000 molecular weight glycosylated form and a 17,000 molecular weight non-glycosylated form. The 17,000 molecular weight form has been cloned and expressed in *Echerichia Coli* (Carter et al., 1990). This recombinant molecule, which has a 26% amino acid sequence homology with IL-1β, binds to cells expressing primarily the 87 kDa type I IL-1 receptor with nearly the same affinity as IL-1 and competes with either IL-1α or IL-1β on these cells without detectable IL-1 agonist effects (Arend et al., 1989; Arend, 1991). The type I IL-1 receptor is expressed in T cells, tissue macrophages, endothelial cells, and bone cells (Seckinger et al., 1990; Dinarello, 1991). IL-Ira also binds, although with a lower affinity, to the type II IL-1 receptor which is expressed mainly in blood neutrophils and B cells (Dinarello, 1991). Since the binding of five molecules of IL-1 per cell is sufficient to induce a full biological response, a 50% IL-1 inhibition in bone cells requires amounts of IL-Ira up to 100 times in excess of the amounts of IL-1α or IL-1β present.

IL-6 stimulates the early stages of osteoclastogenesis in human and murine cultures (Kurihara et al., 1991; Roodman, 1992). IL-6 increases bone resorption in systems rich in osteoclast precursor, such as the mouse fetal metacarpal assay (Lowik et al., 1993), whereas it has no effect in organ cultures where more mature cells predominate, such as murine fetal radii (Kurihara et al., 1991). This suggests that IL-6 increases the formation of osteoclasts from hemopoietic precursors but does not activate mature osteoclasts.

The essential role of CSFs in the proliferation and differentiation of osteoclast precursors is best demonstrated by the presence of osteopetrosis in a natural M-CSF knockout, the op/op mouse (Suda et al., 1992). These mice, which are cured by the administration of M-CSF (Felix et al., 1990), have an extra thymidine inserted within the coding region of the M-CSF gene, a mutation that generates a stop codon within the coding sequence (Yoshida et al., 1990), thereby resulting in the production of a defective M-CSF (Suda et al., 1992). The formation of osteoclasts in bone marrow cultures is also increased by GM-CSF (Macdonald et al., 1986; Lorenzo et al., 1987b). This factor stimulates the early stages of osteoclastogenesis in cooperation with IL-3 (Kurihara et al., 1989, 1990). Although in the mouse osteoclast formation is completely blocked by anti-M-CSF but not anti-GM-CSF antibodies (Tanaka et al., 1993), GM-CSF is critical for the proliferation and differentiation of human osteoclast precursors (Matayoshi et al., 1995).

Although most bone cell-targeting cytokines are produced by either bone and bone marrow cells, mononuclear cells of the monocyte/macrophage lineage are recognized as the major source of IL-1 and TNF (Dinarello, 1991). In contrast, proosteoclastogenic downstream cytokines are mainly produced by stromal cells and osteoblasts (Fibbe et al., 1988; Rifas et al., 1995). Thus, osteoclastogenesis requires the hierarchical interaction of mononuclear cells, stromal cells and/or osteoblasts and hematopoietic osteoclast precursors.

III. EFFECTS OF MENOPAUSE ON CYTOKINE PRODUCTION

The cytokines first recognized to be regulated by estrogen were IL-1 and TNF. This observation was prompted by the finding that monocytes of patients with "high turnover" osteoporosis, the histological hallmark of postmenopausal osteoporosis, secrete increased amounts of IL-1 (Pacifici et al., 1987). Cross-sectional and prospective comparisons of pre- and postmenopausal women revealed that monocytic production of IL-1 and TNF increases after natural and surgical menopause, and is decreased by treatment with estrogen and progesterone (Pacifici et al., 1989, 1990). Subsequent observations showed that the postmenopausal increase in IL-1 activity results from an effect of estrogen on the production of both IL-Iβ and IL-lra (Pacifici et al., 1993). Studies in normal women undergoing ovariectomy (ovx) (Pacifici et al., 1991b; Fiore et al., 1993) revealed that estrogen withdrawal is associated not only with an increased production of IL-1 and TNF, but also of GM-CSF. The changes in these cytokine levels occur in a temporal sequence consistent with a causal role of IL-1, TNF, and GM-CSF in the pathogenesis of ovx induced bone loss (Pacifici et al., 1991b). Moreover, since the increase in GM-CSF production precedes the increase in IL-1 and TNF (Pacifici et al., 1991b), the data suggest that the increased production of GM-CSF is not a result of enhanced secretion of IL-1 and TNF, but rather a direct effect of estrogen withdrawal (Pacifici et al., 1991b).

The *in vitro* production of cytokines from cultured monocytes reflects phenotypic characteristics acquired from local stimuli during their maturation in the bone marrow (Horowitz, 1993). This phenomenon is thought to play an important role in providing the basis for tissue and functional specificity. Consequently, monocyte cytokine secretion is relevant to postmenopausal bone loss as it mirrors cytokine secretion from marrow resident cells of the monocyte macrophage lineage or monocytes that have homed to bone (Horowitz, 1993). This hypothesis was proved correct by studies showing that the secretion of IL-1 from blood monocytes correlates with that from bone marrow mononuclear cells in subjects with Paget's disease and osteoporosis (Pioli et al., 1989; Cohen-Solal et al., 1995) and by observations in rats and mice, where ovx and estrogen replacement were found to regulate the bone marrow mononuclear cell production of IL-1 and TNF (Kitazawa et al., 1994).

It is also important to recognize that monocytes are the major source of IL-1 and TNF in the bone marrow (Dinarello, 1989). Moreover, the anatomical proximity of mononuclear cells to remodeling loci, the capacity to secrete numerous products all recognized for their effects in bone remodeling, and the expression of integrin receptors (Hynes, 1992) which make these cells capable of adhering to the bone matrix, make them likely candidates as participants in skeletal remodeling.

More direct evidence in favor of cause-effect relationship between increased production of IL-1, TNF, and IL-6 and postmenopausal osteoporosis is also provided by the findings of Ralston (1994) demonstrating that IL-1, TNF, and IL-6 mRNAs are expressed more frequently in bone cells from untreated postmenopausal women than in those from women on estrogen replacement. That the increased monocytic production of cytokines plays a direct role in inducing bone resorption was later demonstrated by Cohen-Solal et al. (1993) by examining the bone resorption activity of culture supernatants from monocytes obtained from pre- and postmenopausal women. Using this approach it was found that the culture media of monocytes obtained from postmenopausal women have an increased *in vitro* bone resorption activity which is blocked by the addition of IL-Ira and anti-TNF antibody. Although the molecular mechanism by which estrogen regulates the production of IL-1 and TNF in mononuclear cells remains undetermined, estrogen receptors have been found in mononuclear cells (Weusten et al., 1986) and macrophages (Gulshan et al., 1990; Cutolo et al., 1993). Estrogen has also been shown to decrease the steady-state expression of IL-1 mRNAs in monocytes (Polan et al., 1989). Moreover, in preliminary experiments conducted in human monocytic cell lines cotransfected with estrogen receptors and an IL-I/CAT construct, we have observed that estrogen downregulates IL-1 promoter activity (unpublished data). This effect is likely to be indirect, since the IL-1 promoter does not contains estrogen responsive elements (Dinarello, 1994).

Interestingly, the association between estrogen deficiency and increased IL-1 and TNF activity was confirmed by others when IL-1 activity was measured by bioassay (Ralston et al., 1990; Kaneki et al., 1991; Matsuda et al., 1991; Pioli et al., 1992; Fiore et al., 1993). Conversely, this association was not observed when IL-1 was measured by ELISA or IRMA (Zarrabeitia et al., 1991; Hustmeyer et al., 1993). It should be noted that IL-1 bioactivity reflects the relative amounts of biologically active IL-1 and IL-1 antagonists present in the test sample. Consequently, IL-1 bioassays provide a reliable estimate of target cell response to IL-1. In contrast, ELISAs and IRMAs, although more specific, do not provide information on the amount of biologically active IL-1 which binds to the signal transducing type I IL-1 receptor. The binding of IL-1 to the type I receptor is, in fact, antagonized by IL-Ira (Dinarello, 1991; Thompson et al., 1991), soluble type I (sIL-1 RI) and type

II IL-1 receptor (sIL-1 RII) (Burger et al., 1995; Symons et al., 1995), anti-IL-1α autoantibodies (Hansen et al., 1990), and IL-lβ binding proteins (Simon et al., 1990). Moreover, while sIL-1 RI antagonizes the effects of IL-Ira (Burger et al., 1995), sIL-1 RII binds IL-1β, but does not bind IL-Ira (Burger et al., 1995; Symons et al., 1995). Thus, sIL-1 RII can compete with cell-associated receptors for IL-lβ and potentiate the inhibitory action of IL-Ira. Since estrogen could regulate IL-1 bioactivity by modulating factors which antagonize the binding of IL-1 to its active receptor, investigations on the effects of estrogen on production of IL-Ira, soluble IL-1 receptors and IL-1 binding proteins are likely to provide important information on the effect of estrogen on IL-1 bioactivity.

Subsequent studies conducted to determine if estrogen regulates the production of IL-6 revealed that in murine stromal and osteoblastic cells IL-6 production is inhibited by the addition of estrogen (Girasole et al., 1992) and stimulated by estrogen withdrawal (Passeri et al., 1993). *In vivo* studies also revealed that the production of IL-6 is increased in cultures of bone marrow cells from ovx mice (Jilka et al., 1992). This effect is mediated, at least in the mouse, by an indirect effect of estrogen on the transcription activity of the proximal 225-bp sequence of the IL-6 promoter (Pottratz et al., 1994; Ray et al., 1994)

Interestingly, although studies with human cell lines demonstrated an inhibitory effects of estrogen on the human IL-6 promoter (Rickard et al., 1992), three independent groups have failed to demonstrate an inhibitory effect of estrogen on IL-6 production from human bone cells and stromal cells expressing functional estrogen receptors (Chaudhary et al., 1992; Rickard et al., 1992; Rifas et al., 1995). These data raise the possibility that the production of human IL-6 protein does not increase in conditions of estrogen deficiency. This is further supported by a report that in humans surgical menopause is not followed by an increase in IL-6, although it causes an increase in soluble IL-6 receptor (Girasole et al., 1995).

Another possible intermediate in estrogen action is transforming growth factor β (TGFβ). This growth factor is a multifunctional protein that is produced by many mammalian cells including osteoblasts and has a wide range of biological activities. TGFβ is a potent osteoblast mitogen (Oursler, 1994). In specific experimental conditions TGFβ decreases both osteoclastic resorptive activity and osteoclast recruitment. Oursler et al. (1991b) have reported that estrogen increases the steady-state level of TGFβ mRNA and release of TGFβ protein. This mechanism provides the first example of positive effects of estrogen in bone which may result in decreased bone turnover. Attempts to demonstrate that menopause increases circulating levels of IL-1, TNF, and IL-6 have been, for the most part, unsuccessful (McKane et al., 1993; Khosla et al., 1994), presumably because only a small fraction of the cytokine produced in the bone marrow leaks into the peripheral circulation. The lack of increased serum cytokine levels in estrogen deficient women is also consistent with the notion that since cytokine release requires the adherence of cells to a solid substrate (Pacifici et al., 1991a, 1992), estrogen deficiency is unlikely to stimulate cytokine production from circulating cells.

IV. CYTOKINE KNOCKOUTS AND INFUSION OF CYTOKINE INHIBITORS: TOOLS TO ASSESS THE ROLE OF CYTOKINES IN OVARIECTOMY-INDUCED BONE LOSS

The development of transgenic mice has made it possible to link genes to specific phenotypes and determine the function of numerous proteins. Although this approach has been used to determine which cytokines are causative factors in ovx-induced bone loss, it should be emphasized that postmenopausal osteoporosis results from the impact of estrogen deficiency on a normally developed skeleton. The lack of deactivatable promoters does not allow a gene to "switch off" when the animal reaches maturity. Thus, knockout mice are characterized by bone modeling and remodeling defects which ensue during fetal development and lead to the formation of an abnormal mature skeleton. This is because the absence of a single gene is known to alter expression of other genes and developmental programs (Routtenberg, 1995). Moreover, the phenomenon of "gene compensation" may induce unstable phenotypes and make the recognition of the function of a factor problematic (Routtenberg, 1995). For example, IL-6 deficient mice are characterized by a greater numbers of osteoclasts, a lower bone density, and a greater bone turnover than control mice (Most et al., 1994; Pioli et al., 1994). Moreover, while initial studies conducted using IL-6 deficient mice of the first generation revealed that IL-6 deficient mice do not lose bone in response to ovx (Pioli et al., 1994), more recent data acquired with mice of the third generation which express a stable phenotype revealed that IL-6 deficiency does not protect from the bone loss induced by ovx (Balena et al., personal communication).

Another approach which is not subject to the limitations discussed above is to block the functional activity of cytokines by infusing specific inhibitors such as IL-Ira, the TNF antagonist, TNF binding protein (TNFbp), or anti-IL-6 antibodies in mature animals. This approach has

Table 1. Effect of Cytokine Inhibition on Bone Mass and Bone Turnover in Ovariectomized Rats and Mice in the First Month After Surgery (Early Postovariectomy Period)

Ab	*IL-Ira*	*TNFbp*	*IL-Ira + TNFbp*	*Anti-IL-6*
Prevents ovx-induced bone loss	+	+	+++	–
Blocks osteoclast formation	++	++	++	+
Blocks mature osteoclasts	++	++	+++	–
Stimulates bone formation	+++	+++	+++	–

revealed (Table 1) that in mature rats simultaneous treatment with IL-Ira and the TNFbp is required to completely prevent the bone loss and the increase in bone resorption induced by ovx in the first month after ovx (early post-ovx period) (Kimble et al., 1995a). In contrast, treatment of mature rats with IL-lra only started during the second month after ovx (late post-ovx period) (Table 2) is sufficient to block bone loss and bone resorption and to fully reproduce the effects of estrogen replacement (Kimble et al., 1994a,b). These data demonstrate that IL-1 and TNF play a critical causal role in the pathogenesis of ovx-induced bone loss and that the contribution of individual factors varies as time elapses after ovx. The specific contribution of IL-1 and TNF to ovx-induced bone loss is also species dependent because, in the mouse, treatment with TNFbp alone in the early post-ovx period duplicates the bone-sparing effect of estrogen (Kimble et al., 1997). In contrast, administration of TNFbp alone in the rat decreases, but does not completely block, ovx-induced bone loss (Kimble et al., 1997). The critical role of TNF in the mouse model is further documented by the failure of ovx to induce bone loss in transgenic mice which overexpress soluble TNF receptor (Ammann et al., 1997). Conversely, infusion of anti-IL-6 antibody does not prevent bone loss and the increase in *in vivo* bone resorption induced by ovx (Kimble et al., 1997), although it decreases the formation of osteoclasts in *ex vivo* bone marrow cultures from both estrogen deficient and estrogen replete mice (Jilka et al., 1992; Kitazawa et al., 1994). These data clearly demonstrate that although the production of IL-6 is likely to be regulated by estrogen in the mouse, this cytokine does not mediate the bone sparing effect of estrogen. This is because IL-6 is not required for inducing the production of other upstream estrogen regulated factors, such as IL-1 and TNF. Consequently, in conditions of estrogen deficiency, IL-I and TNF stimulate bone resorption and induce bone loss even in the absence of IL-6.

Studies have also revealed that treatment with either IL-Ira or TNFbp alone blocks the increase in *ex vivo* osteoclast formation induced by ovx (Kitazawa et al., 1994). Thus, IL-1 and TNF must possess either synergistic

Table 2. Effect of Cytokine Inhibition on Bone Mass and Bone Turnover in Ovariectomized Rats and Mice in the Second Month After Ovariectomy (late Postovariectomy Period)

	IL-lra	*TNFbp*	*IL-lra + TNFbp*	*Anti-IL-6Ab*
Prevents ovx-induced bone loss	+++	+++[a]	+++[a]	?
Blocks osteoclast formation	+++	+++[a]	+++[a]	?
Blocks mature osteoclasts	++	++	++	?
Stimulates bone formation	–	–	–	?

Note: [a]Preliminary unpublished data.

or sequential effects on osteoclastogenesis, the primary mechanism underlying long-term elevations of bone resorption (Wronski et al., 1993; Kimble et al., 1994c). In contrast, since inhibition of both IL-1 and TNF is required to prevent the increase in bone resorption observed in the early post-ovx period, IL-1 and TNF are likely to have independent and redundant stimulatory effects on osteoclast activation, the primary mechanism by which bone resorption increases acutely after ovx.

Although the exact mechanism by which IL-1 and TNF promote osteoclastogenesis remains to be elucidated, recent studies have revealed that stromal cells from ovx mice produce higher amounts of M-CSF than cells from estrogen replete mice and that this phenomenon is abolished by *in vivo* treatment with IL-lra and TNFbp (Kimble et al., 1996). These data suggest that the high levels of IL-1 and TNF which characterize the bone environment of ovx mice lead to the selection and expansion of a high M-CSF producing stromal cell population. The increased stromal cell production of M-CSF results, in turn, in increased osteoclast formation. The existence of two sequential regulated steps (monocytic production of cytokines and stromal cell response to these cytokines) explains the marked sensitivity of the osteoclast formation process to the inhibitory effect of estrogen.

V. SUMMARY AND CONCLUSIONS

Although the data discussed above demonstrate that IL-1 and TNF play a prominent causal role in mechanism by which estrogen deficiency induces bone loss, it should be recognized that the bone-sparing effect of estrogen is mediated by numerous cytokines which, by simultaneously stimulating multiple target cells, induce effects which cannot be accounted for by one single factor. The ability of estrogen to regulate some, but not all, the cytokines involved in this process is not inconsistent with this hypothesis because cytokines have potent synergistic effects. Thus, a considerable increase in bone resorption may result from a relatively small increase in the concentration of only few of the bone resorbing factors present in the bone microenvironment. This concept is best illustrated by the study of Miyaura et al. (1995) demonstrating that the concentrations of either IL-1, IL-6, IL-6 receptor and prostaglandins detected in the bone marrow of ovx mice are insufficient to account for the increased bone resorption caused by estrogen withdrawal. In contrast, the increase in bone resorption induced by ovx can be explained by the cumulative effects of these cytokines. Thus, a better understanding of the cooperative effects of cytokines, and a recognition that the contribution of individual cytokines to postmenopausal bone loss varies with the passage of time after the menopause, is necessary to fully understand the mechanism of action of estrogen in bone. Although the relevance of individual bone-targeting cytokines is species-specific, the development of transgenic mice with activatable or deactivatable promoters is likely to result in a further clarification of the integrated action of estrogen regulated cytokines in human bone cells and lay the foundation for the use of cytokine inhibitors in the treatment of postmenopausal osteoporosis.

REFERENCES

Ammann, P., Rizzoli, R., Bonjour, J.P., Bourrin, S., Meyer, J.M., Vassalli, P., and Garcia, I. (1997). Transgenic mice expressing soluble tumor necrosis factor-receptor are protected against bone loss caused by estrogen deficiency. J. Clin. Invest. 99, 1699-1703.

Arend, W.P. (1991). Interleukin 1 receptor antagonist: A new member of the interleukin I family. J. Clin. Invest. 88, 1445-1451.

Arend, W.P., Joslin, F.G., Thompson, R.C., and Hannum, C.H. (1989). An IL-1 inhibitor from human monocytes. J. Immunol. 143, 1851-1858.

Bain, S.D., Bailey, S.C., Celino, D.L., Lantry, M.M., and Edwards, M.W. (1993.) High-dose estrogen inhibits bone resorption and stimulates bone formation in the ovariectomized mouse. J. Bone Min. Res. 8, 435-442.

Bellido, T., Girasole, G., Passeri, G., Yu, X.P., Mocharla, A., Jilka, R.L., Notides, A., and Manolagas, S.C. (1993). Demonstration of estrogen and vitamin D receptors in bone marrow-derived stromal cells: Upregulation of the estrogen receptor by 1,25-dihydroxyvitamin-D3. Endocrinology 133, 553-562.

Bertolini, D.R., Nedwin, G.E., Bringman, T.S., Smith, D.D., and Mundy, G.R. (1986). Stimulation of bone resorption and inhibition of bone formation in vitro by human tumor necrosis factor. Nature 319, 516-518.

Boyce, B.F., Aufdemorte, T.B., Garrett, L.R., Yates, A.J.P., and Mundy, G.R. (1989). Effects of interleukin-1 on bone turnover in normal mice. Endocrinology 125, 1142-1150.

Burger, D., Chicheportiche, R., Giri, J.G., and Dayer, J.M. (1995). The inhibitory activity of human interleukin-I receptor anatgonist is enhanced by type-11 interleukin-I soluble receptor and hindered by type-I interleukin-I soluble receptor . J. Clin. Invest. 96, 38-41.

Canalis, E. (1986). Interleukin-I has independent effects on deoxyribonucleic acid and collagen synthesis in cultures of rat calvariae. Endocrinology 118, 74-81.

Carter, D.B., Deibel, Jr., M.R., Dunn, C.J., C-S., Tomich, C., Laborde, A.L., Slightom, J.L., Berger, A.E., Bienkowski, M.J., Sun, F.F., McEwan, R.N., Harris, P.K.W., Yem, A.W., Waszak, G.A., Chosay, J.G., Sieu, L.C., Hardee, M.M., Zurcher-Neely, H.A., Reardon, M., Heinrikson, R.L., Truesdell, S.E., Shelly, J.A., Eessalu, T.E., Taylor, B.M., and Tracey, D.E. (1990). Purification, cloning, expression, and biological characterization of an interleukin-I receptor antagonist protein. Nature 344, 633-638.

Chaudhary, L.R., Spelsberg, T.C., and Riggs, B.L. (1992). Production of various cytokines by normal human osteoblastlike cells in response to interleukin-1.9 and tumor necrosis factor-α: Lack of regulation by 17b,-estradiol. Endocrinology 130 (2), 528-534.

Cohen-Solal, M.E., Graulet, A.M., Guerris, J., Denne, M.A., Bergot, C., Morieux, C., Sedel, L., Kuntz, D., and de Vernejoul, M.C. (1995). Bone resorption at the femoral neck is dependent on local factors in nonosteoporotic late postmenopausal women: An in vitro-in vivo study. J. Bone Min. Res. 10, 307-314.

Chow, J., Tobias, J.H., Colston, K.W., and Chambers, T.J. (1992). Estrogen maintains trabecular bone volume in rats not only by suppression of bone resorption but also by stimulation of bone formation. J. Clin. Invest. 89, 74-78.

Cohen-Solal, M.E., Craulet, A.M., Denne, M.A., Gueris, J., Baylink, D. and de Vernejoul, M.C. (1993). Peripheral monocyte culture supernatants of menopausal women can induce bone resorption: Involvement of cytokines. J. Clin. Endocrinol. Metab. 77, 1648-1653.

Cutolo, M., Accardo, S., Villaggio, B., Clerico, P., Bagnasco, M., Coviello, D.A., Carruba, G., lo Casto, M., and Castagnetta, L. (1993). Presence of estrogen-binding sites on macrophagelike synoviocytes and CD8+, CD29+, CD45RO+ T-lymphocytes in normal and rheumatoid synovium. Arthritis and Rheumatism 36, 1087-1097.

Dinarello, C.A. (1988). Biology of interleukin 1. FASEB 2, 108-115.

Dinarello, C.A. (1989). Interleukin-I and its biologically related cytokines. Adv. Immunol. 44, 153-205.

Dinarello, C.A. (1991). Interleukin-I and Interleukin-1 antagonism. Blood 777, 1627-1652.

Dinarello, C.A. (1994). The biological properties of interleukin-1. Eur. Cytokine Netw. 5, 517-531.

Elias, J.A. and Lentz, V. (1990). IL-1 and tumor necrosis factor synergistically stimulate fibroblast IL-6 production and stabilize IL-6 messenger RNA. J. Immunol. 145, 161-166.

Eriksen, E.F., Colvard, D.S., Berg, N.J., Graham, M.L., Mann, K.G., Spelsberg, T.C., and Riggs, B.L. (1988). Evidence of estrogen receptors in normal human osteoblastlike cells. Science 241, 84-86.

Ettinger, B., Genant, H.K., and Cann, E.C., (1985). Long-term estrogen replacement therapy prevents bone loss and fractures. Ann. Intern. Med. 102, 319-324.

Felix, R., Fleish, H., and Elford, P.R. (1989). Bone-resorbing cytokines enhance release of macrophage colony-stimulating activity by osteoblastic cell MC3T3-El. Calcif Tissue. Int. 44, 356-360.

Felix, R., Cecchini, M.G., and Fleish, H. (1990). Macrophage colony-stimulating factor restores in vivo bone resorption in the op/op osteopetrotic mouse. Endocrinology 127, 2592-2597.

Fibbe, W.E., Van Damme, J., Billiau, A., Goselink, H.M., Vogt, P.J., Van Eeden, P.R., Altroch, B.W., and Falkenburg, J.H.F. (1988). Interleukin-I induces human marrow stromal cells in long-term culture to produce G-CSF and M-CSF. Blood 71, 431-435.

Fiore, C.E., Falcidia.E, Foti, R., Motta, M., and Tamburino, C. (1993). Differences in the time course of the effects of oophorectomy in women on parameters of bone metabolism and Interleukin-I levels in the circulation. Bone Miner. 20, 79-85.

Gallagher, J.C., Riggs, B.L., and DeLuca, H.F. (1980). Effect of estrogen on calcium absorption and serum vitamin D metabolites in post-menopausal osteoporosis. J. Clin. Endocrinol. Metab. 51, 1359-1364.

Genant, H.K., Cann, C.E., Ettinger, B., and Gordan, G.S. (1982). Quantitative computed bone tomography of vertebral spongiosa: A sensitive method for detecting early bone loss after oophorectomy. Ann. Intern. Med. 97, 699-705.

Girasole, G., Pedrazzoni, M., Giuliani, N., Passeri, G., and Passeri, M. (1995). Increased serum soluble interleukin-6 receptors levels are induced by ovariectomy, prevented by estrogen replacement and reversed by alendronate administration. J. Bone Miner. Res. 10, A86.

Girasole, G., Jilka, R.L., Passeri, G., Boswell, S., Boder, G., Williams, D.C., and Manolagas, S.C. (1992). 17β-estradiol inhibits interleukin-6 production by bone marrow-derived stromal cells and osteoblasts in vitro: A potential mechanism for the antiosteoporotic effect of estrogens. J. Clin. Invest. 89, 883-891.

Gowen, M., Wood, D.D., and Russell, R.G.G. (1985). Stimulation of the proliferation of human bone cells in vitro by human monocyte products with interleukin-I activity. J. Clin. Invest. 75, 1223-1229.

Gowen, M., Wood, D.D., Ihrie, E.J., McGuire, M.K.B., and Russell, R.G.G. (1983). An interleukin-l-like factor stimulates bone resorption in vitro. Nature 306, 378-380.

Gulshan, S., McCruden, A.B., and Stimson, W.H. (1990). Estrogen receptor in macrophages. Scan. J. Immunol. 31, 691-697.

Hannum, C.H.,Wflcox, C.J., Arend, W.P., Joslin, F.G., Dripps, D.J., Heftndal, P.L., Armes, L.G., Sommer, A., Eisenberg, S.P., and Thompson, R.C. (1990). Interleukin-I receptor antagonist activity of a human interleukin-1 inhibitor. Nature 343, 336-340.

Hansen, M.B., Svenson, M., and Bendtzen, K. (1990). Human anti-interleukin lα antibodies. Immunol. Lett. 30, 133-140.

Heaney, R.P., Recker, R.R., and Saville, P.D. (1978). Menopausal changes in bone remodelling. J. Lab. Clin. Med. 92, 964-970.

Horowitz, M. (1993). Cytokines and estrogen in bone: Antiosteoporotic effects. Science 260, 626-627.

Horsman, A., Jones, M., Francis, R., and Nordin, B.E.C. (1983). The effect of estrogen dose on postmenopausal bone loss. New. Eng. J. Med. 309, 1405-1407.

Hughes, D.E., Jilka, R.L., Manolagas, S., Dallas, S.L., Bonewald, L.F., Mundy, G.R., and Boyce, B.F. (1995). Sex steroids promote osteoclast apoptosis in vitro and in vivo. J. Bone Miner. Res. 10, 48 (Abstract.)

Hustmyer, F.G., Walker, E., Yu, X.P., Girasole, G., Sakagami, Y., Peacock, M., and Manolagas, S.C. (1993). Cytokine production and surface antigen expression by peripheral blood mononuclear cells in postmenopausal osteoporosis. J. Bone Miner. Res. 8, 51-59.

Hynes, R.O. (1992). Integrins: Versatility, modulation, and signaling in cell adhesion. Cell 69, 11-25.

Jilka, R.L., Hangoc, G., Girasole, G., Passeri, G., Williams, D.C., Abrams, J.S., Boyce, B., Broxmeyer, H., and Manolagas, S.C. (1992). Increased osteoclast development after estrogen loss: Mediation by Interleukin-6. Science 257, 88-91.

Johnson, R.A., Boyce, B.F., Mundy, G.R., and Roodman, G.D. (1989). Tumors producing human tumor necrosis factor induce hypercalcemia and osteoclastic bone resorption in nude mice. Endocrinology 124, 1424-1427.

Johnston, Jr., C.C., Hui, S.L., Witt, R.M., Appledorn, R., Baker, R.S., and Longcope, C. (1985). Early menopausal changes in bone mass and sex steroids. J. Clin. Endocrinol. Metab. 61, 905-911.

Kaneki, M., Nakamura, T., Masuyama, A., Chen, J.T., Seimiya, Y., Shiraki, M., Ouchi, Y., and Orimo, H. (1991). The effect of menopause on IL-1 and IL-6 release from peripheral blood monocytes. J. Bone Miner. Res. 6, 76 (Abstract.)

Khosla, S., Peterson, J.M., Egan, K., Jones, J.D., and Riggs, L.B. (1994). Circulating cytokine levels in osteoporotic and normal women. J. Clin. Endocrinol. Metab. 79, 707-711.

Kimble, R.B., Matayoshi, A.B., Vannice, J.L., and Pacifici, R. (1994a). Long-term treatment with IL-1 receptor antagonist (IL-Ira) blocks bone loss in ovariectomized rats. J. Bone Min. Res. 9, (Abstract; In press.)

Kimble, R.B., Matayoshi, A.B., Vannice, J.L., Kung, V.T., Williams, C., and Pacifici, R. (1995a). Simultaneous block of Interleukin I and tumor necrosis factor is required to completely prevent bone loss in the early post-ovariectomy period. Endocrinology 136, 3054-3061.

Kimble, R.B.S., Srivastava, S., Bellone, C.J., and Pacifici, R. (1996). Estrogen inhibits IL1 TNF gene expression in transfected murine macrophagic cells. J. Bone Min. Res. 11(Suppl.1), 235 (Abstract.)

Kimble, R.B., Vannice, J.L., Brownfield, C., and Pacifici, R. (1994b). Persistent bone-sparing effect of interleukin-I receptor antagonist: A hypothesis on the role of IL-1 in ovariectomy-induced bone loss. Calcif Tissue Int. 55, 260-265.

Kimble, R.B., Vannice, J.L., Bloedow, D.C., Thompson, R.C., Hopfer, W., Kung, V., Brownfield, C., and Pacifici, R. (1994c). Interleukin-I receptor antagonist decreases bone loss and bone resorption in ovariectomized rats. J. Clin. Invest. 93, 1959-1967.

Kimble, R., Bain, S.D., and Pacifici, R. (1997). The functional block of TMF but not of IL-6 prevent bone loss in ovariectomized mice. J. Bone Miner. Res. 12, 935-941.

Kitazawa, R., Kimble, R.B., Vannice, J.L., Kung, V.T., and Pacifici, R. (1994). Interleukin-I receptor antagonist and tumor necrosis factor binding protein decrease osteoclast formation and bone resorption in ovariectomized mice. Clin. Invest. 94, 2397-2406.

Komm, B.S., Terpening, C.M., Benz, D.J., Graeme, K.A., Gallegos, A., Korc, M., Greene, G.L., O'Mafly, B.W., and Haussler, M.R. (1988). Estrogen-binding, receptor mRNA and biologic response in osteoblastlike osteosarcoma cells. Science 241, 81-83.

Kurihara, N., Chenu, C., Miller, M., Civin, C., and Roodman, G.D. (1990). Identification of committed mononuclear precursors for osteoclastlike cells formed in long-term human marrow cultures. Endocrinology 126, 2733-2741.

Kurihara, N., Civin, C., and Roodman, G.D. (1991). Osteotropic factor responsiveness of highly purified populations of early and late precursors for human multinucleated cells expressing the osteoclast phenotype. J. Bone. Miner. Res. 6, 257-261.

Kurihara, N., Suda, T., Miura, Y., Nakauchi, H., Kodama, H., Hiura, K., Hakeda, Y., and Kumegawa, M. (1989). Generation of osteoclasts from isolated hematopoietic progenitor cells. Blood 74, 1295-1302.

Lindsay, R., Hart, D.M., Forrest, C., and Baird, C. (1980). Prevention of spinal osteoporosis in oophorectomized women. Lancet 2, 1151-1154.

Lorenzo, J.A., Sousa, S.L., Alander, C., Raisz, L.G., and Dinarello, C.A. (1987a). Comparison of the bone-resorbing activity in the supernatants from phytohemaglutinin-stimulated human peripheral blood mononuclear cells with that of cytokines through the use of an antiserum to interleukin 1. Endocrinology 121, 1164-1170.

Lorenzo, J.A., Souss, S.L., Fonseca, J.M., Hock, J.M., and Medlock, E.S. (1987b). Colony-stimulating factors regulate the development of multinucleated osteoclasts from recently replicated cells in vitro. J. Clin. Invest. 160, 164-160.

Lowik, C.W., G.M., Van der Plujim, G., Bloys, H., Hoekeman, K., Bijvoet, O.L.M., Aarden, L.A., and Papapoulos, S.E. (1993). Parathyroid hormone (PTH) and PTH-like protein (PLP) stimulate IL-6 production by osteogenic cells: A possible role of interleukin-6 in osteoclastogensis. Biochem. Biophys. Res. Commun. 162, 1546-1552.

Macdonald, B.R., Mundy, G.R., Clark, S., Wang, E.A., Kuehl, T.J., Stanley, E.R., and Roodman, G.D. (1986). Effects of human recombinant CSF-GM and highly purified CSF-I on the formation of multinucleated cells with osteoclast characteristics in long-term bone marrow cultures. J. Bone Miner. Res. 1, 227-232.

Matayoshi, A., Brown, C., DiPersio, J., Haugh, J., Kuestner, R., and Pacifici, R. (1995). Generation of human osteoclasts from peripheral blood CD34+ Cells. J. Bone Miner. Res. 10, T298 (Abstract.)

Matsuda, T., Matsui, K., Shimakoshi, Y., Aida, Y., and Hukuda, S. (1991). 1-Hydroxyethilidene-1,1-bisphosphonate decreases the postovariectomy-enhanced interleukin I production by peritoneal macrophages in adult rats. Calcif Tissue. Int. 49, 403-406.

McKane, R., Khosla, S., Peterson, J., Egan, K., and Riggs, B.L. (1993). Effect of age and menopause on serum Interleukin-1.9 and Intereleukin-6-levels in women. J. Bone Miner. Res. 8, 162A.

Miyaura, C., Kusano, K., Masuzawa, T., Chaki, O., Onoe, Y., Aoyagi, M., Sasaki, T., Tamura, T., Koishihara, Y., Ohsugi, Y. and Suda, T. (1995). Endogenous bone-resorbing factors in estrogen deficiency cooperative effects of IL-1 and IL-6. J. Bone Miner. Res. 10, 1365-1373.

Most, W., Beek, E.V., Ruwhof, C., Berzooijen, R.V., Aderveen, A., Kopf, M., Papapoulos, S., and Lowik, C. (1994). Osteoclastic resorption in interleukin-6-deficient mice. J. Bone Miner. Res. 9, 48 (Abstract.)

Nguyen, L., Dewhirst, F.E., Hauschka, P.V., and Stashenko, P. (1991). Interleukin-lF, stimulates bone resorption and inhibits bone formation in vivo. Lymphokine Cytokine Res. 10, 15-21.

Nilas, L. and Christiansen, C. (1987). Bone mass and its relationship to age and the menopause. J. Clin. Endocrinol. Metab. 65, 697-702.

Ohta, H., Makita, K., Suda, Y., Ikeda, T., Masuzawa, T., and Nozawa, S. (1992). Influence of oophorectomy on serum levels of sex steroids and bone metabolism and assessment of bone mineral density in lumbar trabecular bone by QCT-C value. J. Bone Min. Res. 7, 659-665.

Ohta, H., Ikeda, T. Masuzawa, T., Makita, K., Suda, Y., and Nozawa, S. (1993). Differences in axial bone mineral density, serum levels of sex steroids, and bone metabolism between postmenopausal and age- and body size-matched premenopausal subjects. Bone 14, 111-116.

Oursler, M.J. (1994). Osteoclast synthesis and secretion and activation of latent transforming growth factor β. J. Bone Miner. Res. 9, 443-452.

Oursler, M.J., Cortese, C., Keeting, P., Anderson, M.A., Bonde, S.K., Riggs, B.L., and Spelsberg, T.C. (1991a). Modulation of transforming growth factor-β production in normal human osteoblastlike cells by 17β-estradiol and parathyroid hormone. Endocrinology 129, 3313-3320.

Oursler, M.J., Pederson, L., Pyfferoen, J., Osdoby, P., Fitzpatrick, L., and Spelsberg, T.C. (1993). Estrogen modulation of avian osteoclast lysosomal gene expression. Endocrinology 132, 1373-1380.

Oursler, M.J., Osdoby, P., Pyfferoen, J., Riggs, B.L., and Spelsberg, T.C. (1991b). Avian osteoclasts as estrogen target cells. Proc. Natl. Acad. Sci. USA 88, 6613-6617.

Pacifici, R., Carano, A., Santoro, S.A., Rifas, L., Jeffrey, J.J., Malone, J.D., McCracken, R., and Avioli, L.V. (1991a). Bone matrix constituents stimulate interleukin-I release from human blood mononuclear cells. J. Clin. Invest. 87, 221-228.

Pacifici, R., Basilico, C., Roman, J., Zutter, M.M., Santoro, S.A., and McCracken, R. (1992). Collagen-induced release of interleukin 1 from human blood mononuclear cells. Potentiation by fibronectin binding to the α5 β1 integrin. J. Clin. Invest. 89, 61-67.

Pacifici, R., Brown, C., Puscheck, E., Friedrich, E., Slatopolsky, E., Maggio, D., McCracken, R., and Avioli, L.V. (1991b). Effect of surgical menopause and estrogen replacement on cytokine release from human blood mononuclear cells. Proc. Nat. Acad. Sci. USA 88, 5134-5138.

Pacifici, R., Brown, C., Rifas, L., and Avioli, L.V. (1990). TNFCa and GM-CSF secretion from human blood monocytes: Effect of menopause and estrogen replacement. J. Bone Miner. Res. 5, 181 (Abstract.)

Pacifici, R., Vannice, J.L., Rifas, L., and Kimble, R.B. (1993). Monocytic secretion of interleukin-I receptor antagonist in normal and osteoporotic women: Effect of menopause and estrogen/progesterone therapy. J. Clin. Endocrinol. Metab. 77, 1135-1141.

Pacifici, R., Rifas, L., McCracken, R., Vered, L., McMurtry, C., Avioli, L.V., and Peck, W.A. (1989). Ovarian steroid treatment blocks a postmenopausal increase in blood monocyte interleukin I release. Proc. Natl. Acad. Sci. USA 86, 2398-2402.

Pacifici, R., Rifas, L., Teitelbaum, S., Slatopolsky, E., McCracken, R., Bergfeld, M., Lee, W., Avioli, L.V., and Peck, W.A. (1987). Spontaneous release of interleukin I from human blood monocytes reflects bone formation in idiopathic osteoporosis. Proc. Natl. Acad. Sci. USA 84, 4616-4620.

Passeri, G., Girasole, G., Jilka, R.L., and Manolagas, S.C. (1993). Increased interleukin-6 production by murine bone marrow and bone cells after estrogen withdrawal. Endocrinology 133, 822-828.

Pfeilschifter, J., Chenu, C., Bird, A., Mundy, G.R., and Roodman, G.D. (1989). Interleukin-1 and tumor necrosis factor stimulate the formation of human osteoclastlike cells in vitro. J. Bone Miner. Res. 4, 113-118.

Pioli, G., Basini, G., Pedrazzoni, M., Musetti, G., Ulietti, V., Bresciani, D., Villa, P., Bacci, A., Hughes, D., Russell, G., and Passeri, M. (1992). Spontaneous release of Interleukin-I and interleukin-6 by peripheral blood monocytes after ovariectomy. Clin. Scien. 83, 503-507.

Pioli, G., Girasole, G., Pedrazzoni, M., Sansoni, P., Davoli, L., Ciotti, G., Mantovani, A., and Passeri, M. (1989). Spontaneous release of Interleukin I (IL-1) from medullary mononuclear cells of Pagetic subjects. Calcif Tissue. Int. 45, 257-259.

Polan, M.L., Loukides, J., Nelson, P., Carding, S., Diamond, M., Walsh, A., and Bottomly, K. (1989). Progesterone and estradiol modulate interleukin-1]R messenger ribonucleic acid levels in cultured human peripheral monocytes. J. Clin. Endocrinol. Metab. 69, 1200-1206.

Poli, V., Balena, R., Fattori, E., Markatos, A., Yamamoto, M., Tanaka, H., Ciliberto, G., Rodan, G.A., and Costantini, F. (1994). Interleukin-6-deficient mice are protected from bone loss caused by estrogen depletion. EMBO J. 13, 1189-1196.

Pottratz, S.T., Bellido, T., Mocharia, H., Crabb, D., and Manolagas, S. (1994). 17β,-Estradiol inhibits expression of human interleukin-6 promoter-reporter constructs by a receptor-dependent mechanism. J. Clin. Invest. 93, 944-950.

Prince, R.L. (1994). Counterpoint: Estrogen effects on calcitropic hormones and calcium homeostasis. (Review.) Endocr. Rev. 15, 301-309.

Raisz, L.G. (1988). Local and systemic factors in the pathogenesis of osteoporosis. N. Eng. J. Med. 318, 818-828.

Ralston, S.H. (1994). Analysis of gene expression in human bone biopsies by polymerase chain reaction: Evidence for enhanced cytokine expression in postmenopausal osteoporosis. J. Bone Miner. Res. 9, 883-890.

Ralston, S.H., Russell, R.G.G., and Gowen, M. (1990). Estrogen inhibits release of tumor necrosis factor from peripheral blood mononuclear cells in postmenopausal women. J. Bone Miner. Res. 5, 983-988.

Ray, A., Prefontaine, K.E., and Ray, P. (1994). Downmodulation of interleukin-6 gene expression by 17β-estradiol in the absence of high affinity DNA binding by the estrogen receptor. J. Biol. Chem. 269, 12940-12946.

Rickard, D., Russell, G., and Gowen, M. (1992). Oestradiol inhibits the release of tumour necrosis factor but not interleukin 6 from adult human osteoblasts in vitro. Osteoporosis Int. 2, 94-102.

Rifas, L., Kenney, J.S., Marcelli, M., Pacifici, R., Dawson, L.L., Cheng, S., and Avioli, L.V. (1995). Production of interleukin-6 in human osteoblasts and human bone marrow stromal cells: Evidence that induction by interleukin-I and tumor necrosis factor-α is not regulated by ovarian steroids. Endocrinology 136, 4056-4067.(Abstract.)

Riggs, B.L and Melton, L.J. (1986a). Medical progress: Involutional osteoporosis. New Eng J. Med. 314, 1676-1684.

Riggs, B.L. and Melton, L.J. (1986b). Evidence for two distinct syndromes of involutional osteoporosis. Am. J. Med. 75, 899-901.

Riggs, B.L., Wahner, H.W., Dunn, W.L., Mazess, R.B., Offord, K.P., and Melton, L.J. (1981). Differential changes in bone mineral density of the appendicular and the axial skeleton with aging. J. Clin. Invest. 67, 328-335.

Riggs, B.L., Jowsey, J., Goldsmith, S., Kelly, P.J., Hoffman, D.L., and Arnaud, C.C. (1972). Short- and long-term effects of estrogen and synthetic anabolic hormones in postmenopausal osteoporosis. J. Clin. Invest. 51, 1649-1663,

Roodman, G.D. (1992). Interleukin-6: An osteotropic factor? J. Bone Miner. Res. 7, 475-478.

Routtenberg, A. (1995). Knockout mouse fault lines [letter]. Nature 374, 314-315.

Sabatini, M., Boyce, B., Aufdemorte, T., Bonewald, L., and Mundy, G.R. (1988). Infusions of recombinant human interleukin-1α and -1β cause hypercalcemia in normal mice. Proc. Nat. Acad. Sci. USA 85, 5235-5239.

Schneider, G.B. and Relfson, M. (1989). Pluripotent hemopoietic stem cells give rise to osteoclasts in vitro: Effect of RGM-CSF. Bone. Min. 5, 129-138.

Seckinger, P., Klein-Nulend, J., Alander, C., Thompson, R.C., Dayer, J-M. and Raisz, L.G. (1990). Natural and recombinant human IL-1 receptor antagonists block the effects of IL-1 on bone resorption and prostaglandin production. J. Immunol. 145, 4181-4184.

Simon, J.A., Eastgate, J.A., and Duff, G.W. (1990). A soluble binding protein specific for interleukin IF, is produced by activated mononuclear cells. Cytokine 2, 190-198.

Slemenda, C., Hui, S.L., Longcope, C., and Johnston, C.C. (1987). Sex steroids and bone mass. A study of changes about the time of menopause. J. Clin. Invest. 80, 261-1269.

Srivastava, S., McHugh, K., Kimble, R., Ross, F.P., and Pacifici, R. (1995). Estrogen down regulates the expression of M-CSF mRNA in bone marrow stromal cells by stimulating the production of the transcription factor EGR-1. J. Bone Miner. Res. 10 (Suppl.1), S18.

Stashenko, P., Dewhirst, F.E., Rooney, M.L., Desjardins, L.A., and Heeley, J.D. (1987). Interleukin-1F is a potent inhibitor of bone formation in vitro. J. Bone Miner. Res. 2, 559-565.

Stein, B. and Yang, M.X. (1995). Repression of the interleukin-6 promoter by estrogen receptor is mediated by NFκB and C/EBP-β. Mol. Cell. Biol. 15, 4971-4979.

Stock, J.L., Coderre, J.A., and Mallette, L.E. (1985). Effects of a short course of estrogen on mineral metabolism in postmenopausal women. J. Clin. Endocrinol. Metab. 61, 595-600.

Suda, T., Takahashi, N., and Martin, T.J. (1992). Modulation of osteoclast differentiation. Endocr. Rev. 13, 66-80.

Symons, J.A., Young, P.R., and Duff, G.W. (1995). Soluble type-il interleukin-1 (il-1) receptor binds and blocks processing of il-l-β precursor and loses affinity for il-1 receptor antagonist. Proc. Nat. Acad. Sci. USA 92, 1714-1718.

Takahashi, N., Udagawa, N., Akatsu, T., Tanaka, H., Shionome, M., and Suda, T. (1991). Role of colony-stimulating factors in osteoclast development. J. Bone Miner. Res. 6, 977-985.

Tanaka, S., Takahashi, N., Udagawa, N., Tamura, T., Akatsu, T., Stanley, E.R., Kurokawa, T., and Suda, T. (1993). Macrophage colony-stimulating factor is indispensable for both proliferation and differentiation of osteoclast progenitors. J. Clin. Invest. 91, 257-263.

Thompson, R.C., Dripps, D.J., and Eisenberg, S.P. (1991). IL-Ira: properties and uses of an interleukin-1 receptor antagonist. Agents Actions Supp. 35, 41-49.

Thomson, B.M., Mundy, G.R., and Chambers, T.J. (1987). Tumor necrosis factor α and F, induce osteoblastic cells to stimulate osteoclastic bone resorption. J. Immunol. 138, 775-779.

Thomson, B.M., Saklatvala, J., and Chambers, T.J. (1986). Osteoblasts mediate interleukin I stimulation of bone resorption by rat osteoclasts. J. Exp. Med. 164, 104-112.

Tiegs, R.D., Body, J.J., Wahner, H.W., Barta, J., Riggs, B.L., and Heath, H.I. (1985). Calcitonin secretion in postmenopausal osteoporosis. N. Eng. J. Med. 312, 1097-1101.

Weusten, J.J., Blankenstein, M.A., Gmelig-Meyling, F.H., Schuurman, H.J., Kater, L., and Thijssen, J.H. (1986). Presence of estrogen receptors in human blood mononuclear cells and thymocytes. Acta Endocrinol. 112, 409-414.

Wronski, T.J., Dann, L.M., Qi, H., and Yen, C.F. (1993). Skeletal effects of withdrawal of oestrogen and diphosphonate treatment in ovariectomized rats. Calcif. Tissue Int. 53, 210-216.

Yoshida, H.S., Hayashi, S., Kunisada, T., Ogawa, M., Nishikawa, S., Okamura, H., Sudo, T., and Schultz, L.D. (1990). The murine mutation osteopetrosis is in the coding region of macrophage colony stimulating factor gene. Nature 345, 442-444.

Zarrabeitia, M.T., Riancho, J.A., Amado, J.A., Napal, J., and Gonzales-Macias, J. (1991). Cytokine production by blood cells in postmenopausal osteoporosis. Bone Miner. 14, 161-167.

PAGET'S DISEASE OF BONE

G. David Roodman

I. INTRODUCTION

Paget's disease is the second most common metabolic bone disease in the world, only surpassed by osteoporosis (Singer, 1985). The prevalence of the disease rises with age with about 8% of the population over the age of 75 years being affected by Paget's disease. Paget's disease has been reported in young people, although it is rarely seen in patients under the age of 45. It af-

Advances in Organ Biology
Volume 5C, pages 661-675.

ISBN: 0-7623-0390-5

fects about 3% of the population over the age of 40 in the Western hemisphere and Australia, and about 2 to 3 million patients in the United States (Collins, 1956). Paget's disease has a marked geographical distribution, and is seen most prevalently in Lancashire, England with about 7% of the population over the age of 55 years being affected. In contrast, Paget's disease is an extremely rare disease in Scandinavia, Africa, the Far East, and the Middle East. There is a familial incidence to Paget's disease, with up to 20% of patients having an affected sibling or relative, and vertical transmission of the disease has been reported.

Paget's disease of bone was first described more than 100 years ago by Sir James Paget (1877), a British surgeon, and the description of the disease—with deformity and enlargement of the skull, clavicles or long bones, which may be mono-ostotic or polyostotic, as well as significant morbidity from bone pain, fracture, neurologic complications due to nerve root compression, dental abnormalities and deafness—has changed little since his initial description. An interesting feature of Paget's disease is that new lesions rarely, if ever, develop after diagnosis, but rather lesions that are present continue to increase in size unless treated. Thus, Paget's disease is a highly localized process, which is relatively common in the Western hemisphere.

The primary pathologic abnormality in patients with Paget's disease is increased bone resorption by abnormal osteoclasts. In Paget's disease the osteoclasts are increased in number and size, and have increased activity. Bone resorption is coupled to new bone formation, so that there is constant destruction and formation of bone with a loss of normal bone architecture. The bone that is formed is of poor quality, which leads to fractures, degenerative arthritis, and neurologic impairment (Ryan, 1983). The osteoblasts in lesions of patients with Paget's disease appear to be normal (Hosking, 1981), although they are increased in number. These repeated cycles of bone resorption, followed by new bone formation, result in deposition of cement lines which produce the mosaic pattern that is a characteristic feature seen in histologic sections of pagetic bone. There is loss of definition between the cortical and trabecular bone, with areas of trabecular thickening and other areas in which the trabeculae are completely resorbed (Meunier et al., 1980). These trabeculae are often numerous and thick, and trabecular bone volume is markedly increased in patients with Paget's disease compared to normal.

Importantly, in patients with Paget's disease, bone not clinically involved with Paget's disease also appears to show increased bone remodeling. Approximately 45% of uninvolved bone shows significantly increased trabecular bone resorption and a modest increase in the number of osteoclasts (Meunier et al., 1980). Siris and co-workers (1989) and Meunier and co-workers (1980) attributed the increased bone remodeling in uninvolved bone from Paget's patients due to secondary hyperparathyroidism, rather than subclinical involvement with Paget's disease. However, a minority of patients (less than 20%) have increased parathyroid hormone levels, and these levels of parathyroid hormone are not dramatically increased.

II. OSTEOCLAST MORPHOLOGY IN PAGET'S DISEASE

Transmission electron microscopy studies have revealed several unique features that differentiate pagetic osteoclasts from normal osteoclasts. Rebel and co-workers (1976) were the first to report several cytological abnormalities present in osteoclasts from patients with Paget's disease. The osteoclasts were irregular in shape with multiple extensions and invaginations. The plasma membrane showed an external coat of dense particles consistent with membranes involved in active transport. Occasionally masses of small electron dense glycogen-like particles were found in the cytoplasm, and the cytoplasm also contained numerous microfilaments of various dimensions. The osteoclasts had typical ruffled borders with a clear zone. The most striking feature of these cells was their nuclei. There were large numbers of nuclei that were polymorphic. Some were smooth and ovoid while others were badly deformed with multiple indentations. The paranuclear space was dilated and contained clear vesicles. The nuclei had a peripheral distribution of dense chromatin and large nucleoli. In each of these osteoclasts several nuclei contained microcylindrical inclusion bodies that were filamentous structures of about 150 Å thick. Transverse sections of these inclusions showed a clear center surrounded by dense structures of about 50 Å in diameter. Occasionally these filaments were closely packed in a paracrystalline array with the interspace reduced to about 50 Å. The filaments were organized in bundles. These nuclear inclusions in Paget's osteoclasts were present in all cells and were never seen in normal osteoclasts. The nuclear inclusions present in pagetic osteoclasts were similar to nuclear inclusions seen in glial cells from patients with progressive multifocal leukoencephalopathy, a disease most probably due to a papovavirus (Zu Rhein, 1969; Dayan, 1974). In addition, patients who had subacute sclerosing panencephalitis virus infections also showed similar nuclear inclusions (Raine et al., 1974). These investigators suggested that Paget's disease may be due to an "external agent." Mills and Singer (1976) confirmed that nuclear inclusions were present in osteoclasts of patients with Paget's disease. These investigators studied 18 patients with Paget's disease. Nuclear inclusions similar to those reported by Rebel and co-workers (1976) were present in 20–40% of osteoclasts and were present in about one-fourth of the nuclei. Nuclear inclusions were present in all biopsy specimens from patients with Paget's disease. These nuclear inclusions differed from nuclear bodies which are nuclear organelles associated with cellular hyperactivity (Bouteille et al., 1967). They described these nuclear inclusions as most closely resembling the viral nucleocapsids of measles type virus. In addition to the nuclear inclusions, these pagetic osteoclasts also contained cytoplasmic inclusions which were similar to nuclear inclusions from patients with measles virus. Harvey and co-workers (1982) found nuclear inclusions in 56 to 100% of osteoclasts present in Paget's bone biopsy sections viewed by transmission electron microscopy. These nuclear inclusions occupied 15 to 75% of the nuclear cross-sectional area. Intracytoplasmic inclusions also were seen in 30 to 40% of osteoclasts in Paget's bone biopsies. Of interest is that treatment of patients with calcitonin or bisphosphonates, although reducing the number of osteoclasts, did not affect the morphology or prevalence of the nuclear inclusions present in these cells.

Howatson and Fornasier (1982) compared the nucleocapsids of measles virus and respiratory syncytial virus with the nuclear inclusion microfilaments associated with Paget's disease of bone. These authors confirmed that only osteoclasts contained these nuclear inclusions, and these inclusions were present in all biopsy specimens from Paget's patients. In eight of 10 patients cytoplasmic inclusions were also detected. The distribution of microfilament inclusions in pagetic osteoclasts closely paralleled that of measles nucleocapsids in monkey cells, but the dimensions of the pagetic microtubules were significantly different from those of measles virus. They also noted that the microtubules in the nuclear inclusions of Paget's osteoclasts were indistinguishable in dimension from the nucleocapsids from respiratory syncytial virus, but were dissimilar in the confirmation and location. These data supported the hypothesis that Paget's disease of bone was a slow viral disease and suggested that the respiratory syncytial virus rather than measles virus may be the cause of Paget's disease. Similar ultrastructural studies by Gherardi and co-workers (1980) further suggested a viral etiology for Paget's disease.

III. PAGET'S DISEASE: A POTENTIAL SLOW VIRUS DISEASE

The ultrastructural features of pagetic osteoclasts noted above, which demonstrate that they contain paramyxoviral-like nuclear and cytoplasmic inclusions, have suggested that Paget's disease may be a slow viral disease. However, to date no one has isolated or cloned the Paget's virus, nor has anyone demonstrated that they can infect normal cells with viral material isolated from pagetic osteoclasts. Thus, the identification and characterization of the virus, if present, is still unclear. Basle et al. (1979) and Rebel et al. (1980, 1981), using indirect immunofluorescence techniques with antimeasles polyclonal antibodies, showed reactivity of pagetic osteoclasts with these antibodies. These investigators showed that most of the osteoclasts, but none of the other bone cells (e.g., osteoblasts or osteocytes), reacted with the antimeasles antibodies. These positive reactions with the polyclonal antimeasles antibodies could be abolished with specific absorption of the antibodies on measles virus infected vero cells, but not by absorption on uninfected vero cells. No reactivity of normal osteoclasts with antimeasles antibodies could be detected. Similar results were obtained either by immunofluorescence or immunoperoxidase techniques. Mills and co-workers (1981, 1984) reported evidence for both respiratory syncytial virus and measles virus antigens in osteoclasts from patients with Paget's disease of bone. Using indirect immunofluorescent antibody assays, bone biopsies specimens from 28 of 29 Paget's patients showed reactivity with antibodies against respiratory syncytial virus, and 11 of 22 patients had positive immunofluorescence for measles virus. Using antiviral monoclonal antibodies, Basle et al. (1985) showed that antigens of measles virus, simian virus 5 (SV5) and human parainfluenza virus type 3 could be detected in pagetic osteoclasts. Measles and SV5 nucleoprotein and hemagglutinin-neuraminidase antigens were also present in all cases of Paget's bone examined. These studies further supported the hypothesis that a paramyxovirus may be responsible for Paget's disease of bone. Recently other investigators suggested canine distemper virus rather than other paramyxoviruses may play a role in the etiology of Paget's disease (Barker and Detheridge, 1985; Gordon et al., 1992). *In situ* hybridization studies carried out by Basle and co-workers (1987) have shown that measles nucleocapsid protein mRNA was expressed in the cytoplasm and nuclei of 80–90% of pagetic osteoclasts. In addition, the tritiated labeled DNA probes also hybridized with 30–40% of mononuclear cells including osteoblasts, osteocytes, fibroblasts, and lymphocytes. *In situ* hybridization studies with hepatitis B cDNA were negative. These data suggest that a wide range of cells contained measles virus mRNA, but only in the osteoclast was this message being translated into mature virus. Thus, the presence of viral inclusions in all cases of Paget's disease has led to the suggestion that Paget's disease is a slow viral disease analogous to subacute sclerosing panencephalitis and canine distemper. However, investigators have failed to demonstrate complement fixing antiviral antibodies for measles in serum from patients with Paget's disease (Winfield and Sutherland, 1981). In contrast, these antimeasles antibodies are present in the serum of patients with subacute sclerosing panencephalitis. In addition, attempts to isolate and transfer the virus has also been unsuccessful (Singer and Mills, 1983; Mills and Singer, 1987). Furthermore, the geographic prevalence of acute measles infection does not match the geographic distribution of Paget's disease. For example, Paget's disease is very rare in Scandinavia but is common in Lancashire, although the two regions have a similar incidence of measles. Potentially other etiologic factors may be active in Paget's disease in addition to the virus. Furthermore, viral inclusions have been found in some patients with giant cell tumors of bone (Mirra et al., 1981) as well as in rare patients with osteopetrosis (Mills et al., 1988), suggesting that these viral inclusions are not unique to Paget's disease.

We have used bone marrow culture techniques with marrow isolated from bones involved with Paget's disease to further characterize the viral transcripts present in Paget's patients. We have previously reported that long-term cultures of marrow from involved bones from patients with Paget's disease form multinucleated cells that have many of the characteristics of pagetic osteoclasts (Kukita et al., 1990). These multinucleated cells form more rapidly than normal osteoclast-like cells in these marrow cultures, have increased numbers of nuclei, have elevated levels of tartrate-resistant acid phosphatase, and the cultures formed between 10 to 100 times more osteoclast-like cells than formed in normal marrow cultures (Kukita et al., 1990). However, viral-like nuclear inclusions were not seen on ultrastructural studies of these osteoclast-like multinucleated cells formed in marrow from Paget's patients. Mills and her co-workers (1994) have tested pagetic marrow derived multinucleated cells formed in long-term marrow cultures for expression of viral antigens for paramyxoviruses. These investigators found that Paget's marrow derived multinucleated cells expressed the nucleocapsid antigens for both respiratory syncytial virus and measles virus, while normal marrow derived multinucleated cells did not express these viral proteins. Thus, the osteoclast-like cells express measles virus nucleocapsid protein, but the mature nucleocapsids are not formed in these cells.

To determine if evidence of measles virus could be confirmed in these osteoclast-like cells, polymerase chain reaction (PCR) analysis of RNA isolated from highly purified populations of these osteoclast-like cells derived from marrow from Paget's patients was tested using reverse-transcriptase PCR (RT-PCR) techniques for the presence of measles virus nucleocapsid transcripts. Measles virus nucleocapsid transcripts were consistently found in marrow cultures from patients with Paget's disease. Because of the findings of Mills and co-workers (1994) that mononuclear cells, as well as multinucleated cells, formed in the long-term marrow cultures from patients with Paget's disease, also expressed nucleocapsid protein, we then determined if mononuclear cells from marrow samples freshly isolated from patients with Paget's disease also expressed measles virus nucleocapsid protein. RT-PCR analysis of freshly isolated marrow mononuclear cells from these patients demonstrated measles virus nucleocapsid mRNA in 9 of 12 patients studied (Reddy et al., 1995a). Interestingly, many of these measles virus transcripts had point mutations at a highly localized area in the region of mRNA that coded for the carboxy-terminus of the nucleocapsid protein. All of these mutations were sense mutations resulting in amino acid substitutions in the nucleocapsid protein. This region of the nucleocapsid mRNA does not bind to the RNA polymerase, so that this region of the nucleocapsid mRNA can be mutated without affecting the viability of the virus. These data suggest that osteoclast precursors, which are increased in pagetic marrow samples from marrow from involved bones from patients with Paget's disease (Demulder et al., 1993), may also express measles virus nucleocapsid transcripts. Therefore, we cultured bone marrow mononuclear cells in the presence of granulocyte-macrophage colony-stimulating factor (GM-CSF) to form colony-forming unit granulocyte-macrophage-(CFU-GM) derived cells, which are the earliest recognizable precursor for the osteoclast, and then treated these CFU-GM-derived cells with 1,25-dihydroxyvitamin D_3 to induce commitment to the osteoclast lineage (Kurihara et al., 1990a). We found that both the early and the committed precursors for osteoclast-like cells from marrow samples from involved bones of patients with Paget's disease expressed measles virus nucleocapsid transcripts. In addition, we have shown that the measles virus hemagglutinin gene is also expressed by these cells. These data suggest that in addition to the osteoclast, osteoclast precursors also express measles virus transcripts and may be the primary site of infection of measles virus in these patients (Figure 1).

Since CFU-GM circulate in the peripheral blood and, in addition to forming osteoclasts, are also the precursors for monocytes and granulocytes that also can circulate, we then examined peripheral blood samples from patients with Paget's disease for the presence of measles virus nucleocapsid transcripts by RT-PCR. We found that peripheral blood mononuclear cells expressed measles virus nucleocapsid transcripts (Reddy et al., 1995b). We confirmed the presence of measles virus transcripts in peripheral blood monocytes using *in situ* hybridization techniques. In 10 normal blood samples, no measles virus transcripts could be detected by

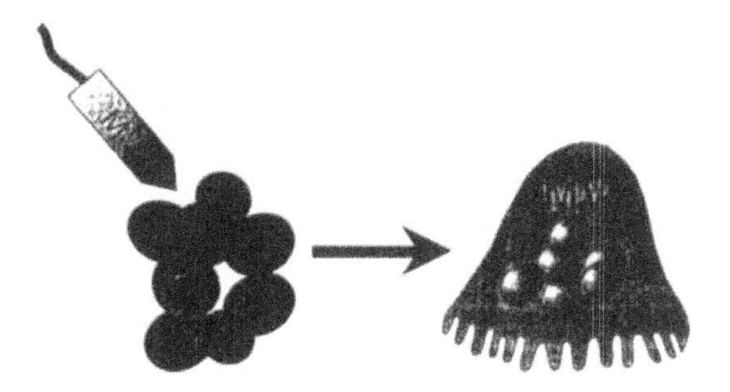

Figure 1. Infection of osteoclast precursors by paramyxoviruses may be the primary pathologic event in Paget's disease. In this model, osteoclast precursors are the initial site of infection by measles virus, which then differentiate and fuse to form multinucleated osteoclasts that also express measles virus. This contrasts with the mature osteoclast being the primary site of infection.

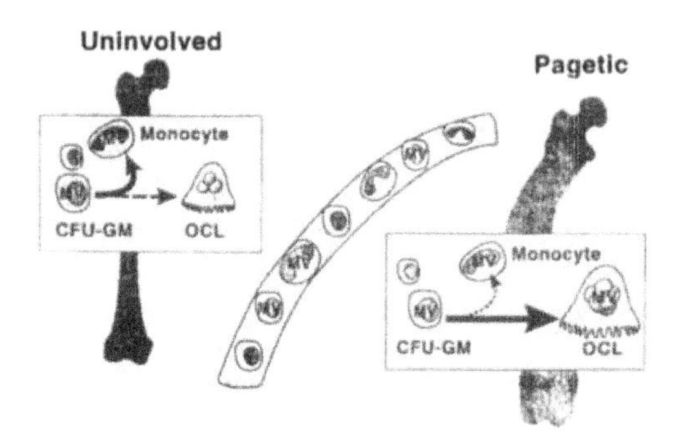

Figure 2. The role of the marrow microenvironment in Paget's disease. In pagetic lesions the marrow microenvironment preferentially induces osteoclast formation, while in uninvolved bone the marrow microenvironment preferentially induces CFU-GM, the earliest identifiable osteoclast precursor, to differentiate to the granulocyte-macrophage lineage, rather than to osteoclasts. Circulating cells in the peripheral blood contain measles virus nucleocapsid transcripts, but when they home to the marrow do not form osteoclasts that contain measles virus transcripts in uninvolved bone, but preferentially do so in involved bone.

RT-PCR or by *in situ* hybridization studies. These data demonstrate that in Paget's patients, circulating cells that contain measles virus transcripts are present and brings into question why Paget's disease remains such a highly localized process throughout the life of the patient. These data suggest that other factors such as the marrow microenvironment in pagetic lesions may play an important role in maintaining the localized nature of Paget's disease (Figure 2).

IV. ABNORMALITIES IN OSTEOCLAST PRECURSORS IN THE MARROW MICROENVIRONMENT IN PATIENTS WITH PAGET'S DISEASE

As noted above, osteoclast precursors and circulating peripheral blood cells contain measles virus transcripts, suggesting that early cells in the osteoclast lineage are infected by the measles virus in Paget's patients. In addition to containing viral transcripts, osteoclast precursors of Paget's patients are abnormal in several other important ways. Demulder and colleagues (1993) have shown that osteoclast precursors are increased in marrow aspirates from patients with Paget's disease compared to normals. When these osteoclast precursors were purified from marrow stromal elements, using a monoclonal antibody that identifies the CD34 antigen that is present on hematopoietic precursors and not on the marrow stromal cells, the absolute number of osteoclast precursors in pagetic and normal marrow aspirates were similar. Coculture of pagetic osteoclasts with marrow stromal cells from Paget's patients or with normal cells enhance growth of these osteoclast precursors above expected levels (Figure 3), suggesting that the osteoclast precursor is hyperresponsive to the marrow microenvironment. When osteoclast precursors derived from normal marrow were cocultured with pagetic stromal cells, there was enhanced growth of these osteoclast precursors, suggesting also that the marrow microenvironment is also abnormal in patients with Paget's disease and can induce increased osteoclast formation. In addition to increased responsivity to the marrow microenvironment, pagetic osteoclast precursors from marrow obtained from involved bones from Paget's patients are hyperresponsive to 1,25-dihydroxyvitamin D_3 and will form osteoclast-like cells in long-term marrow cultures treated with concentrations of 1,25-dihydroxyvitamin D_3 that are one-tenth that of normal. Thus, there are several abnormalities present in the pagetic osteoclast precursor which distinguishes it from the normal osteoclast precursor, including hypersensitivity to 1,25-dihydroxyvitamin D_3, hyperresponsivity to the marrow microenvironment, and the presence of measles virus nucleocapsid transcripts (Figure 4).

The studies described above suggest that the marrow microenvironment is abnormal in patients with Paget's disease. To further characterize abnormalities in the marrow microenvironment in patients with Paget's disease, we have produced marrow stromal cell lines from normal bone marrow and marrow obtained from involved bones from patients with Paget's disease. The stromal cell lines were produced by infecting marrow stromal cells with a recombinant SV40 adenovirus construct. These immortalized cell lines

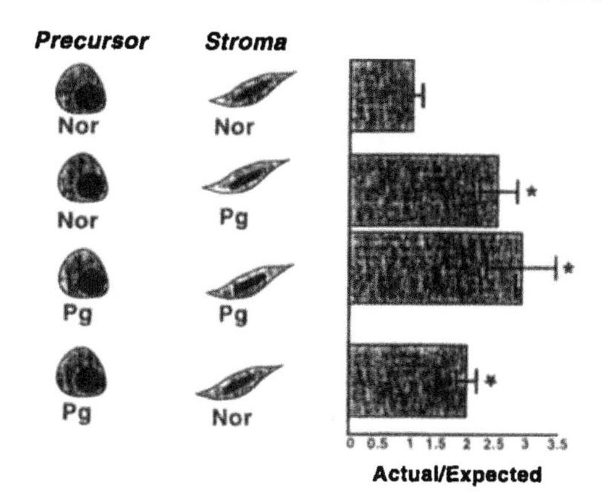

Figure 3. Coculture studies using highly purified osteoclast precursors from normal and pagetic patients and stromal cells from normal and pagetic patients. When pagetic osteoclast precursors are cocultured with normal or pagetic stromal cells, the number of CFU-GM-derived colonies formed is significantly enhanced above the expected level. When highly purified normal osteoclast precursors are cocultured with normal stromal cells, the expected number of CFU-GM-derived colonies are formed. In contrast, when normal osteoclast precursors are cocultured with pagetic marrow stromal cells, enhanced CFU-GM colony formation occurs. These data suggest that the growth of normal and pagetic osteoclast precursors is enhanced by pagetic marrow stroma, and that the pagetic stroma can also enhance the growth of normal osteoclast precursors above expected levels. The results are presented as the ratio of actual over the expected number of CFU-GM colonies formed.

have the characteristics of normal marrow stromal cells. Silverton and colleagues (1994) have shown that stromal cell lines from patients with Paget's disease enhance osteoclast-like cell formation in long-term marrow cultures, and that these osteoclast-like cells further enhance bone resorption. Furthermore, in contrast to the coculture studies with normal human marrow stromal cell lines in which cell to cell contact is required to enhance osteoclast-like cell formation (Takahashi et al., 1995), cell to cell contact between stromal cells and osteoclast precursors is not absolutely required for enhanced osteoclast formation seen when pagetic marrow stromal cells are used in these assays. Conditioned media from pagetic marrow stromal cell lines can enhance osteoclast-like cell formation in normal human marrow cultures, while conditioned media from normal marrow stromal cell lines do not by themselves enhance osteoclast-like cell formation (Takahashi et al., 1995). The pagetic marrow stromal cell lines express high levels of cytokines including interleukin-6 (IL-6), interleukin-1 (IL-1), and GM-CSF, but the mechanisms by which they enhance osteoclast-like cell formation in long-term human marrow cultures are not clear at present.

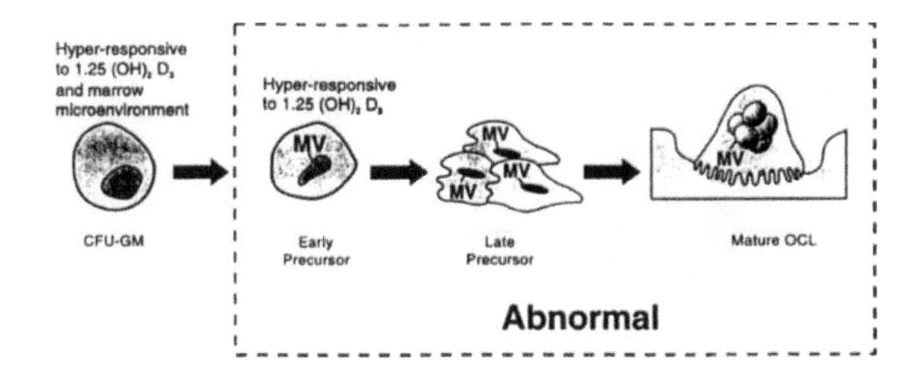

Figure 4. Abnormalities in osteoclast precursors and mature osteoclasts in Paget's disease. In this model system, the early osteoclast precursors contain measles virus transcripts and demonstrate hyperresponsivity to 1,25-dihydroxyvitamin D_3 and the marrow microenvironment. These cells then differentiate to form committed osteoclast precursors which then fuse and form multinucleated osteoclasts that express measles virus transcripts.

V. INTERLEUKIN-6 AS AN AUTOCRINE/PARACRINE FACTOR IN PAGET'S DISEASE

Recent studies by our group (Roodman et al., 1992) have suggested that IL-6 may play a role in the increased osteoclast formation seen in Paget's patients. Bone marrow samples were obtained from patients with Paget's patients and placed in long-term marrow culture. Conditioned media from these long-term pagetic marrow cultures were then added to normal marrow cultures and tested for their capacity to increase osteoclast-like multinucleated cell formation in normal marrow cultures. The Paget's marrow culture conditioned media increased normal multinucleated cell formation. One of the factors present in this conditioned media was IL-6, because antibodies to IL-6 but not IL-1, GM-CSF, or TNF-α neutralized the stimulating activity present in the Paget's conditioned media. Furthermore, *in situ* hybridization studies showed that the multinucleated cells present in the Paget's marrow cultures were actively transcribing IL-6 mRNA. When bone marrow plasma samples from patients with Paget's disease were assayed for IL-6, increased levels of IL-6 were found in 9 of 10 patients, while elevated

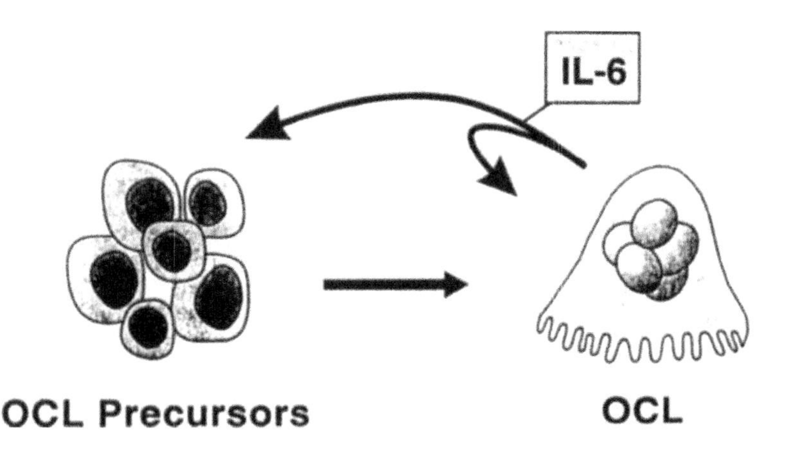

Figure 5. Role of IL-6 in Paget's disease. IL-6 appears to be an autocrine/paracrine factor in Paget's disease that enhances both osteoclast formation and osteoclastic bone resorption.

levels of IL-6 were not found in 10 of 10 normal marrow plasma samples. In addition, increased levels of IL-6 were found in the peripheral blood plasma in 17 of 27 patients with Paget's disease while elevated levels of IL-6 were not found in normal marrow plasma from a similar number of controls. These data suggest that IL-6 may be an autocrine/paracrine factor in Paget's disease of bone (Figure 5). Studies by Hoyland and co-workers (1994) support this hypothesis. Their *in situ* hybridization studies of pagetic bone demonstrated that IL-6 and IL-6 receptor mRNAs were expressed at high levels by osteoclasts isolated from Paget's bone. Ishimi and co-workers (1980) and Lowik et al. (1989) have shown that IL-6 may play a critical role in osteoclast formation and activity. IL-6 increased formation of osteoclasts in bone organ culture systems and also increased osteoclastic bone resorption in newborn mouse calvarial cultures. In addition, Kurihara et al. (1990b) have shown that IL-6 stimulates osteoclast-like multinucleated cell formation in normal human marrow cultures and increases the percentage of the cells reacting with the 23c6 monoclonal antibody that preferentially reacts with osteoclasts.

VI. CONCLUSION

The molecular and cell biology of the osteoclast in Paget's disease is just beginning to be understood. The presence of measles virus transcripts and canine distemper viral transcripts in osteoclasts and osteoclast precursors in patients with Paget's disease suggest a possible pathogenetic role for paramyxoviruses in Paget's disease of bone. However, until the virus is isolated, cloned, and normal osteoclasts or osteoclast precursors are infected by the virus and display a pagetic phenotype, the role of the paramyxoviruses in Paget's disease is still unclear. Abnormalities in the osteoclast precursor and the marrow microenvironment may contribute to the enhanced osteoclast formation and osteoclastic bone resorption seen in patients with Paget's disease and further amplify the initial pathologic event that induces increased osteoclast formation in these patients. The presence of circulating cells which contain measles virus transcripts in patients with Paget's disease further suggests that the marrow microenvironment may play an important role in maintaining the highly localized nature of Paget's disease. Studies to identify cytokines induced by osteoclasts and marrow stroma which can enhance osteoclast formation or osteoblastic bone formation, should provide important insights into the pathogenesis and pathophysiology of Paget's disease.

REFERENCES

Barker, D.J. and Detheridge, F.M. (1985). Dogs and Paget's disease (letter), Lancet 30, 1245.

Basle, M.F., Rebel, A., Pouplard, A., Kouyoumdjian, S., Filmon, R., and Lepatezour, A. (1979). Mise en evidence d'antigenes viraux de rougeole dans les osteoclastes de la maladie osseuse de Paget. C.R. Acad. Sci. Paris 289, 225-228.

Basle, M.F., Russell, W.C., Goswami, K.A., Rebel, A., Giraudon, P., Wild, R., and Filmon, R. (1985). Paramyxovirus antigens in osteoclasts from Paget's bone tissue detected by monoclonal antibodies. J. Gen. Virol. 66, 2103-2110.

Basle, M.F., Rebel, A., Fournier J.G., Russell, W.C., and Malkani, K. (1987). On the trail of paramyxoviruses in Paget's disease of bone. Clin. Ortho. Rel. Res. 217, 9-15.

Bouteille, M., Kalifat, S.R., and Delarue, J. (1967). Ultrastructural variations of nuclear bodies in human diseases. J. Ultrastruct. Res. 19, 474-486.

Collins, D.M. (1956). Paget's disease of bone, incidence, and subclinical forms. Lancet 2, 51-57.

Dayan, A.D. (1974). An unusual intranuclear structure (? viral nucleocapsid) in the brain in subacute sclerosing panencephalitis. J. Neurol. Neurosurg. Psych. 37, 201-206.

Demulder, A., Takahashi, S., Singer, F.R., Hosking, D.J., and Roodman, G.D. (1993). Abnormalities in osteoclast precursors and the marrow accessory cells in Paget's disease. Endocrinology 133 (5), 1978-1982.

Gherardi, G., Cascio, V.L.O., and Bonucci, E. (1980). Fine structure of nuclei and cytoplasm of osteoclasts in Paget's disease of bone. Histopath. 4, 63-74.

Gordon, M., Mee, A.P., Sharpe, P., and Anderson, D. (1992). Canine distemper virus transcripts sequenced from pagetic bone. Bone and Miner. 19, 159-174.

Harvey, L., Gray, T., Beneton, M.N.C., Douglas, D.L., Kanis, J.A., and Russell, R.G.G. (1982). Ultrastructural features of the osteoclasts from Paget's disease of bone in relation to a viral etiology. J. Clin. Pathol. 35, 771-779.

Hosking, D.J. (1981). Paget's disease of bone. Br. Med. J. 283, 686-688.

Howatson, A.F. and Fornasier, V.L. (1982). Microfilaments associated with Paget's disease of bone: Comparison with nucleocapsids of measles virus and respiratory syncytial virus. Intervirol. 18, 150-159.

Hoyland, J.A., Freemont, A.J., and Sharpe, P.T. (1994). Interleukin-6, IL-6 receptor, and IL-6 nuclear factor gene expression in Paget's disease. J. Bone Miner. Res. 9(1), 75-80.

Ishimi, Y., Miyaura, C., Jin, C.H., Akatsu, T., Abe, E., Nakamura, Y., Yamaguchi, A., Yoshiki, S., Matsuda, T., Hirano, T., Kishimoto, T., and Suda, T. (1980). IL-6 is produced by osteoblasts and induces bone resorption. J. Immunol. 145, 3297-3303.

Kukita, A., Chenu, C., McManus, L.M., Mundy, G.R., and Roodman, G.D. (1990). Atypical multinucleated cells form in long term marrow cultures from patients with Paget's disease. J. Clin. Invest. 85, 1280-1286.

Kurihara, N., Chenu, C., Civin, C.I., and Roodman, G.D. (1990a). Identification of committed mononuclear precursors for osteoclastlike cells formed in long-term marrow cultures. Endocrinology 126, 2733-2741.

Kurihara, N., Bertolini, D., Suda, T., Akiyama, Y., and Roodman, G.D. (1990b). Interleukin-6 stimulates osteoclastlike multinucleated cell formation in long term human marrow cultures by inducing IL-1 release. J. Immunol. 144, 426-430.

Lowik, C.W.G.M., van der Pluijm, G., Bloys, S., Hoekman, K., Bivoet, O.L.M., Aarden, L.A., and Papapoulos, S.E. (1989). Parathyroid hormone (PTH) and PTH-like protein (PLP) stimulate interleukin-6 production by osteogenic cells: A possible role of interleukin-6 in osteoclastogenesis. Biochem. Biophys. Res. Commun. 162, 1546-1552.

Meunier, P.J., Coindre, J.M., Edouard, C.M., and Arlot, M.E. (1980). Bone histomorphometry in Paget's disease. Arthr. Rheum. 23, 1095-1103.

Mills, B.G. and Singer, F.R. (1976). Nuclear inclusions in Paget's disease of bone. Science 194, 201-202.

Mills, B.G., Singer, F.R., Weiner, L.P., and Holst, P.A. (1981). Immunohistological demonstration of respiratory syncytial virus antigens in Paget's disease of bone. Proc. Natl. Acad. Sci. U.S.A., 78, 1209-1213.

Mills, B.G., Singer, F.R., Weiner, L.P., Suffin, S.C., Stabile, E., and Holst, P. (1984). Evidence for both respiratory syncytial virus and measles virus antigens in the osteoclasts of patients with Paget's disease of bone. Clin. Orthop. Rel. Res. 183, 303-311.

Mills, B.G., and Singer, F.R. (1987). Critical evaluation of viral antigen data in Paget's disease of bone. Clin. Ortho. Rel. Res. 217, 16-25.

Mills, B.G., Yabe, H., and Singer, F.R. (1988). Osteoclasts in human osteopetrosis contain viral-nucleocapsid-like nuclear inclusions. J. Bone Miner. Res. 3, 101-106.

Mills, B.G., Frausto, A., Singer, F.R., Ohsaki, Y., Demulder, A., and Roodman, G.D. (1994). Multinucleated cells formed in vitro from Paget's bone marrow express viral antigens. Bone 15 (4), 443-448.

Mirra, J.M., Bauer, F.C.H., and Grant, T.T. (1981). Giant cell tumor with viral-like intranuclear inclusions associated with Paget's disease. Clin. Ortho. Rel. Res. 158, 243-251.

Paget, J. (1877). On a form of chronic inflammation of bones (osteitis deformans). Med. Chir. Trans. 60, 37-63.

Raine, C.S., Feldman, L.A., Sheppard, R.D., Barbosa, L.H., and Bornstein, M.B. (1974). Subacute sclerosing panencephalitis virus. Lab. Invest. 31, 42-53.

Rebel, A., Malkani, K., Basle, M., and Bregeon, C.H. (1976). Osteoclast ultrastructure in Paget's disease. Calcif. Tiss. Res. 20, 187-199.

Rebel, A., Basle, M., Pouplard, A., Malkani, K., Filmon, R., and Lepatezour, A. (1980). Bone tissue in Paget's disease of bone. Arthr. Rheum. 23, 1104-1114.

Rebel, A., Basle, M., Pouplard, A., Malkani, K., Filmon, R., and Lepatezour, A. (1981). Toward a viral etiology for Paget's disease of bone. Metab. Bone Dis. Rel. Res. 4, 235-238.

Reddy, S.V., Singer, F.R., and Roodman, G.D. (1995a). Bone marrow mononuclear cells from patients with Paget's disease contain measles virus nucleocapsid mRNA that have mutations in a specific region of the sequence. J. Clin. Endo. Metab. 80 (7), 2108-2111.

Reddy, S.V., Singer, F.R., Mallette, L., and Roodman, G.D. (1995b). Detection of measles virus transcripts in peripheral blood mononuclear cells and marrow mononuclear cells from patients with Paget's disease. J. Bone Miner. Res. 10(1), 61, S154.

Roodman, G.D., Kurihara, N., Ohsaki, Y., Kukita, A., Hosking, D., Demulder, A., Smith, J.F., and Singer, F.R. (1992). Interleukin-6: A potential autocrine/ paracrine factor in Paget's disease of bone. J. Clin. Invest. 89(1), 46-52.

Ryan, W.G. (1983). Pathophysiology and modern management of Paget's disease, Comprehens. Ther. 9, 64-69.

Silverton, S.F., Takahashi S., Bunin, N., and Howard, D.F. (1994). Pagetic cell lines require cell contact with human bone marrow cultures for optimal expression of TRAP positive cells in vitro. J. Bone Miner. Res. 9 (1), A532, S229.

Singer, F.R. and Mills, B.G. (1983). Evidence for a viral etiology of Paget's disease of bone. Clin. Ortho. Rel. Res. 178, 245-251.

Singer, F.R. (1985). Paget's disease of bone. In: Cecil Textbook of Medicine, 17th edn. (Wyngaarden, J.B., Smith, L.H., and Plum, F., Eds), pp. 1461-1463. W.B. Saunders, Philadelphia.

Siris, E.S., Clemens, T.P., McMahon, D., Gordon, A., Jacobs, T.P., and Canfield, R.E. (1989). Parathyroid function in Paget's disease of bone. J. Bone Miner. Res. 4, 75-79.

Takahashi, S., Reddy, S.V., Dallas, M., Devlin, R., Chou, J.Y., and Roodman, G.D. (1995). Development and characterization of a human marrow stromal cell line that enhances osteoclastlike cell formation. Endocrinology 136 (4), 1441-1449.

Winfield, J. and Sutherland, S. (1981). Measles antibody in Paget's disease. Lancet 1, 891.

Zu Rhein, G.M. (1969). Association of papova-virions with a human demyelinating disease, progressive multifocal leukoencephalopathy. Prog. Med. Virol. 11, 185-247.

INHERITED AND ACQUIRED DISORDERS OF THE EXTRACELLULAR $CA^{2+}{}_{O}$-SENSING RECEPTOR

Edward M. Brown and Steven C. Hebert

Advances in Organ Biology
Volume 5C, pages 677-707.

ISBN: 0-7623-0390-5

I. INTRODUCTION

Calcium (Ca^{2+}) ions are of critical importance for a wide variety of vital bodily functions within both the extracellular and intracellular compartments (Pietrobon et al., 1990; Brown, 1991). The cytosolic free calcium concentration (Ca^{2+}_i), by acting as an intracellular second messenger and as a cofactor for various enzymes, plays a central role in coordinating and controlling cellular processes as diverse as muscular contraction, secretion and glycogen metabolism as well as cellular proliferation, differentiation, and motility (Pietrobon et al., 1990). Although Ca^{2+}_i is usually at much lower levels than the extracellular ionized calcium concentration (Ca^{2+}_o), it can rapidly undergo large fluctuations due to release from intracellular stores and/or influx through the plasma membrane. Extracellular Ca^{2+}, in contrast, remains nearly constant, varying by only a few percent under normal circumstances (Parfitt and Kleerekoper, 1980; Aurbach et al., 1985; Brown, 1991; Kurokawa, 1994). Ca^{2+}_o also participates in numerous key processes, including the promotion of blood clotting, regulation of neuromuscular excitability, and maintenance of the integrity of the skeleton.

Ca^{2+}_o is maintained at a nearly invariant level through a sophisticated homeostatic system that has been well-characterized in tetrapods and comprises two central elements (Figure 1) (Parfitt and Kleerekoper, 1980; Aurbach et al., 1985; Brown, 1991; Kurokawa, 1994). The first are Ca^{2+}_o-sensing, calciotropic hormone-secreting cells, such as parathyroid cells, thyroidal C-cells, and renal proximal tubular cells. In response to changes in Ca^{2+}_o these cells alter their secretion of calciotropic hormones, namely parathyroid hormone (PTH), calcitonin (CT) and 1,25-dihydroxyvitamin D (1,25$(OH)_2$D), respectively, in ways that are designed to restore the extracellular ionized calcium concentration to normal. These calciotropic hormones act on the second element of the system, the effector tissues (i.e., bone, intestine, and kidney), which alter their transport of calcium and/or phosphate [and to a lesser extent magnesium (Mg^{2+})] ions into or out of the extracellular fluid (ECF) so as to normalize Ca^{2+}_o.

A great deal of progress has been made in recent years toward understanding both the afferent (e.g., Ca^{2+}_o-sensing) and efferent (i.e., effector)

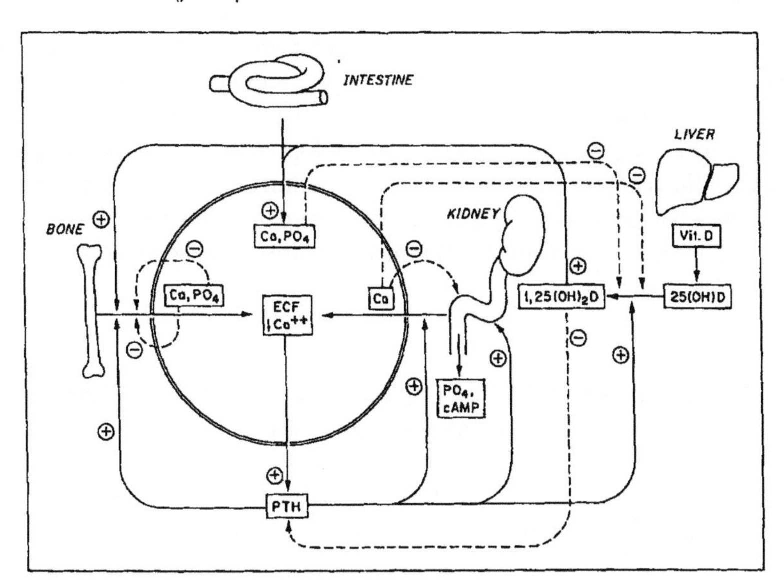

Figure 1. Schematic diagram illustrating the regulatory system maintaining extracellular Ca^{2+} homeostasis. The solid arrows and lines show the effects of PTH and $1,25(OH)_2D_3$; the dotted arrows and lines demonstrate examples of how extracellular Ca^{2+} or phosphate ions exert direct actions on target tissues. Abbreviations are as follows: Ca^{2+}, extracellular calcium; PO_4, phosphate; ECF, extracellular fluid, PTH, parathyroid hormone; $1,25(OH)_2D$, 1,25-dihydroxyvitamin D; 25(OH)D, 25-hydroxyvitamin D; minus signs indicate inhibitory actions while plus signs show positive effects. Reproduced with permission from Brown et al., 1994.

arms of this system. Initially progress was more rapid in characterizing the latter, with the isolation and structural characterization of first the calciotropic hormones in the 1960s and 1970s (Aurbach et al., 1985) and then, more recently, their receptors during the 1980s and 1990s (Abou-Samra et al., 1992; Lin et al., 1991). In contrast, it was for many years unclear how cells could respond directly to Ca^{2+}_{0} as an extracellular, "first" messenger (Brown, 1991; Brown et al., 1995). Within the past five years, however, the application of molecular techniques has enabled cloning of an extracellular Ca^{2+}-sensing receptor (CaR), initially from parathyroid (Brown et al., 1993). Subsequently, the same CaR has been isolated from kidney (Aida et al., 1995; Riccardi et al., 1995), C-cells (Garrett et al., 1995c), and brain (Ruat et al., 1995). Moreover, the availability of the cloned CaR made it possible to document that several inherited diseases of calcium homeostasis resulted from mutations in the CaR (Brown et al., 1995). The hypercalcemic disorders, familial hypocalciuric hypercalcemia (FHH) and neonatal severe hyperparathyroidism (NSHPT), are caused in most cases by mutations that reduce the activity of the CaR, while a hypocalcemic syndrome, familial autosomal dominant hypocalcemia, can result from activating mutations in the CaR. This chapter will review these developments briefly outlining the cloning of the receptor and how it functions within the calcium homeostatic system as a background for describing in more detail how mutations in the CaR produce the clinical and biochemical features of these inherited disorders of Ca^{2+}_0-sensing.

II. ROLE OF CA^{2+}_0-SENSING IN EXTRACELLULAR CALCIUM HOMEOSTASIS

Although the Ca^{2+}_0-sensing capacities of parathyroid and C-cells have long been recognized to be a critical element in the afferent arm of the mineral ion homeostatic system (Austin and Heath, 1981; Parfitt and Kleerekoper, 1980; Aurbach et al., 1985; Brown, 1991; Kurokawa, 1994), there are several other cells that also sense Ca^{2+}_0 and respond with changes in cellular function relevant to Ca^{2+}_0 homeostasis (Brown, 1991; Kurokawa, 1994). The kidney shows several such responses. In the proximal tubule, elevated levels of Ca^{2+}_0 directly inhibit the 1-hydroxylation of vitamin D (Weisinger et al., 1989). The thick ascending limb (TAL) of Henle's loop and the collecting duct also show the capacity to sense Ca^{2+}_0. Tubules from the TAL perfused *in vitro* show a marked reduction in reabsorption of Ca^{2+} and Mg^{2+} when the basolateral but not the luminal surface of the cells are exposed to elevated extracellular levels of either Ca^{2+} or Mg^{2+} (Quamme, 1989). Moreover, exposure of cells from either the medullary TAL (MTAL) or cortical TAL (CTAL) to elevated Ca^{2+}_0 produces a pertussis toxin-sensitive inhibition of the stimulation of cAMP accumulation by vasopressin, PTH, or a variety of other hormones acting on that segment of the nephron (Takaichi et al., 1986; Takaichi and Kurokawa, 1988). It is possible the Ca^{2+}_0-sensing in the CTAL enables local regulation of renal tubular reabsorption of calcium and magnesium by directly increasing their excretion in response to increments in systemic levels of Ca^{2+}_0 or Mg^{2+}_0. Elevated levels of Ca^{2+}_0 are thought to inhibit the actions of vasopressin on both the MTAL and the collecting ducts, although the physiological relevance of these actions has been obscure (Suki et al., 1969).

Osteoclasts and osteoblasts likewise sense Ca^{2+}_0 (Raisz and Kream, 1983; Malgaroli et al., 1989; Zaidi et al., 1989; Quarles, 1997). When isolated osteoclasts are exposed to elevated levels of Ca^{2+}_0, there is a rapid increase in $Ca^{2+}{}_{i}$ followed by inhibition of bone resorption probably related, at least in part, to reduced cellular adhesion to the bone surface and inhibition of the release of hydrolytic enzymes (Malgaroli et al., 1989; Zaidi et al., 1989). Therefore, $Ca^{2+}{}_{0}$-sensing could play a physiologically relevant role in controlling osteoclastic activity at a local level within bone. The proliferation of some osteoblastic cells is enhanced by increases in the concentrations of $Ca^{2+}{}_{0}$ and other polyvalent cations (e.g., Al^{3+}) (Quarles et al., 1997), an effect that could potentially provide a mechanism for contributing to the disposal of systemic calcium loads.

III. CLONING AND CHARACTERIZATION OF A $CA^{2+}{}_{0}$-SENSING RECEPTOR

The recent application of the tools of molecular biology has greatly illuminated how $Ca^{2+}{}_{0}$-sensing takes in parathyroid and kidney cells. Racke et al. (1993) and Shoback and co-workers (Chen et al., 1994) showed that injection of *Xenopus laevis* oocytes with parathyroid mRNA renders them responsive to polycationic agents known to modulate the $Ca^{2+}{}_{0}$-sensing mechanism in parathyroid cells, such as extracellular Ca^{2+} or Gd^{3+}. Brown and co-workers (1993) then utilized a similar assay system to screen a bovine parathyroid cDNA library and were able to isolate a single full length, functional clone of the CaR (BoPCaR = bovine parathyroid $Ca^{2+}{}_{0}$-sensing receptor) (Brown et al., 1993) (Figure 2).

The deduced amino acid sequence of BoPCaR predicts three principal structural domains. The first is a large hydrophilic extracellular domain (ECD) at the amino-terminus which consists of 613 amino acids (Brown et al., 1993). A second, hydrophobic transmembrane domain follows, comprising 250 amino acids and containing seven membrane-spanning segments characteristic of the superfamily of G protein-coupled receptors (GPCRs). The third domain is a putatively cytoplasmic, carboxy-terminal tail of 222 amino acids. There are nine predicted N-linked glycosylation sites within the ECD, which likely account for the substantial content of carbohydrate found in the native receptor (Brown et al., 1993; Garrett et al., 1995a,c). Within its intracellular loops and carboxy-terminal tail, BoPCaR contains four predicted protein kinase C (PKC) phosphorylation sites. It is possible that PKC-mediated phosphorylation of the CaR contributes to the inhibition of the activation of phospholipase C (PLC) by high $Ca^{2+}{}_{0}$ that is observed following treatment of bovine parathyroid cells with activators of PKC (Kifor et al., 1990). Subsequently, several more

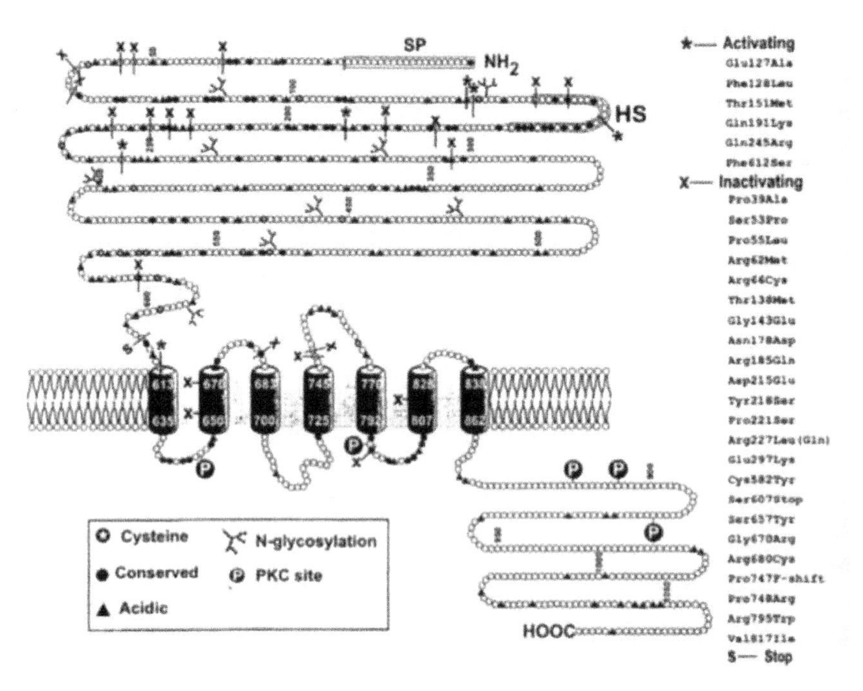

Figure 2. Schematic illustration of the predicted structure of the CaR protein, showing activating and inactivating mutations. Symbols are given in key. Additional abbreviations include the following: SP, predicted signal peptide; HS, hydrophobic segment. Reproduced with kind permission from Academic Press, San Diego, California.

CaRs have been isolated from additional tissues and mammalian species, including human parathyroid (Garrett et al., 1995a), kidney (Aida et al., 1995), as well as rat kidney (Riccardi et al., 1995), and brain (Ruat et al., 1995). These all share a high level of sequence homology (>90% identity in amino acid sequence) and almost certainly represent the various species homologues of the same ancestral gene.

Among the superfamily of GPCRs, the CaR shares amino acid sequence homology with three other families of receptors, the metabotropic glutamate receptors (mGluRs) (Nakanishi, 1992), the $GABA_B$ receptors (Kaupman et al., 1997), and putative pheromone receptors isolated from rat vomeronasal organ (Ryba et al., 1997). The mGluRs are expressed at high levels pre- and postsynaptically in the central nervous system and respond to glutamate, the major excitatory neurotransmitter in the brain, as their principal ligand *in vivo*. The degree of homology between the CaRs and mGluRs is modest (18-24% identity in amino acid sequence) (Brown et al., 1993; Garrett et al., 1995a). There is, however, striking topological similarity between the two classes of receptors. Both possess large (-600 amino acid) amino-terminal ECDs as well as a total of 20 strictly conserved cysteine residues (17 within the ECD and 3 in transmembrane segments or extracellular loops) (Brown et al., 1993; Garrett et al., 1995a). The construction of chimeric receptors containing the ECD of one class of receptor fused to the transmembrane and carboxy-terminal tail of the other has established that the ECD plays a key role in determining agonist specificity. For example, if the ECD is derived from the CaR, the chimeric receptor responds to CaR agonists, whereas an mGluR amino-terminus confers specificity to glutamatergic agonists when fused to the CaR transmembrane and cytoplasmic domains (Hammerland et al., 1995). The key residues within the ECD of the CaR that interact with its polycationic agonists, however, are currently unknown. It is possible that the conserved cysteines within the ECD of the CaR and mGluRs contribute to organizing this domain of the receptor into a structure with a binding pocket that interacts with small charged ligands (e.g., glutamate or Ca^{2+} and other polyvalent cations for the mGluRs and CaR, respectively). O'Hara and co-workers (1993) and Conklin and Bourne (1994) have postulated that the mGluRs and BoPCaR bear an evolutionary relationship to the bacterial periplasmic binding proteins (PBPs) (Adams and Oxander, 1989; Sharff et al., 1992). The latter sense extracellular ligands which either activate chemotaxis or cellular uptake of the ligand by the bacteria and function, in effect, as cell surface receptors. In Conklin and Bourne's (1994) formulation, the CaR represents the product of the fusion of an extracellular "sensing" domain, which originated from the PBPs, with a seven membrane-spanning serpentine motif to create a hybrid receptor molecule capable of transducing information in the extracellular space into alterations in the intracellular signaling pathways regulated by the GPCRs.

IV. TISSUE DISTRIBUTION AND PHYSIOLOGICAL ROLES OF THE CAR

Northern blot analysis has shown that transcripts for the CaR are present in diverse tissues, not all of which play obvious roles in mineral ion homeostasis (Brown et al., 1993; Garrett et al., 1995a; Riccardi et al., 1995; Ruat et al., 1995). These include the parathyroid cells, calcitonin-secreting thyroidal C-cells, kidney, intestine, lung, and various regions of brain. The use of *in situ* hybridization as well as immunohistochemistry with antibodies produced against synthetic peptides based on predicted sequences within the ECD of the CaR has enabled more detailed localization of the CaR in these tissues (Riccardi et al., 1996, 1998; Ruat et al., 1995). Moreover the localization of the CaR, combined with additional physiological and biochemical data, have elucidated significantly the roles of the CaR in the tissues expressing it.

A. Role of the CaR in Regulating Parathyroid Function

In the parathyroid, the CaR is thought to couple both to activation of PLC and inhibition of adenylate cyclase (Brown, 1991; Brown et al., 1995). The intracellular mechanism(s) through which the CaR inhibits rather than stimulates hormonal secretion (as is the case with most GPCRs activating PLC), however, remains an important unsolved problem. Nevertheless, the documentation that there is markedly attenuated high Ca^{2+}_{o}-mediated inhibition of PTH secretion in patients with homozygous inactivating mutations in the CaR (Cooper et al., 1986; Marx et al., 1986) or in mice homozygous for targeted deletion of the CaR gene (Ho et al., 1995) (see below) provides strong support for the central role of the CaR in Ca^{2+}_{o}-regulated PTH release. Moreover, the use of specific "calcimimetic," allosteric activators of the CaR in the presence of submaximal levels of Ca^{2+}_{o} has shown that the CaR likely also mediates the high Ca^{2+}_{o}-induced reduction in the levels of preproPTH mRNA (Garrett et al., 1995b). Finally, there is marked chief cell hyperplasia in both NSHPT as well as in mice homozygous for targeted disruption of the CaR gene (Randall and Lauchlan, 1963; Spiegel et al., 1977; Ho et al., 1995), suggesting that the receptor is directly or indirectly involved in suppressing parathyroid cell proliferation.

B. Role of the CaR in Regulating Renal Function

As assessed by reverse transcription-polymerase chain reaction (RT-PCR) of RNA isolated from individually microdissected nephron segments combined with immunohistochemistry and *in situ* hybridization, the CaR is expressed within nearly all nephron segments, including the glomerulus; proximal convoluted and straight tubules; MTAL and CTAL; distal convoluted tubule (DCT); and cortical, outer medullary, and inner medullary collecting ducts (Riccardi et al., 1996, 1998). In the MTAL and CTAL, where the CaR mRNA and protein are expressed at the highest levels, the receptor is located predominantly on the basolateral surface of the cells (Riccardi et al., 1998), where it presumably senses systemic (e.g., blood) levels of $Ca^{2+}{}_{o}$. In the inner medullary collecting duct (IMCD), on the other hand, it is localized on the apical (e.g., luminal) surface of the cell enabling it to monitor urinary levels of $Ca^{2+}{}_{o}$ (Sands et al., 1997). There is coexpression of the CaR and PTH receptor along the CTAL and DCT, allowing for interactions between the effects of the two receptors on distal renal tubular reabsorption of calcium (Morel et al., 1982; Riccardi et al., 1996).

Of the effects of $Ca^{2+}{}_{o}$ on renal function, several can be tentatively linked to the CaR. In the MTAL, by analogy with the parathyroid cell, the pertussis toxin-sensitive inhibition of vasopressin-stimulated cAMP accumulation likely reflects a CaR-mediated process (Takaichi et al., 1986; Takaichi and Kurokawa, 1988). High $Ca^{2+}{}_{o}$ also inhibits NaCl reabsorption in the TAL (Hebert and Andreoli, 1984), which could reduce the vasopressin-mediated generation of the countercurrent gradient in the MTAL and contribute to the impaired urinary concentrating ability in some hypercalcemic individuals. This effect on NaCl reabsorption could potentially be related, at least in part, to the concomitant diminution in vasopressin-stimulated cAMP accumulation. In addition to inhibiting reabsorption of water indirectly by decreasing the magnitude of the countercurrent gradient, high $Ca^{2+}{}_{o}$ may also directly reduce transepithelial water flow in the collecting duct. As noted above, recent studies have localized the CaR to the apical surface of the collecting duct, where it has been postulated to decrease the availability and/or activity of water channels, perhaps by modulating the trafficking of water channel-containing, subapical endosomes (Sands et al., 1997).

In the CTAL, the PTH-stimulated accumulation of cAMP (Morel et al., 1982) and resultant activation of NaCl reabsorption via the apical $Na^{+}/K^{+}/2Cl$ cotransporter generates a lumen positive transepithelial potential difference which drives passive paracellular reabsorption of NaCl, Ca^{2+}, and Mg^{2+} (Di Stefano et al., 1993; Hebert and Brown, 1996). Therefore, high $Ca^{2+}{}_{o}$-elicited suppression of PTH-stimulated cAMP accumulation in this nephron segment (Takaichi and Kurokawa, 1988), presumably caused by the CaR, could be an important factor in the previously documented reduction in the tubular reabsorption of both Ca^{2+} and Mg^{2+} by high peritubular $Ca^{2+}{}_{o}$. CaR-regulated intracellular mediators other than cAMP may also contribute to $Ca^{2+}{}_{o}$-evoked changes in solute transport in the CTAL. Wang and co-workers (1996) have recently shown that a metabolite of AA, probably 20-HETE, appears to mediate the high $Ca^{2+}{}_{o}$-elicited inhibition of an apical potassium (K^{+}), channel in the CTAL. The latter action, by preventing recycling of K^{+}, transported intracellularly by the apical $Na^{+}/K^{+}/2Cl$ cotransporter back into the lumen, effectively inhibits NaCl reabsorption by limiting the concentration of luminal potassium ions available to the co-transporter (Hebert and Brown, 1996). Additional studies will be needed to determine the role of the CaR, if any, in mediating other known effects of $Ca^{2+}{}_{o}$ on renal function, such as reducing 1-hydroxylase activity in the proximal tubule (Weisinger et al., 1989), lowering glomerular filtration rate (Humes et al., 1978) and reducing renal blood flow (Edvall, 1958).

C. Role of the CaR in Other Tissues

C-cells likewise contain abundant CaR transcripts (Garrett et al., 1995c) and the receptor expressed in C-cells appears to be identical to that present in parathyroid cells as assessed by actual sequencing of PCR-amplified, reverse-transcribed transcripts from the C-cell derived rMTC44-2 cell line (Garrett et al., 1995c). It is striking, therefore, that the same receptor that inhibits PTH secretion is apparently capable of simulating calcitonin secretion. While there conceivably could be an additional receptor(s) in the C-cell that regulates CT secretion, TT cells, a C-cell line that does not express the CaR, is also unresponsive to changes in $Ca^{2+}{}_{o}$ (Garrett et al., 1995c). Moreover, high $Ca^{2+}{}_{o}$ likewise stimulates the release of ACTH from the murine, pituitary-derived AtT20 cell line, which expresses the mouse homologue of the CaR, in association with high $Ca^{2+}{}_{o}$-elicited increases in $Ca^{2+}{}_{i}$ and inositol phosphates (Emanuel et al., 1996).

Additional cells expressing the CaR that are involved in systemic $Ca^{2+}{}_{o}$ homeostasis include intestinal epithelial cells that absorb dietary calcium and other nutrients (Gama et al., 1997; Chattopadhyay et al., 1998), bone-resorbing osteoclasts (Kameda et al., 1998), the bone-forming osteoblasts (Yamaguchi et al., 1998). Whether the CaR mediates known actions of extracellular calcium on the functions of these various cell types (for review, see Brown, 1991) and these actions contribute to the maintenance of $Ca^{2+}{}_{o}$ within its normally narrow limits remains to be proven.

In brain, the CaR protein is located throughout the central nervous system (CNS) with particular abundance in the cerebellum and hippocampus, hypothalamus, thalamus, basal ganglia, olfactory bulbs, ependymal zones of the cerebral ventricles, and cerebral arteries (Rogers et al., 1997; Ruat et al., 1995). In some cases, CaRs within the brain may be directly or indirectly involved in systemic fluid and electrolyte metabolism. For example, the subfornical organ, which is an important thirst center, has a high density of CaRs. Therefore, the increase in thirst in some hypercalcemic individuals may result not only indirectly from defective urinary concentration but also from a direct action on the subfornical organ, which could serve to anticipate concomitant losses of water as a consequence of the direct actions of high $Ca^{2+}{}_{0}$ on water handling by the kidney. In other regions of the brain, it is more likely that the CaR responds to local changes in $Ca^{2+}{}_{0}$ that are known to occur as a result of alterations in the level of brain cell activity, owing to uptake of calcium through various types of plasma membrane Ca^{2+} channels (Heinemann et al., 1977). CaRs in the brain could potentially also respond to ligands other than $Ca^{2+}{}_{0}$ (e.g., polycations, such as spermine, which are abundant in the CNS).

V. INHERITED DISORDERS OF $CA^{2+}{}_{0}$-SENSING

A. Familial Hypocalciuric Hypercalcemia

Clinical and biochemical features of FHH

FHH, originally named familial benign hypercalcemia or FBH (Foley et al., 1972), is an autosomal dominant trait with greater than 90% penetrance, characterized by lifelong hypercalcemia that is moderate (usually <12 mg/dl) in degree and largely asymptomatic (Marx et al., 1981a; Law and Heath, 1985; Heath, 1989). Unlike individuals with primary hyperparathyroidism (PHPT), who often have a similar level of hypercalcemia, those with FHH generally exhibit rates of urinary calcium excretion that are within the normal range and are inappropriately low given their elevated serum calcium concentration (the urinary calcium to creatinine clearance ratio is usually <0.01 in FHH). Moreover, particularly when measured using assays that are specific for the intact form of circulating PTH [PTH (I-84)] (Chou et al., 1992), individuals with FHH exhibit levels of the hormone that are generally normal and may even reside within the lower half of the normal range (Marx et al., 1981a; Law and Heath, 1985; Heath, 1989; Chou et al., 1992). In PHPT, in contrast, at least 90% of patients have frankly elevated PTH levels, although confusion may arise in differentiating FHH and PHPT in cases where the level of intact PTH is in the upper part of the normal range (Heath, 1989; Gunn and Wallace, 1992). Hypophosphatemia is variably present, and mild increases in serum magnesium concentration may coexist with the hypercalcemia, which are caused by excessively avid renal tubular reabsorption in Mg^{2+} in addition to that for Ca^{2+} (Marx et al., 1981a; Law and Heath, 1985; Heath, 1989). Serum levels of the vitamin D metabolites, 25(OH)D and $1{,}25(OH)_2D$, are usually within the normal range (Law et al., 1984; Kristiansen et al., 1985).

Most patients with FHH do not develop the typical complications encountered in PHPT and other chronic hypercalcemic conditions, such as nephrolithiasis and renal dysfunction, although patients with FHH seemingly have a higher prevalence of chondrocalcinosis than in the normal population (Marx et al., 1981a). While patients have been encountered in FHH kindreds with pancreatitis, it is currently thought that pancreatitis is not a direct consequence of FHH (Law and Heath, 1985; Stuckey et al., 1990). A family from Oklahoma has exhibited an unusual variant of this condition with progressive, age-related elevations in PTH levels along with hypophosphatemia and osteomalacia (McMurtry et al., 1992). The generally benign clinical course of FHH dictates conservative medical management and observation with periodic monitoring of serum levels of calcium and PTH. Establishing the diagnosis is clinically important, however, in order to avoid unnecessary and fruitless attempts at surgical correction of the hypercalcemia which invariably recurs unless total parathyroidectomy is performed. The diagnosis can generally be made by demonstrating an autosomal dominant pattern of inheritance of mild hypercalcemia in affected family members with a urinary calcium to creatinine clearance ratio of less than 0.01 (Marx et al., 198la). As discussed below, more direct genetic approaches to diagnosis are now possible in selected cases with the identification of inactivating mutations in the CaR as the cause of most cases of FHH.

The inappropriately normal (i.e., nonsuppressed) level of PTH in the setting of overt hypercalcemia and relative hypocalciuria in individuals with FHH indicates that there is abnormal regulation of both PTH secretion and renal calcium handling by Ca^{2+}_{0} (Marx et al., 198la; Law and Heath, 1985). Moreover, abnormal Ca^{2+}_{0}-sensing by parathyroid and kidney in FHH has been confirmed by direct biochemical testing. During intravenous infusion of calcium, patients with FHH require supranormal levels of Ca^{2+}_{0} to achieve a degree of suppression of circulating PTH levels equivalent to that seen in normals with increases in Ca^{2+}_{0} within the upper part of the normal range (Auxwerx et al., 1984; Khosla et al., 1993). That is, the parathyroid glands of patients with FHH exhibit an increase in set-point (i.e., the level of Ca^{2+}_{0} half-maximally suppressing PTH release). The enhanced renal tubular reabsorption of calcium in FHH is not the result of the altered PTH dynamics, as it persists even after total parathyroidectomy (Attie et al., 1983; Davies et al., 1984). The studies of Attie and co-workers (1983) suggested that the abnormal renal calcium handling in FHH is localized in the TAL, since a loop diuretic, ethacrynic acid, was the only calciuric stimulus that increased urinary calcium excretion in these parathyroidectomized patients. Recall that this is the nephron segment where basolaterally situated CaRs are thought to mediate the inhibitory actions of elevated peritubular levels of Ca^{2+} and Mg^{2+} on their own reabsorption. In effect, the TAL, like the parathyroid cell, is "resistant" to $Ca^{2+}{}_{o}$ in FHH, limiting the increase in urinary calcium excretion that can be achieved in the face of an increase in $Ca^{2+}{}_{o}$. Therefore, the abnormal $Ca^{2+}{}_{o}$ sensing/handling in both parathyroid and kidney provided an important clue that FHH could be an inherited disorder of the CaR, providing a strong impetus to identify the FHH gene.

Genetics of FHH

Over the past few years, several studies have shown that in most FHH families the disease gene is linked to a locus on the long arm of chromosome 3 (Chou et al., 1992). This was initially shown by linkage analysis of four large kindreds with FHH, which mapped the gene to band 3q21-24 (Chou et al., 1992). In one family, however, a phenotypically similar disorder maps to band p13.3 on the short arm of chromosome 19 (Heath et al., 1993). In addition, the variant of FHH in the Oklahoma kindred described above is not linked to either chromosome 3 or 19, confirming that FHH is genetically heterogeneous (Trump et al., 1995). Pollak et al. (1993) then utilized a ribonuclease A (RNAse A) protection assay to identify unique missense mutations (i.e., point mutations in which a change in a single nucleotide base substitutes a new amino acid for the one originally coded for) in affected but not in unaffected individuals from three unrelated FHH families. Two of the three families had mutations (Arg185Gln and Glu297Lys) in the amino-terminal extracellular domain of the CaR, which is known to play a key role in ligand binding. The third family had a mutation (Arg795Trp) within the third intracellular loop that could potentially disrupt CaR function in at least two ways: the intracellular loops of GPCRs serve an important role in coupling the receptors to G protein(s), and the replacement of the positively charged arginine residue (basic residues are frequently found in this location in other GPCRs) with the bulky, hydrophobic tryptophan could interfere substantially with G protein coupling. Indeed, when expressed in *X. laevis* oocytes by injection of synthetic CaR mRNA engineered to encode the Arg795Trp mutation, the receptor showed markedly blunted responsiveness to Ca^{2+}, Gd^{3+}, and neomycin, affording strong evidence that this mutation impaired $Ca^{2+}{}_{o}$-sensing (Pollak et al., 1993). This study, therefore, provided the initial direct confirmation that FHH results from inactivating CaR mutations, which render target cells, in effect, resistant to the $Ca^{2+}{}_{o}$ signal. In the parathyroid, this increases the set-point for $Ca^{2+}{}_{o}$-regulated PTH release, while in the kidney, most likely in the TAL, it impairs the normal high $Ca^{2+}{}_{o}$-evoked reduction in renal tubular reabsorption of Ca^{2+}. Both abnormalities contribute to the initiation and maintenance of hypercalcemia in FHH. Subsequent studies (Heath et al., 1994; Chou et al., 1995; Janicic et al., 1995; Pearce et al., 1995) have confirmed and extended these findings. With few exceptions, each FHH kindred harbors its own unique missense mutation, which reduces the function of the receptor when expressed in either *X. laevis* oocytes or mammalian expression systems. Mutations other than missense mutations that have been identified to date include the following: (1) a nonsense mutation at codon 607 within the ECD that predicts a truncated receptor protein lacking all of its membrane-spanning segments (Pearce et al., 1995); (2) a C to T transition accompanied by deletion of a single nucleotide within codon 747 that changes the reading frame, resulting in premature termination of translation that would be expected to produce a receptor protein lacking its last three transmembrane domains (Pearce et al., 1995); and (3) insertion of an Alu-repetitive element within the predicted carboxy-terminal tail of the CaR (Janicic et al., 1995). The latter introduces an unrelated sequence, which includes a stretch of 29 phenylalanines encoded by the poly A tract within the Alu-repetitive element, followed by premature truncation due to the presence of an in-frame stop codon. It should be noted that about one-half to two-thirds of families showing linkage to the chromosome 3 locus have no detectable mutations within the coding sequence of the CaR gene (Chou et al., 1992; Heath et al., 1994). Such families presumably harbor mutations in regions of the gene that interfere with its transcription or translation or the stability of its mRNA.

Even within kindreds showing linkage to chromosome 3, there is some degree of biochemical heterogeneity. Some families show minimal hypercalcemia or levels of serum calcium that are within the upper part of the normal range (Pollak et al., 1993). Others may show considerably more severe, albeit generally asymptomatic hypercalcemia, with levels of serum calcium that average 2–3 mg/dl above that of unaffected family members. Some of the reasons for this variability are becoming apparent as a variety of mutations are encountered and with the recent report of the development of a mouse model of FHH with targeted disruption of the FHH gene (Ho et al., 1995) (described in more detail in below). In general, mutations that would be predicted to totally abolish the activity of the receptor and prevent its expression on the cell surface (e.g., the presence of the stop codon within the ECD) show the mildest hypercalcemia, which is similar in degree to that observed in mice heterozygous for disruption of the CaR gene (Ho et al., 1995;

Pearce et al., 1995). In these cases, there is probably simply a reduction in the number of functionally intact receptors because they are being produced by only one allele, which apparently does not upregulate its level of expression sufficiently to maintain normocalcemia. It is likely, on the other hand, that missense mutations associated with more severe hypercalcemia in the heterozygous state produce a dominant negative effect. That is, the abnormal receptor interferes in some manner with the function of the normal receptor, perhaps by interfering with the biosynthesis and/or posttranslational processing of the latter, competing with them for a limited pool of G protein, or forming oligomeric complexes of inactive and active CaRs that prevent activation of the normal receptors (Bai et al., 1996, 1997).

Individuals with FHH show mild hypermagnesemia due, in large part, to excessive magnesium reabsorption in the distal nephron (Marx et al., 1981a). Elevated peritubular levels of not only Ca^{2+}, but also Mg^{2+}, inhibit reabsorption of both ions in the TAL (Quamme, 1989). Moreover, both the cloned parathyroid (BoPCaR) (Brown et al., 1993) and renal (rat kidney CaR = RaKCaR) (Riccardi et al., 1995) CaRs respond to $Mg^{2+}{}_{0}$ when expressed in *X. laevis* oocytes, albeit at somewhat higher concentrations than for $Ca^{2+}{}_{0}$. It is likely, therefore, that the CaR could play an important, although as yet unproven, role in the regulation of renal Mg^{2+} reabsorption by the TAL and, in turn, of $Mg^{2+}{}_{0}$. Of interest, mice heterozygous or homozygous for disruption of the CaR gene both show statistically significant increases in $Mg^{2+}{}_{0}$ providing additional evidence for a role of the CaR in $Mg^{2+}{}_{0}$ homeostasis (Ho et al., 1995).

One of the well-described effects of hypercalcemia on the kidney is to decrease maximal urinary concentrating ability, leading to hyposthenuria and, in some cases, overt nephrogenic diabetes insipidus (NDI) (Suki et al., 1969). This could occur in two ways: (1) elevations in extracellular calcium are known to inhibit NaCl absorption in the TAL (Hebert and Andreoli, 1984) where the CaR is located (Riccardi et al., 1996, 1998). NaCl reabsorption in the TAL is crucial for generation of the countercurrent gradient that is required for hydrosmotic water flow to occur in the collecting duct under the influence of vasopressin. (2) Hypercalcemia likewise exerts an inhibitory effect on the action of vasopressin on water transport in the collecting duct (Jones et al., 1988). Recent data indicate that the CaR is present on the apical surface of cells of the collecting ducts, and, therefore, this effect on vasopressin action might also be CaR-mediated (Sands et al., 1997). The excretion of a dilute urine in the setting of hypercalciuria may be a protective mechanism that reduces the risk of developing calcium-containing renal stones and/or nephrocalcinosis when a concentrated urine is being elaborated because of dehydration. Persons with FHH, on the other hand, concentrate their urine normally despite being hypercalcemic (Marx et al., 1981b). This may occur because of reduced activity of mutant CaRs in the MTAL and/or collecting duct. Impaired Ca^{2+}_{o}-sensing in the MTAL could lead to a reduced or absent inhibitory action of hypercalcemia on NaCl reabsorption and attendant generation of the countercurrent gradient. Alternatively, or in addition, the presence of CaRs with reduced activity in the collecting duct could block the inhibitory action of hypercalcemia on vasopressin action in this nephron segment, likewise mitigating the development of NDI.

B. Neonatal Severe Primary Hyperparathyroidism

NSHPT is a state of symptomatic hypercalcemia with hyperparathyroid bone disease that occurs in children less than six months of age (Heath, 1994). Most cases present at birth or within the first week of life. Failure to thrive, anorexia, constipation, and respiratory distress are the most common presenting clinical features. Respiratory complications, resulting from chest wall deformity or multiple rib fractures due to the associated bone disease, are the major cause of morbidity (Marx et al., 1982, 1985; Heath, 1994). Radiographs show striking demineralization, often with fractures of the ribs and long bones, as well as subperiosteal bone resorption, metaphyseal widening, and, in occasional cases, rickets (Marx et al., 1982; Gaudelus et al., 1983). The degree of hypercalcemia is often severe, ranging from 14 to 20 mg/dl or higher, and values as high as 30.8 mg/dl have been reported (Eftekhari and Yousufzadeh, 1982). Strikingly, relative hypocalciuria has been documented in some cases, even without a family history of FHH (Corbeel et al., 1968). Serum PTH levels are usually high, although the degree of the elevation can be modest (Corbeel et al., 1969; Lutz et al., 1986; Marx et al., 1986; Harris and D'Ercole, 1989). On histological examination the bones can show typical signs of severe hyperparathyroidism, with classical osteitis fibrosa cystica (Matsuo et al., 1982).

NSHPT can be fatal if vigorous medical and surgical management are delayed. In some cases particularly within the past 15 years, however, the disorder runs an apparently self-limited course, with eventual clinical stabilization with a milder degree of hypercalcemia by six to seven months of age with only conservative medical therapy (Marx et al., 1982; Page and Haddow, 1987; Harris and D'Ercole, 1989; Fujitomo et al., 1990; Pomeranz et al., 1992). The initial management of patients with NSHPT should include rapid hydration and aggressive respiratory support, particularly if the integrity of the chest wall is compromised. In very severe cases or in those with clinical deterioration despite vigorous medical therapy, total parathyroidectomy with autotransplantation of part of one of the resected glands is generally recommended during the first month of life (Heath, 1994). Some authors recommend total parathyroidectomy with lifelong medical management of the resultant hypoparathyroidism, since some degree of hypercalcemia will generally persist if parathyroid tissue is left *in situ* (Matsuo et al., 1982; Marx et al., 1986). There is usually a dramatic improvement following parathyroidectomy, with prompt healing of skeletal lesions, even if hypercalcemia recurs following autotransplantation or a less than total parathyroidectomy (Heath, 1994; Matsuo et al., 1982). The parathyroid glands in NSHPT encountered at the time of surgery are uniformly enlarged, sometimes being up to 10 times the mass of those in normal infants, and show chief cell or water-clear cell hyperplasia (Garcia-Banniel et al., 1974; Matsuo et al., 1982; Page and Haddow, 1987; Heath, 1994).

Since the initial descriptions of FHH and NSHPT, several authors have noted the occurrence of NSHPT in families with other members showing the typical features of FHH (Thompson et al., 1978; Marx et al., 1981a, 1985; Matsuo et al., 1982). In 1981 Marx and co-workers reported that among their 15 kindreds with FHH there were three infants from two families with NSHPT. They suggested that NSHPT may be the homozygous form of FHH (Marx et al., 198la). These observations lent credence to the hypothesis that NSHPT represents the homozygous form of FHH and, furthermore, that the apparently sporadic occurrence of NSHPT could in some cases result from failure to identify very mild hypercalcemia in family members who were heterozygous for the abnormality. After the gene for FHH was linked to chromosome 3q, Pollak and co-workers (1994b) showed that among 11 families with FHH mapping to chromosome 3q2, four with consanguineous marriages produced offspring with NSHPT. By analyzing genetic markers closely linked to the FHH gene, these authors showed in these cases that NSHPT represents the homozygous form of FHH. The cloning of the CaR then made it possible to confirm in these kindreds that patients with NSHPT have two copies of the mutated CaR gene (Pollak et al., 1993; Chou et al., 1995; Janicic et al., 1995). As a result, these patients show much more severe hypercalcemia, with marked elevations in PTH and striking parathyroid cell hyperplasia.

Indeed, in two cases, it has been possible to study the function of parathyroid tissues from infants with NSBPT *in vitro* (Cooper et al., 1986; Marx et al., 1986). Both showed markedly increased set-point for $Ca^{2+}{}_{o}$-regulated PTH secretion, further supporting a key role of the CaR in $Ca^{2+}{}_{o}$-sensing by the parathyroid cell. Recent studies, however, have directly documented that heterozygous *de novo* CaR mutations were present in two sporadic cases of neonatal hyperparathyroidism that were sufficiently severe to merit the diagnosis of NSHPT (Pearce et al., 1995). Both showed milder elevations in serum calcium concentration (~12 mg/dl) than generally encountered in the homozygous form of NSHPT but had evidence of hyperparathyroid bone disease. Such cases may provide an explanation for the previously observed clinical spectrum in neonatal severe hyperparathyroidism, ranging as it does from typically severe, life-threatening manifestations (probably resulting in most cases from the homozygous form of the disorder) to a milder, even self-limited form of the disease (i.e., in patients with *de novo* heterozygous CaR mutations) (Page and Haddow, 1987; Harris and D'Ercole, 1989; Pomeranz et al., 1992).

There are several possible explanations for why some neonates heterozygous for inactivating mutations in the CaR present with a phenotype that is more severe than is the norm for FHH. First, if the affected child is born of a father with FHH and a normal mother, the exposure of the fetal parathyroid glands (which have an elevated set-point for Ca^{2+}_{o}) to the relative hypocalcemia of the mother's intrauterine environment may cause some degree of "secondary" hyperparathyroidism similar to that encountered in normal children born of hypoparathyroid mothers (Bai et al., 1997). Secondly, the abnormal CaR allele could produce an unusually severe dominant negative effect on the normal CaR, thereby producing a more pronounced abnormality in Ca^{2+}_{o}-sensing than encountered in most FHH patients, particularly if the child is the product of a normal mother (Bai et al., 1997).

C. Mice with Targeted Disruption of the CaR Gene: An Animal Model for FHH and NSHPT

Ho et al. (1995) reported the development of a mouse with a "knockout" of the CaR gene using homologous recombination. The heterozygous mice show mild hypercalcemia and mild hypermagnesemia with a slight elevation in their serum levels of PTH but are otherwise phenotypically normal. Mice homozygous for deletion of the CaR gene, on the other hand, have more marked hypercalcemia of 14–15 mg/dl, a slightly higher level of serum magnesium than the heterozygotes, and several-fold elevations in serum levels of PTH. They also show markedly reduced growth in the postnatal period and die within the first several weeks after birth. The availability of these animal models of FHH and NSHPT should enable additional studies directed at understanding how the CaR regulates a variety of tissues that express it as well as the consequences of the loss of one or both alleles of the CaR.

D. Autosomal Dominant Hypocalcemia

Autosomal dominant hypocalcemia (ADH) is a familial syndrome characterized by mild hypocalcemia with few or no symptoms and detectable levels of PTH (Pollak et al., 1994a). Since these patients generally tolerate their hypocalcemia well, it had previously been hypothesized that they might have a reduction in the set-point for $Ca^{2+}{}_o$-regulated PTH release (Estep et al., 1981; Pollak et al., 1994a). This would reset their calcium homeostatic system to maintain a subnormal level of $Ca^{2+}{}_o$, an alteration that is the converse of the abnormality in $Ca^{2+}{}_o$-sensing in FHH. Indeed, in one such family with ADH, EDTA infusion provoked PTH secretion, consistent with a downward shift in the set-point of the parathyroid gland (Estep et al., 1981), prompting Pollak et al. (1994a) to search for activating mutations in this family's CaR gene. In one of the probands as well as in all other hypocalcemic family members, they demonstrated a missense mutation (Glu127Ala) in the extracellular domain of the CaR. When expressed in *X. laevis* oocytes, a mutant receptor containing this amino acid substitution produced several-fold higher levels of IP_3 at both low and high $Ca^{2+}{}_o$ than the wild-type receptor. Thus *in vivo* this mutation presumably increases CaR activity inappropriately even at frankly low levels of $Ca^{2+}{}_o$, thereby suppressing PTH secretion and causing hypocalcemia.

Subsequently, several other families with "autosomal dominant hypoparathyroidism" as well as occasional sporadic cases of hypoparathyroidism (Baron et al., 1996; DeLuca et al., 1997) have been reported to have missense mutations in the extracellular domain of the CaR (Finegold et al., 1994; Davies et al., 1995; Pearce et al., 1996; Perry et al., 1995). Pearce et al. (1996) recently studied six families with ADH. In addition to having asymptomatic hypocalcemia with detectable PTH levels, affected family members also had hypomagnesemia and hyperphosphatemia. The reason for classifying these families as having ADH rather than familial isolated hypoparathyroidism was the unusual response of affected members to vitamin D treatment, which led to nephrocalcinosis and renal impairment in several cases (Pearce et al., 1996). Some also complained of thirst and polyuria with normalization of their serum calcium concentrations, possibly due to the development of nephrogenic diabetes insipidus. Five of these families demonstrated mutations in the extracellular domain of the CaR that apparently produce "hypercalcemic" manifestations during normalization of their serum calcium concentration with vitamin D by activating the receptor at normal or even low levels of $Ca^{2+}{}_{0}$. These mutations within the extracellular domain of the CaR may enhance its affinity for $Ca^{2+}{}_{0}$ or mimic the ligand-bound state of the receptor, thereby initiating intracellular signaling at subnormal levels of extracellular calcium or perhaps even in the total absence of calcium. The clinical distinction of these patients from those with sporadic hypoparathyroidism or other familial forms of hypoparathyroidism is of great clinical importance, since patients with ADH may end up with irreversible renal damage if they are treated with sufficient calcitriol to normalize their serum calcium concentration.

VI. ACQUIRED DISORDERS OF $CA^{2+}{}_{0}$-SENSING

The cloning of the CaR and the development of specific anti-CaR antibodies (Garrett et al., 1995c) has made it possible to examine the potential involvement of the CaR in various forms of primary and secondary hyperparathyroidism. Hosokawa et al. (1995) sought mutations similar to those causing FHH in the coding region of the CaR in some 40 parathyroid tumors from patients with various types of hyperparathyroidism, including parathyroid adenoma, carcinoma, as well as various forms of primary and uremic, secondary/tertiairy parathyroid hyperplasia. None had detectable mutations in the CaR indicating that the production of a CaR with an abnormal primary structure must be an uncommon cause of hyperparathyroidism. More recently, Kifor and co-workers (1996) and Gosuser and colleagues (1997) used anti-CaR antibodies to show that there is a substantial (50–60% on average) reduction in the immunoreactivity of the CaR in parathyroid adenomas relative to normal parathyroid glands from the same patients biopsied at the time of parathyroidectomy. A similar finding was observed in hyperplastic glands from individuals with marked uremic secondary/tertiary hyperparathyroidism (Kifor et al., 1996; Gognsev et al., 1997). The basis for this apparent reduction in the level of expression of the CaR in these circumstances is not known, although it has recently been shown that there can be loss of heterozygosity in the region of chromosome 3 where the CaR gene is located in some parathyroid tumors (Thompson et al., 1995). Although loss of one allele of the CaR gene per se has not yet been shown directly in these cases, such allelic loss could reduce the overall level of expression of the CaR, similar to the situation in the mice heterozygous for knockout of the CaR.

VII. DIAGNOSTIC IMPLICATIONS

Detection of mutations in the CaR in patients with sporadic, asymptomatic hypercalcemia could clearly be helpful in diagnosing *de novo* cases of FHH, although the large size of the CaR's coding sequence makes this an arduous undertaking if direct sequencing is carried out. More rapid procedures for detecting point mutations (i.e., the use of denaturing gradient gel electrophoresis) could facilitate mutational analysis in such cases (Pearce et al., 1996). Failure to identify a mutation does not rule out FHH, however, as not all patients with this disorder show CaR mutations, even when the disorder is linked to the locus on chromosome 3. Thus diagnosing FHH will likely continue to involve the time-tested approach of demonstrating an autosomal dominant pattern of inheritance of asymptomatic hypercalcemia in family members in addition to the proband, which is accompanied by relative hypocalciuria (calcium/creatinine clearance ratio of <0.01) (Marx et al., 1981a). Screening of all first degree relatives of individuals with a provisional diagnosis of FHH for hypercalcemia should be carried out to the extent possible, since even borderline hypercalcemic patients may harbor mutations. Such affected persons should be aware of their diagnosis and that they need not undergo surgery for a mistaken diagnosis of mild primary hyperparathyroidism. Combined with family screening for hypercalcemia, analysis for mutations should also be considered in the evaluation of cases presenting with neonatal hyperparathyroidism, especially when there is no family history of hypercalcemia.

The identification of cases of ADH due to activating mutations of the CaR is important, because many of these individuals are at potential risk for well-meaning but ill-advised overtreatment with calcium/vitamin D, which can have deleterious and occasionally irreversible renal consequences if the disorder is not identified. Recognition of these cases necessitates careful clinical and genetic evaluation. ADH should be suspected in cases with a presumed diagnosis of sporadic or familial hypoparathyroidism who tolerate treatment with vitamin D poorly and develop substantial hypercalciuria, so that complications such as nephrocalcinosis and renal failure can be avoided.

VIII. THERAPEUTIC IMPLICATIONS

The recognition that parathyroid cells recognize and respond to changes in $Ca^{2+}{}_{o}$ through a cell surface, G protein-linked receptor led to attempts to develop drugs targeted at the receptor. Calcimimetics, drugs that mimic the effects of high Ca^{2+}_{o} (Fox et al.; 1993; Steffey et al., 1993; Nemeth et al., 1998) on the receptor, for instance, have been developed recently. One such agent, NPS R-568, is currently in the process of undergoing clinical trials for the treatment of primary and secondary hyperparathyroidism (Heath et al., 1995; Silverberg et al., 1997). The latter compound is a small organic molecule which inhibited PTH secretion *in vitro* and produced sustained hypocalcemia in preclinical studies in normal rats. Recent short-term studies in humans have shown similar results (Silverberg et al., 1997). In addition, in rats with mild chronic renal insufficiency due to partial nephrectomy, treatment for four weeks with NPS R-568 prevented the development of secondary hyperparathyroidism, indicating its potential utility in this circumstance in humans. (Wada et al., 1997). Thus NPS R-568 could represent a significant advance in the treatment of both primary and secondary hyperparathyroidism. In addition to inhibiting PTH secretion, it is conceivable that calcimimetic agents might also reduce the level of expression of the PTH gene and even prevent or reverse parathyroid cellular proliferation, if it turns out that the CaR regulates the latter two processes.

IX. SUMMARY

The cloning of a G protein-coupled CaR directly documents that a variety of cells can directly recognize and respond to small changes in Ca^{2+}_{o} through a receptor-mediated mechanism similar to that through which cells respond to a wide variety of hormones, neurotransmitters, and other extracellular messengers (i.e., PTH or CT). Thus Ca^{2+}_{o} acts as an extracellular, first messenger in addition to serving its better recognized function as a nearly universal intracellular second messenger. Several tissues that express the CaR are key components of the mineral ion homeostatic system that have been known to sense Ca^{2+}_{o} for many years (e.g., parathyroid and C-cells). The presence of the CaR on several cell types within the kidney, however, strongly suggests that several of the poorly understood direct effects of Ca^{2+}_{o} on kidney function might be mediated by the CaR. These actions include the increase in the urinary excretion of calcium and magnesium ions in the setting of hypercalcemia, which enhances the reduction of renal tubular calcium reabsorption that results from high $[Ca^{2+}_{o}]$-mediated suppression of PTH secretion. The decrease in maximal renal concentrating ability observed in some hypercalcemic patients likely indicates a functionally important relationship between the homeostatic systems regulating renal calcium and water handling, whose purpose is to reduce the risk of abnormal renal deposition of calcium salts in circumstances where excess urinary calcium loads must be disposed of. Further support for the role of the receptor in regulating renal function and for coordinating renal handling of calcium and water has come from studying human syndromes of $Ca^{2+}{}_{o}$ resistance or overresponsivenss due to loss-of-function or gain-of-function mutations in the CaR, respectively. There is still a great deal to be learned, however, about the role of the CaR in locations such as the brain, where it appears that it responds to changes in the local rather than the systemic level of $Ca^{2+}{}_{o}$. The development of drugs that activate or inhibit the CaR have great potential utility in treating various conditions in which the receptor is either under- or overactive. Finally, there may well be additional receptors for $Ca^{2+}{}_{o}$ (Malgaroli et al., 1989; Zaidi et al., 1989; Lundgren et al., 1994; Saito et al., 1994; Quarles et al., 1997) or for other ions (indeed, the CaR may also function as a physiologically important $Mg^{2+}{}_{o}$-receptor). Such ion receptor/sensors, in turn, could malfunction in certain disease states and be amenable to pharmacological manipulation with appropriate therapeutics.

ACKNOWLEDGMENTS

The authors gratefully acknowledge the generous grant support provided by the USPHS [DK41415, 44588, 46422, and 52005 (to E.M.B.) and 48330 (to E.M.B. and S.C.H.)], the St. Giles Foundation and NPS Pharmaceuticals, Inc., Salt Lake City, UT.

REFERENCES

Abou-Samra, A.B., Juppner, H., Force, T., Freeman, M.W., Kong, X.F., Schipani, E., Urena, P., Richards, J., Bonventre, JV, Potts, Jr., J.T. and Kronenberg, H.M. (1992). Expression cloning of a common receptor for parathyroid hormone and parathyroid hormone-related peptide from rat osteoblast-like cells: a single receptor stimulates intracellular accumulation of both cMP and inositol triphosphates and increases intracellular free calcium. Proc. Natl Acad. Sci. U.S.A. 89, 2732-2736.

Adams, M.D. and Oxender, D.L. (1989). Bacterial periplasmic binding protein tertiary structures. J. Biol. Chem. 264, 15739-15742.

Aida, K., Kois, S., Tawata, M. and Onaya, T. (1995). Molecular cloning of a putative Ca^{2+}-sensing receptor cDNA from human kidney. Biochem. Biophys. Res. Commun. 214, 524-529.

Attie, M.F., Gill, Jr., J.R., Stock, J.L., Spiegel, A.M., Downs, Jr., R.W., Levine, M.A., and Marx, S.J. (1983). Urinary calcium excretion in familial hypocalciuric hypercalcemia. J. Clin. Invest. 72, 667-676.

Aurbach, G.D., Marx, S.J., and Spiegel, A.M. (1985). Parathyroid hormone, calcitonin, and the calciferols. In: Textbook of Endocrinology, 7th ed. (Wilson, J.D. and Foster, D.W., Eds.), pp. 1137-1217. Saunders, Philadelphia, PA.

Austin, L.A. and Heath, H. (1981). Calcitonin: Physiology and Pathophysiology. N. Engl. J. Med. 304, 269-278.

Auxwerx, J., Demedts, M., and Bouillon, R. (1984). Altered parathyroid set point to calcium in familial hypocalciuric hypercalcemia. Acta Endocrinol. 106, 215-218.

Bai, M., Quinn, S., Trivedi, S., Kifor, O., Pearce, S.H.S., Pollak, M.R., Krapcho, K., Hebert, S.C., and Brown, E.M. (1996). Expression and characterization of inactivating and activating mutations of the human Ca^{2+}_{o}-sensing receptor. J. Biol. Chem. 271, 19537-19545.

Bai, M., Pearce, S.H.S., Kifor, O., Trivedi, S., Stauffer, U.G., Thakker, R.V., Brown, E.M., and Steinmann, B. (1997). In vivo and in vitro characterization of neonatal hyperparathyrodism resulting from a de novo, heterozygous mutations in the Ca^{2+}-sensing receptor gene: Normal maternal calcium homeostasis as a cause of secondary hyperparathyroidism in familial benign hypocalciuric hypercalcemia. J. Clin. Invest. 99, 88-96.

Baron, J., Winer, K.K., Yanovski, J.A., Cunningham, A.W., Laue, L., Zimmerman, D., and Cutler, Jr., G.B. (1996). Mutations in the Ca^{2+}-sensing receptor gene cause autosomal dominant and sporadic hypoparathyroidism. Human Mol. Genet. 5, 601-606.

Brown, E.M. (1991). Extracellular Ca^{2+} sensing, regulation of parathyroid cell function, and role of Ca^{2+} and other ions as extracellular (first) messengers. Physiol. Rev. 71, 371-411.

Brown, E.M., Gamba, G., Riccardi, D., Lombardi, D., Butters, R., Kifor, O., Sun, A., Hediger, M.A., Lytton, J, and Hebert, S.C. (1993). Cloning and characterization of an extracellular Ca^{2+}-sensing receptor from bovine parathyroid. Nature 366, 575-580.

Brown EM, Pollak M, Hebert SC 1994 Cloning and characterization of extracellular Ca^{2+}-sensing receptors from parathyroid and kidney: Molecular physiology and pathophysiology of Ca^{2+}-sensing. The Endocrinologist 4:419-426.

Brown, E.M., Pollak, M., Seidman, C.E., Seidman, J.G., Chou, Y.-H.W., Riccardi, D., and Hebert, S.C. (1995). Calcium-ion-sensing cell-surface receptors. New Eng. J. Med. 333, 234-240.

Chattopadhyay, N., Cheng, I., Rogers, K., Riccardi, D., Hall, A., Diaz, R., Hebert, S.C., Soybel, D.I., and Brown, E.M. (1998). Identification and localization of extracellular Ca^{2+}-sensing receptor in rat intestine. Am. J. Physiol. 274, G122-G130.

Chen, T.-H., Pratt, S.A., and Shoback, D.M. (1994). Injection of bovine parathyroid poly(A), RNA into *Xenopus* oocytes confers sensitivity to extracellular calcium. J. Bone Mineral Res. 9, 293-300.

Chou, Y.-H.W., Brown, E.M.. Levi, T., Crowe, G., Atkinson, A.B,, Amquist, H.J., Toss, G., Fuleihan, G.E., Seidman, J.G., and Seidman, C.E. (1992). The gene responsible for familial hypocalciuric hypercalcernia maps to chromosome 3 in four unrelated families. Nature Genet. 1, 295-300.

Chou, Y.-H.W., Pollak, M.R., Brandi, M.L., Toss, G., Arnqvist, H., Atkinson, A.B., Papapoulos, S.E., Marx, S.J., Brown, E.M., Seidman, J.G., and Seidman, C.E. (1995). Mutations in the human calcium-sensing receptor gene. Am. J. Hum. Genet. 56, 1075-1079.

Conklin, B.R. and Bourne, H.R. (1994). Marriage of the flytrap and the serpent. Nature 367, 22.

Cooper, L., Wertheimer, J., Levey, R., Brown, E., LeBoff, M., Wilkinson. R., and Anast, C. (1986). Severe primary hyperparathyroidism in a neonate with two hypercalcemic parents: Management with parathyroidectomy and heterotopic autotransplantation. Pediatrics 78, 263-268.

Corbeel, L., Casaer, P., Malvaux, P., Lormans, J., and Bourgeois, N. (1968). Hyperparathyroidie congeritale. Arch. Fr. Pediatr. 25, 879-891.

Davies, M., Adams, P.H., and Lumb, G.A. (1984). Familial hypocalciuric hypercalcemia: Evidence for continued enhanced renal tubular reabsorption of calcium following total parathyroidectomy Acta Endocrinol. 106, 499-504.

Davies, M., Mughal, Z., Selby, P.L., Tymms, D.J., and Mawer, E.B. (1995). Familial benign hypocalcemia. J. Bone Mineral Res. 10 (Suppl. 1), S507 (Abstract.)

De Luca, F., Ray, K., Mancilla, E.E., Fan, G.-F., Winer, K.K., Gore, P., Spiegel, A.M., and Baron, J. (1997). Sporadic hypoparathyroidism caused by de novo gain-of-function mutations in the Ca^{2+}-sensing receptor. J. Clin. Endocrinol. Metab. 82, 2710-2715.

Di Stefano, A., De Rouffignac, C., and Wittner, M. (1993). Transepithelial Ca^{2+} and Mg^{2+} transport in the cortical thick ascending limb of Henle's loop of the mouse is a voltage-sensitive process. Renal Physiol. Biochem. 16. 157-166.

Edvall, C.A. (1958). Renal function in hyperparathyroidism: A clinical study of 30 cases with special reference to selective renal clearance and renal vein catheterization. Acta Chir. Scand. 229, 1-54.

Eftekhari, F. and Yousufzadeh, D.K. (1982). Primary infantile hyperparathyroidism: Clinical, laboratory, and radiographic features in 21 cases. Skel. Radiol. 8, 201-208.

Emanuel, R.L., Adler, G.K., Krapcho, K., Fuller, F., Quinn, S.J., and Brown, E.M. (1996). Calcium-sensing receptor expression and regulation in AtT-20 pituitary cell line. Molec. Endocrinol. 10, 555-565.

Estep. H.L., Mistry, Z., and Burke, P.K. (1981). Familial idiopathic hypocalcemia. Program and Abstracts, 63rd Annual Meeting of The Endocrine Society, Cincinnati, OH. p. 275 (Abstract #750.)

Finegold, D.N., Armitage, M.M., Galiani, M., Matise, T.C., Pandian, MR., Perry, Y.M., Deka, R., and Ferrell, R.E. (1994). Preliminary localization of a gene for autosomal dominant hypoparathyroidism to chromosome 3ql3. Pediatir. Res. 36, 414-417.

Foley, T.P., Harrison, H.C., Arnaud, C.D., and Harrison, H.E. (1972). Faniilial benign hypercalcemia. J. Pediatr. 81, 1060-1067.

Fox, J., Petty., B.A., and Nemeth, E.F. (1993). A first generation calciniimetic compound (NPS R-568) that acts on parathyroid cell calcium receptor: A novel approach for hyperparathyroidism. J. Bone Mineral Res. 8 (Suppl. 1), S181(Abstract.)

Fujitomo, Y., Hazama, H., and Oku, K. (1990). Severe primary hyperparathyroidism in a neonate having a parent with hypercalcemia: Treatment by total parathyroidectomy and simultaneous heterotopic autotransplantation. Surgery 108, 933-938.

Gama, L., Baxenlale-Cox, L.M., and Breitwieser, G.E. (1997). Ca^{2+}-sensing receptor in intestinal epithelium. Am. J. Physiol. 273, C1168-C1175.

Garcia-Banniel, R., Kutchemeshgi, A., and Brandes, D. (1974). Hereditary hyperparathyroidism. The fine structure of the parathyroid gland. Arch. Pathol. 97, 399-403.

Garrett, J.E., Capuano, I.V., Hammerland, L.G., Hung, B.C.P., Brown, E.M., Hebert, S.C., Nemeth, E.F., and Fuller, F. (1995a). Molecular cloning and characterization of the human parathyroid calcium receptor. J. Biol. Chem. 270. 12919-12925.

Garrett, J.E., Steffey, M.E., and Nemeth, E.F. (1995b). The calcium receptor agonist R-568 suppresses PTH mRNA levels in cultured bovine parathyroid cells. J. Bone Mineral Res. 10, S387 (Abstract, M539.)

Garrett, J.E., Tamir, H., Kifor, O., Simin, R.T., Rogers, K.V., Mithal, A., Gagel, R.F., and Brown, E.M. (1995c). Calcitonin-secreting cells of the thyroid gland express an extracellular calcium-sensing receptor gene. Endocrinology 136, 5202-5211.

Gaudelus, J., Dandine, M., Nathanson, M., Perelman, R., and Hassan, M. (1983). Rib cage deformity in neonatal hyperparathyroidism. Am. J. Dis. Child 137, 408-409.

Gogusev, J., Duchambon, P., Hory, B., Giovannini, M., Goureau, Y., Sarfati E., and Drueke, T. (1997). Depressed expresson of calcium receptor in parathyroid gland tissue of patients with hyperparathyroidism. Kidney Int. 51, 328-336.

Gunn, I.R. and Wallace, J.R. (1992). Urine calcium and serum ionized calcium, total calcium and parathyroid hormone concentrations in the diagnosis of primary hyperparathyroidism and familial benign hypercalcemia. Ann. Clin. Biochem. 29, 52-58

Hammerland, L.G., Krapcho, K.J., Alasti, N., Garrett, J.E., Capuano, I.V., Hung, B.C.P., and Fuller, F.H. (1995). Cation binding determinants of the calcium receptor revealed by functional analysis of chimeric receptors and a deletion mutant. J. Bone Mineral Res. 10, S156 (Abstract, 69.)

Harris, S.S. and D'Ercole, A.J. (1989). Neonatal hyperparathyroidism: The natural course in the absence of surgical intervention. Pediatrics 83, 53-56.

Heath, D.A. (1994). Familial hypocalciuric hypercalcemia. In: The Parathyroids. (Bilezikian, J.P., Marcus, R., and Levine, M.A., Eds.), pp. 699-710. Raven Press, New York.

Heath, III, H., Odelberg, S., Brown, D., Hillm, V.M., Robertson, M., Jackson, C.E., Teh, B.J., Hayward, N., Larsson, C., Buist, N, Garrett, J., and Leppert, M. (1994). Sequence analysis of the parathyroid cell calcium receptor (CaR) gene in familial benign hypercalcemia (FBH): A Multiplicity of Mutations? J. Bone Min. Res. 9 (Suppl. 1), S414 (Abstract.)

Heath, III., H., Jackson, C., Otterud, B., and Leppert, M. (1993). Genetic linkage analysis of familial benign (hypocalciuric) hypercalcemia: Evidence for locus heterogeneity. Am. J. Hum. Genet. 53, 193-200.

Heath, III, H. (1989). Familial benign (hypocalciuric) hypercalcemia. A troublesome mimic of primary hyperparathyroidism. Endocrinol. Metab. Clin. North Am. 18, 723-740.

Heath, III, H., Sanguinetti, E.L., Oglesby, S., and Marriott, T.B. (1995). Inhibition of human parathyroid hormone secretion in vivo by NPS R-568, a calcimimetic drug that targets the parathyroid-cell surface calcium receptor. Bone 16, (Suppl. 1), 85S.

Hebert, S.C. and Andreoli, T.E. (1984). Regulation of NaCl transport in the thick ascending limb. Am. J. Physiol 246, F745-F756.

Hebert. S.C. and Brown, E.M. (1996). The scent of an ion-calcium-sensing and its roles in health and disease. Curr. Opin. Nephrol. Hypertens. 5, 45-52.

Heinemann, U., Lux, H.D., and Gutnick, M.J. (1977). Extracellular free calcium and potassium during paroxysmal activity in the cerebral cortex of the rat. Expt. Brain Res. 27, 237-243.

Ho, C., Conner, D.A., Pollak, M., Ladd, D.J., Kifor, O., Warren, H., Brown, E.M., Seidman, C.E., and Seidman, J.G. (1995). A mouse model for familial hypocalciuric hypercalcemia and neonatal severe hyperparathvroidism. Nature Genet. 11, 389-394.

Hosokawa, Y., Pollak, M.R., Brown, E.M., and Arnold, A. (1995). Mutational analysis of the extracellular Ca^{2+}-sensing receptor gene in human parathyroid tumors. J. Clin. Endocrinol. Metab. 80, 3107-3110.

Humes, H.D., Ichikawa, I., Troy, JL., and Brenner, B.M. (1978). Evidence for a parathyroid hormone-dependent influence of calcium on the glomerular ultrafiltration coefficient. J. Clin. Invest. 61, 32-40.

Janicic, N., Pausova, Z., Cole, D.E.C., and Hendy, G.N. (1995). Insertion of an alu sequence in the Ca^{2+}-sensing receptor gene in familial hypocalciuric hypercalcemia and neonatal severe hyperparathyroidism. Am. J. Hum. Genet. 56, 880-886.

Jones, S.M., Frindt, G., and Windhager, E. (1988). Effect of peritubular [Ca] or ionomycin on hydrosmotic response of CCD to ADH or CAMP. Am. J. Physiol. 254, F240-F253.

Kameda, T., Mano, H., Yamada, Y., Takai, H., Amizuka, N., Kobori, M., Izumi, N., Kawashima, H., Ozawa, H., Ikeda, K., Kameda, A., Hakeda, Y., and Kumegawa, M. (1998). Calcium-sensing receptor in mature osteoclasts, which are bone resorbing cells. Biochem. Biophys. Res. Commun. 245, 419-422.

Kaupman, K., Huggel, K., Heid, J., Flor, P.J., Bischoff, S., Kickel, S.J., McMaster, C., Angst, C., Bittiger, H., Froestl, W., and Bettler, B. (1997). Expression cloning of $GABA_B$ receptors uncovers similarity to metabotropic glutamate receptor. Nature 386, 239-246.

Khosla, S., Ebeling, P.R., Firek, A.F., Burritt, M.M., Kao, P.C., and Heath, H. III, (1993). Calcium infusion suggests a "set-point" abnormality of parathyroid gland function in familial benign hypercalcemia and more complex disturbances in primary hyperparathyroidism. J. Clin. Endocrinol. Metab. 76, 715-720.

Kifor, O., Congo, D., and Brown, E.M. (1990). Phorbol esters modulate the high $Ca^{2+}{}_0$-stimulated accumulation of inositol phosphates in bovine parathyroid cells. J. Bone Miner. Res. 5, 1003-1011.

Kifor, O., Moore, Jr., F.D., Wang, P., Goldstein, M., Vassilev, P., Kifor, I., Hebert, S.C., and Brown, E.M. (1996). Reduced immunostaining for the extracellular Ca^{2+}-sensing receptor in primary and uremic secondary hyperparathyroidism. J. Clin. Endocrinol. Metab. 81, 1598-1606.

Kristiansen, J.H., Rodbro, P., Christiansen, C., Brochner-Mortensen, J., and Carl, J. (1985). Familial hypocalciuric hypercalcemia II: Intestinal calcium absorption and vitamin D metabolism. Clin. Endocrinol. 23, 511-515.

Kurokawa, K. (1994). The kidney and calcium homeostasis. Kidney Int. 45 (Suppl.44), S97-S105.

Law, Jr., W.M., Bollman, S., Kumar, R., and Heath, III, H. (1984). Vitamin D metabolism in familial benign hypercalcemia (hypocalciuric hypercalcemia) differs from that in primary hyperparathyroidism. J. Clin. Endocrinol. Metab. 58, 744-747.

Law, Jr., W.M. and Heath, III, H. (1985). Familial benign hypercalcemia (Hypocalciuric hypercalcemia). Clinical and pathogenic studies in 21 families. Ann. Int. Med. 102, 511-519.

Lin, H.Y., Harris, T.L.. Flannery, M.S., Aruffo, A., Kaji, F.H., Gom, A., Kolakowski, Jr., L.F., Lodish, H.F., and Goldring, S.R. (1991). Expression cloning of an adenylate cyclase-coupled calcitonin receptor. Science 254, 1022-1024.

Lundgren, S., Hjalm, G., Hellman, P., Ek, B., Juhlin, C., Rastad, J., Klareskog, L., Akerstrom, G., and Rask, L. (1994). A protein involved in calcium sensing of the human parathyroid and placental cytotrophoblast cells belongs to the LDL-receptor protein superfamily. Exper. Cell Res. 212, 344-350.

Lutz, P., Kane, O., Pfersdorff, A., Seiller, F., Sauvage, P., and Levy, J.M. (1986). Neonatal primary hyperparathyroidism. Total parathyroidectomy with autotransplantation of cryopreserved parathyroid tissue. Acta Pediatr. Scand. 75, 179-182.

Malgaroli, A.,. Meldolesi, J., Zambone-Zallone, A., and Teti, A. (1989). Control of cytosolic free calcium in rat and chicken osteoclasts. The role of extracellular calcium and calcitonin. J. Biol. Chem. 264, 14342-14349.

Marx, S.J., Attie, M.F., Levine, M.A., Spiegel, A.M., Downs, Jr., R.W., and Lasker, R.D. (1981a). The hypocalciuric or benign variant of familial hypercalcemia: Clinical and biochemical features in fifteen kindreds. Medicine (Baltimore) 60, 397-412.

Marx, S.J., Attie, M.F., Stock, J.L., Spiegel, A.M., and Levine, M.A. (1981b). Maximal urine-concentrating ability: Familial hypocalciuric hypercalcemia versus typical primary hyperparathyroidism. J. Clin. Endocrinol. Metab. 52, 736-740.

Marx, S.J., Attie, M.F., Spiegel, A.M., Levine, M.A., Lasker, R.D., and Fox, M. (1982). An association between neonatal severe primary hyperparathroidism and familial hypocalciuric hypercalcemia in three kindreds. N. Eng. J. Med. 306, 257-284.

Marx, S.J., Fraser, D., and Rapoport, A. (1985). Familial hypocalciuric hypercalcemia. Mild expression of the disease in heterozygotes and severe expression in homozygotes. Am. J. Med. 78, 15-22.

Marx, S., Lasker, R., Brown, E., Brown, E., LeBoff, M., Wilkinson, R., and Anast, C. (1986). Secretory dysfunction in parathyroid cells from a neonate with severe primary hyperparathyroidism. J. Clin. Endocrinol. Metab. 62, 445-449.

Matsuo, M., Okita, K., Takene, H., and Fujita, T. (1982). Neonatal primary hyperparathyroidism in familial hypocalciuric hypercalcemia. Am. J. Dis. Child 136, 728-731.

McMurtry, C.T., Schranck, F.W., Walkenhorst, D.A., Murphy, W.A., Kocher, D.B., Teitelbaum, S.L., Rupich, R.C., and Whyte, M.P. (1992). Significant developmental elevation in serum parathyroid hormone levels in a large kindred with familial benign (hypocalciuric) hypercalcemia. Am. J. Med. 93, 247-258.

Morel, F., Chabardes, D., Imbert-Teboul, M., Le Bouffant, F., and Hus Citharel, A. (1982). Multiple hormonal control of adenylate cyclase in distal segments of the rat kidney. Kid. Intern. (Suppl. II), 555-557.

Nakanishi, S. (1992). Molecular diversity of glutamate receptors and implications for brain function. Science 258, 597-603.

Nemeth, E.F. (1995). Ca2+ receptor–dependent regulation of cellular functions. News in Physiol. Sci. 10, 1-5.

O'Hara, P.J., Sheppard, P.D., Thogerson, T., Venezia, D., Haldeman, B.A., McGrane, V., Houamed, K.M., Thomsen, C., Gilbert, T.L., and Mulvihill, E.R. (1993). The ligand binding domain in metabotropic glutamate receptors is related to bacterial periplasmic binding proteins. Neuron 11, 41-52.

Page, L.A. and Haddow, J.E. (1987). Self-limited neonatal hyperparathyroidism in familial hypocalciuric hypercalcemia. J. Pediatr. 111, 261-264.

Parfitt, A.M. and Kleerekoper, M. (1980). The divalent ion homeostatic system: Physiology and metabolism of calcium, phosphate, magnesium, and bone. In: Clinical Disorders of Fluid and Electrolyte Metabolism, 3rd edn., (Maxwell, M.H. and Kleeman, C.R., Eds.), pp. 269-398, MacGraw-Hill, New York.

Pearce, S.H.S., Trump, D., Wooding, C., Besser, G.M., Chew, S.L., Grant, D.B., Heath, D.A., Hughes, I.A., Petterson, C.R., Whyte, M.P., and Thakker, R.V. (1995).

Calcium-sensing receptor mutations in familial benign hypercalcemia and neonatal hyperparathyroidism. J. Clin. Invest. 96, 2683-2692.

Pearce, S.H.S., Williamson, C., Kifor, O., Bai, M., Coulthard, M.G., Davies, M., Lewis-Barned, N., McCredie, D., Powell, H., Kendall-Taylor, P., Brown, E.M., and Thakker, R.V. (1996). A familial syndrome of hypocalcemia with hypercalciuria due to mutations in the calcium-sensing receptor. N. Engl. J. Med. 335, 1115-1122.

Perry, Y.M., Finegold, D.N., Armitage, M.M., and Ferrell, R.E. (1995). A missense mutation in the Ca-sensing receptor gene causes familial autosomal dominant hypoparathyroidism. Am. J. Hum. Genet. 55 (Suppl.), (Abstract #79.)

Pietrobon, D., DiVirgilio, F., and Pozzan, T. (1990). Structural and functional aspects of calcium homeostasis in eukaryotic cells. Eur. J. Biochem. 120, 599-622.

Pollak, M., Brown, E.M., Chou, Y.-H.W., Hebert, S.C., Marx, S.J., Steinmann, B., Levi, T., Seidman, C.E., and Seidman, J.G. (1993). Mutations in the human Ca^{2+}-sensing receptor gene cause familial hypocalciuric hypercalcemia and neonatal severe hyperparathyroidism. Cell 75, 1297-1303.

Pollak, M., Brown, E.M., Kifor, O., Estep, H., Seidman, C., and Seidman, J.G. (1994a). Autosomal dominant hypocalcemia due to an activating mutation in the human extracellular Ca^{2+}-sensing receptor gene. Nature Genet. 8, 303-308.

Pollak, M., Chou, Y.-H.W., Marx, S.J., Steinmann, B., Cole, D.E.C., Brandi, M.L., Papapoulos, S., Menko, F., Hendy, G.N., Brown, E.M,, Seidman, C.E., and Seidman, J.G. (1994b). Familial hypocalciuric hypercalcemia and neonatal severe hypercalcemia: The effects of mutant gene dosage on phenotype. J. Clin. Invest. 93, 1108-1112.

Pomeranz, A., Wolacti. B., Raz. A., and Ben Ari, Y. (1992). Neonatal hyperparathyroidism. Conservative treatment with intravenous and oral rehydration solutions. Child Nephrol. Urol. 12, 55-58.

Quamme, G.A. (1989). Control of magnesium transport in the thick ascending limb. Am. J. Physiol. 256, F197-F210.

Quarles, L.D. (1997). Cation-sensing receptors in bone: A novel paradigm for regulating bone remodling? J. Bone Miner. Res. 12, 1971-1974.

Racke, F.K., Hammerland, J.G., Dubyak, G.R., and Nemeth, E.F. (1993). Functional expression of the parathyroid calcium receptor in *Xenopus* oocytes. FEBS Lett. 333, 132-136.

Raisz, L.G. and Kream, B.E. (1983). Regulation of Bone Formation. N. Engl. J. Med. 309, 35-39.

Randall, C. and Lauchlan, S.C. (1963). Parathyroid hyperplasia in an infant. Am. J. Dis. Child. 105, 364-367.

Riccardi, D., Hall, A.E., Chattopadhyay, N., Xu, J. Brown, E.M. and Hebert, S.C. (1998). Localization of the extracellular CA^{2+}/(polyvalent cation)-sensing receptor protein in rat kindey. Am. J. Physiol. 274, F611-F622.

Riccardi, D., Lee, W.-S., Lee, K., Segre, G.V., Brown, E.M., and Hebert, S.C. (1996). Localization of the extracellular Ca^{2+}-sensing receptor and PTH/PTHrP receptor in rat kidney. Am. J. Physiol. 271, F951-F956.

Riccardi, D., Park, J., Lee, W.-S., Gamba, G., Brown, E.M., and Hebert, S. (1995a). Cloning and functional expression of a rat kidney extracellular calcium-sensing receptor. Proc. Natl. Acad. Sci. U.S.A. 92, 131-135.

Rogers, K.V., Dunn, C.E., Brown, E.M., and Hebert, S.C. (1997). Localization of calcium receptor mRNA in the adult rat central nervous system by in situ hybridization. Brain. Res. 744, 47-56.

Ruat, M., Molliver, M.E., Snowman, A.M., and Snyder, S.H. (1995). Calcium-sensing receptor: Molecular cloning in rat and localization to nerve terminals. Proc. Natl. Acad. Sci. U.S.A. 92, 3161-3165.

Ryba, N.J.P. and Trindell, R. (1997). A new multigene family of putative pheromone receptors. Neuron 19, 371-379.

Saito, A., Pietromonaco, S., Loo, A.K., and Farquhar, M.G. (1994). Complete cloning and sequencing of rat gp330/"megalin", a distinctive member of the low density lipoprotein receptor gene family. Proc. Natl. Acad. Sci. U.S.A. 91, 9725-9729.

Sands, J.M., Naruse, M., Baum, M., Jo, I., Hebert, S.C., Brown, E.M., and Harris, W.H. (1997). Apical extracellular calcium/polyvalent cation-sensing receptor regulates vasopressin-elicited water permeability in rat kidney inner medullary collecting duct. J. Clin. Invest. 99, 1399-1405.

Sharff, A.J., Rodseth, L.E., Spurlino, J.C., and Quiocho, F.A. (1992). Crystallographic evidence for a large ligand-induced tinge-twist motion between the two domains of the maltodextrin binding protein involved in active transport and chemotaxis. Biochemistry 31, 10657-10663.

Silverberg, S.J., Bone, III, H.G., Marriott, T.B., Locker, F.G., Thys-Jacobs, S., Dziem, G., Kaatz, S., Sanguinetti, E.L., and Bilezikian, J.P. (1997). Short-term inhibition of parathyroid hormone secretion by a calcium-receptor agonist in patients with primary hyperparathyroidism. N. Engl. J. Med. 337, 1506-1510.

Spiegel, A.M., Harrison, H.E., Marx, S.J., Brown, E.M., and Aurbach, G.D. (1977). Neonatal primary hyperparathyroidism with autosomal dominant inheritance. J. Pediatr. 90, 269-272.

Steffey, M.E., Fox, J., VanWagenen, B.C. et al. (1993). Calcimimetics: Structurally and mechanistically novel compounds that inhibit hormone secretion from parathyroid cells. J. Bone Min. Res. 8 (Suppl. 1), S175 (Abstract.)

Stuckey, B.G.A., Kent, G.N., Gutteridge, D.H., and Reed, W.D. (1990). Familial hypocalciuric hypercalcemia and pancreatitis: No causal link proven. Aus. N.Z. J. Med. 20, 718-719.

Suki, W.M., Eknoyan, G., and Rector, F.C. (1969). The renal diluting and concentrating mechanism in hypercalcemia. Nephron 6, 50-61.

Takaichi, K., Uchida, S., and Kurokawa, K. (1986). High Ca^{2+} inhibits AVP-dependent cAMP production in the thick ascending limb. Am. J. Physiol. 250, F770-776.

Takaichi, K. and Kurokawa, K. (1988). Inhibitory guanosine triphosphate-binding protein-mediated regulation of vasopressin action in isolated single medullary tubules of mouse kidney. J. Clin. Invest. 82, 1437-1444.

Thompson, D.B., Samawitz, W., Odelberg, S., Szabo, J., and Heath, III, H. (1995). Genetic abnormalities in sporadic parathyroid adenomas: Loss of heterozygosity for chromosome 3q markers flanking the calcium receptor locus. J. Clin. Endocrinol. Metab. 80, 3377-3380.

Thompson, N.W., Carpenter, L.C., Kessler, D.L., and Nishiyama, R.H. (1978). Hereditary neonatal hyperparathyroidism. Arch. Surg. 113, 100-103.

Trump, D., Whyte, M.P., Wooding, C., Pang, J.T., Pearce, S.H.S, Kocher, D.B., and Tbakker, R.V. (1995). Linkage studies in a kindred from Oklahoma with familial benign (hypocalciuric) hypercalcemia and developmental elevations in serum parathyroid hormone levels, indicate a third locus for FHH. Human Genet. 96, 183-187.

Wada, M., Furuya, Y., Sakiyama, J.-i., Kobayashi, N., Miyata, S., Ishii, H., and Hagano, N. (1997). The calcimimetic compound NPS R-568 suppresses parathyroid cell prolification in rats with renal insufficiency. J. Clin. Invest. 100, 2977-2983.

Wang, W.-H., Lu, M., and Hebert, S.C. (1996). Cytochrome P-450 metabolites mediate extracellular Ca^{2+}-induced inhibition of apical K^{+} channels in the TAL. Am. J. Physiol. 271, C103-C111.

Weisinger, J.R., Favus, M.J., Langman, C.B., and Bushinsky, D. (1989). Regulation of 1,25-dihydroxyvitarnin D by calcium in the parathyroidectomized, parathyroid hormone-replete rat. J. Bone. Min. Res. 4, 929-935.

Yamaguchi, T., Kifor, O., Chattopadhyay, N., and Brown, E.M. (1998). Expression of extracellular calcium ($Ca^{2+}{}_{0}$)-sensing receptor in the clonal osteoblastlike cell lines, UMR-106 and SAOS-2. Biochem. Biophys. Res. Commun. 243, 753-757.

Zaidi, M., Datta, H.K., Patchell, A., Moonga, B., and MacIntyre, L. (1989). "Calcium-activated" intracellular calcium elevation: A novel mechanism of osteoclast regulation. Biochem. Biophys. Res. Commun. 183, 1461-1465.

BONE DISEASE IN MALIGNANCY

Brendan F. Boyce, Toshiyuki Yoneda, and

Theresa A. Guise

Advances in Organ Biology
Volume 5C, pages 709-738.

ISBN: 0-7623-0390-5

I. INTRODUCTION

A firm grasp of the sequence of cellular events involved in the process of bone remodeling is essential for understanding the effects of tumor cells on the skeleton. These have been described in detail earlier and will not be repeated here. However, since we will be describing a number of models in which growing animals are used to examine the effects of tumors on bone, bone modeling and aspects of bone remodeling relevant to the effects of malignancy will be covered briefly.

Long bones increase in length through the process of endochondral ossification in which cancellous and surrounding cortical bone are laid down at the epiphyseal growth plates near the ends of the bones. They increase in width by periosteal apposition of matrix by osteoblasts, and the bone marrow space is increased in diameter in proportion to the length of the growing bone by endosteal resorption of cortical bone by osteoclasts. New cancellous bone matrix laid down at sites of resorption of calcified cartilage immediately adjacent to the growth plate is quickly remodeled and then removed by fairly aggressive osteoclastic resorption as the bone grows. Defects in osteoclast generation or function result in build-up of cancellous bone within the medullary cavity—the hallmark of osteopetrosis.

After epiphyseal closure in humans, trabecular bone remains at the ends of long bones and, like cancellous bone in the axial skeleton, undergoes remodeling throughout life. Hematopoietic tissue also persists at these sites and within the medullary cavities of vertebral bodies and likely cooperates with bone cells in regulating remodeling in these bones. As will be seen later, the intense resorption of new bone matrix near growth plates and the more leisurely resorption of remodeling bone elsewhere are associated with the release of growth factors into the local microenvironment which can promote the proliferation of metastatic tumor cells preferentially at these sites. In contrast, the cavity of the shafts of long bones in normal humans is devoid of trabecular bone, is filled with fatty marrow, and is much less frequently the site of metastatic tumor deposits than the ends.

II. LOCAL EFFECTS OF TUMOR CELLS ON BONE

Cancers such as breast, lung, and prostate exhibit a special predisposition to spread to bone and typically cause osteolytic (breast and lung), osteoblastic (prostate), or mixed osteolytic and osteoblastic bone metastases (Mundy and Martin, 1993). In addition to these solid tumors, myeloma also causes extensive bone destruction and hypercalcemia (Mundy, 1995). Despite some studies indicating that breast cancer cells can resorb bone fragments *in vitro* (Eilon and Mundy, 1978), most investigators now agree that the bone is resorbed by osteoclasts, rather than tumor cells (Figure 1) (Francini et al., 1993; Taube et al., 1994; Yoneda et al., 1994; Mundy and Yoneda, 1995).

Bone involvement by metastatic cancer can result in diverse complications, including bone pain, pathologic fractures (associated with lytic metastases), nerve compression syndromes (especially in myeloma) and hypercalcemia. Most of these are due to the effects of cytokines, growth factors and hormones released into the bone marrow around them and these act in a variety of ways to increase osteoclast numbers and activity (Table 1). For example, interleukin-6 (IL-6), which is produced by myeloma cells, stimulates the proliferation of early osteoclast precursors, but does not promote fusion of these to form osteoclasts and has only weak osteoclast stimulating activity. In contrast, IL-1 which is produced by some solid tumors, stimulates not only the generation of osteoclasts, but also their activity (Uy et al., 1995a) in part by prolonging their life spans through prevention of apoptosis (Hughes et al., 1994). Parathyroid hormone related protein (PTHrP) which is released by many tumor cell types has effects on osteoclasts similar to those of IL-1, although it does not stimulate granulocyte macrophage colony joining units (De La Mata et al., 1995). Its release is often associated with hypercalcemia because of its additional effects on renal tubular calcium reabsorption.

Our knowledge of the mechanisms involved in tumor metastasis and growth in bone is still limited. Here we shall describe our current understanding of the

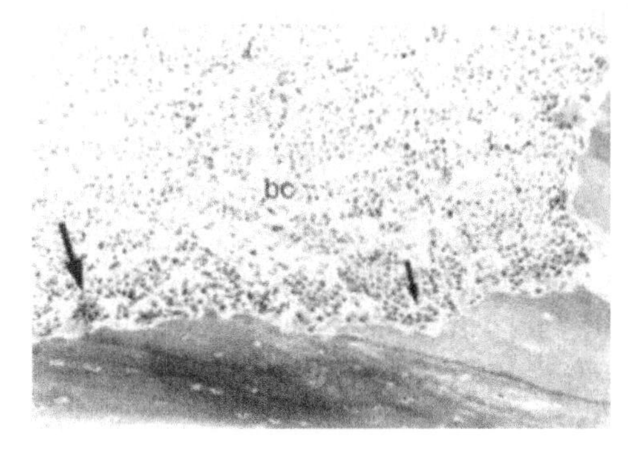

Figure 1. Tumor-induced bone resorption. This high magnification photomicrograph shows multinucleated osteoclasts (large arrow) resorbing the bone matrix in a metastatic deposit of the human breast cancer cell line, MDA-MB-231 (bc). Some cancer cells are in close contact with the bone surface (small arrow), but these are considered to have moved into sites of previous osteoclastic resorption. Hematoxylin and eosin, magnification x 17.

Table 1. Agents Released by TumorCells that Stimulate Bone Resorption

Parathyroid hormone (PTH)
Parathyroid hormone-related protein (PTHrP)
Interleukin-1 (IL-1)
Tumor necrosis factor α (TNFα)
Tumor necrosis factor β (TNFβ, lymphotoxin)
Transforming growth factors α and β (TGFα, TGFβ)
Epidermal growth factor (EGF)
Platelet-derived growth factor (PDGF)
Prostaglandins
Oxygen-derived free radicals
Colony stimulating factors (CSFs)

mechanisms underlying the predilection of certain types of cancers for bone, their growth in the bone microenvironment and interactions between them and bone cells, mainly osteoclasts. We shall focus on metastatic breast cancer as a representative example, although the general principles apply to other metastatic cancers and to myeloma.

A. Breast Cancer Predilection for Metastasis to Bone

The predilection of breast cancer for bone has been described in a number of clinical studies. Walther (1948) reported that 64% of 186 patients who died of breast cancer had bone metastases at autopsy: data which likely represent the frequency of metastasis to bone without the influence of chemotherapy. More recently, Cifuentes and Pickren (1979) and Weiss (1992) reported that 71% of 707 and 62% of 1,060 breast cancer patients, respectively, had bone metastases at autopsy, suggesting that chemotherapy has not influenced the predilection for metastasis to bone. In a study of the clinical course, including the incidence, prognosis, morbidity, and response to treatment of 587 patients dying of breast cancer, Coleman and Rubens (1987) found that 69% had bone metastases, that bone was the most common site of first distant relapse and 10% had hypercalcemia with widespread skeletal metastases. Furthermore, the response of bone lesions to endocrine and chemotherapy was apparently less than that of non-osseous lesions.

B. Mechanisms for Preferential Metastasis of Breast Cancer to Bone

Two major factors control the dissemination of cancers to distant organs: the biological properties of the cancer cells and the environment at the metastatic site.

Specific Properties of Breast Cancer for Metastasis to Bone

Multiple and complex steps are involved in the metastasis to and colonization of distant organs by tumor cells (Fidler, 1990; Liotta, 1992). Cancer cells with high metastatic potential for bone must possess properties not found in cells that do not spread to bone, and these could include production of autocrine growth-stimulating factors, angiogenetic factors, proteolytic enzymes, expression of aberrant numbers of growth factor receptors, temporal and spatial expression of cell adhesion molecules (CAMs), and resistance to host immune surveillance. However, these properties are generally found in all metastatic cancers and unlikely to account for the preferential colonization of bone by breast cancer.

Circulating breast cancer cells enter bone mainly through nutrient arteries which communicate with the sinusoidal network in the bone marrow (Figure 2) (DeBruyn, 1981), rather than with a capillary system found in most solid organs. Thus, they interact with the sinus endothelium which is

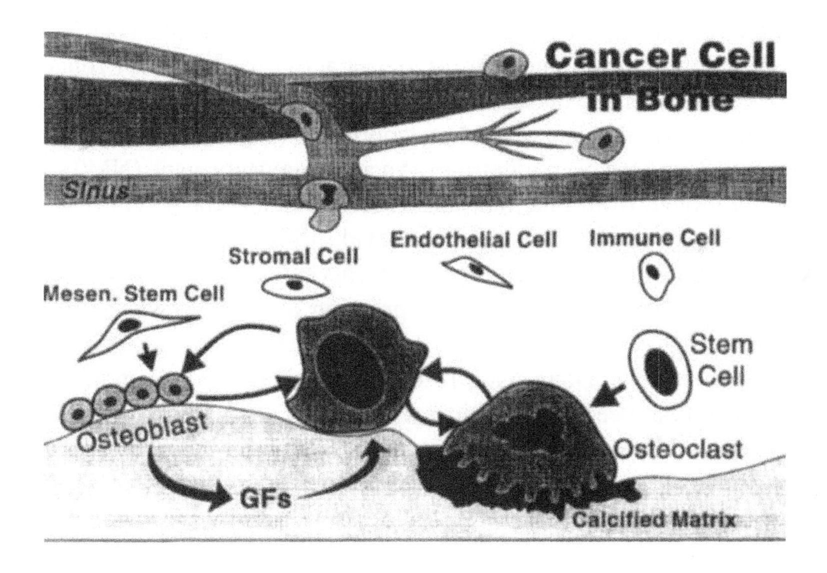

Figure 2. Specific steps involved in cancer metastasis to bone. Metastatic cancer cells enter the bone through nutrient arteries and pass into the marrow space through sinusoids which are lined by endothelial cells. Here tumor cells interact with a variety of host cells whose local production of cytokines may promote their growth. In addition, growth factors (GFs) released by osteoblasts or from bone matrix during resorption by osteoclasts may stimulate cancer cells to proliferate further and to release factors which promote osteoclastic resorption to produce lytic lesions or promote osteoblast proliferation and thus ostesclerotic lesions.

thin but not discontinuous (DeBruyn, 1981) and pass through it to migrate into the bone marrow compartment. To achieve this, they express CAMs to establish cell-cell contact with sinus endothelial cells and subsequently secrete proteolytic enzymes to degrade the endothelial wall. Cancer cells preferentially attach to endothelial cells in their target organs (Auerbach et al., 1987), suggesting a contributory role of endothelial cells in organ preference. Similarly, protease secretion by cancer cells is also influenced by the organ environment (Nakajima et al., 1990). The cell-cell adhesion molecules and proteases which are involved as breast cancer cells accomplish these steps in bone are not known and need to be identified.

In the bone marrow, there are diverse types of cells which play a role in the maintenance of host homeostasis by producing a variety of cytokines and form complex cytokine networks (Figure 2). Breast cancer cells migrating into the bone marrow are exposed to these cytokines and growth factors and it is likely that they interact with one another to enhance tumor cell growth and activity. Establishment of functional interactions between metastatic breast cancer cells and osteoclasts is necessary for the progression of metastases and this could be mediated by direct cell-cell contact and/or production of soluble stimulators of osteoclast activity (Figure 2) (Yoneda et al., 1994; Mundy and Yoneda, 1995).

Recently, several genes which may contribute to the metastatic potential of breast cancer have been identified. An anti-metastatic gene (NM23) was originally cloned from low metastatic murine melanoma cells using subtractive hybridization techniques (Steeg et al., 1988) and high expression of it is associated with a good prognosis in breast cancer patients (Hennessy et al., 1991; Hirayama et al., 1991). Expression of a metastasis suppressor gene called KA11 has been shown to be decreased in metastatic human prostate cancer cells (Dong et al., 1995) while a metastasis-promoting gene named MTA1 has been cloned from rat mammary adenocarcinoma cells using differential hybridization (Toh et al., 1994). Since the protein products of these genes have not yet been characterized, it is not known if they play a role in breast cancer metastasis to and organ selectivity for bone.

Bone Microenvironment

Clinically, it is well-recognized that most cancers exhibit target organ preference when they disseminate. This was first reported by Paget (1889) who found in autopsy records of 735 women who died of breast cancer that the highest numbers of metastases were in the ovaries, followed by the skeleton. He proposed the "seed and soil" theory that the microenvironment of the organs to which cancer cells spread may serve as a fertile soil for their growth. This hypothesis has been widely accepted and remains a basic principle in the field of cancer metastasis (for review, see Rusciano and Burger, 1992). We believe it is particularly relevant to bone.

Bone stores a variety of growth factors that are laid down in bone matrix by osteoblasts during bone formation (Hauschka et al., 1986) These are released in active form into the marrow when bone matrix is degraded during osteoclastic resorption (Pfeilschifter and Mundy, 1987) and thus could stimulate the growth of metastatic cancer cells in the marrow (Figure 2). Among these, transforming growth factor-β (TGFβ) has been shown to stimulate the proliferation of Walker 256 carcinosarcoma cells which metastasize to bone (Orr et al., 1995). Moreover, insulin-like growth factors (IGFs), whose concentrations in bone are higher than other growth factors (Hauschka et al., 1986), promote the growth of human breast cancer MDA-MB-231 cells (Yoneda et al., 1995). Neutralizing antibodies to the IGF-I receptor decreased the mitogenic activity on MDA-MB-231 cells of the culture supernatants harvested from resorbing bone (Yoneda et al., 1995).

Additional evidence that the host environment influences the metastatic potential of cancers is the observation that some cancers increase their metastatic and organ-preferential properties by successive *in vivo* passages in target organs. Murine melanoma B16 cells with low metastatic potential become highly metastatic to lung (B16F10) and liver (B16L8) after repeated selection and culture from pulmonary (Hart and Fidler, 1981) or hepatic (Tao et al., 1979) metastatic foci, respectively. B16L8 cells respond specifically to growth factors from hepatocytes, whereas B16F10 do not (Sargent et al., 1988). In similar experiments, low metastatic human colon cancer cells developed high metastatic ability and organ selectivity for the liver (Morikawa et al., 1988) and then expressed greater numbers of functional receptors for TGF-α and hepatocyte growth factor than low metastatic cancer cells (Fidler, 1995). Although it is unclear whether these changes result from enrichment of a highly metastatic and organ-preferential subpopulation of cancer cells or from the acquisition of metastatic ability and organ preference, they demonstrate that the metastatic behavior of cancer cells can be altered by specific organ environments.

Chemotaxis could also contribute to the bone preference of breast cancer. Culture supernatants of resorbing bone (Orr et al., 1979) and type I collagen and its fragments (Mundy et al., 1981) released during bone resorption stimulate chemotaxis of breast cancer cells in a Boyden chamber assay. Thus, breast cancer cells in bone marrow might preferentially migrate to adjacent resorbing bone surfaces in response to increased local levels of bone products, such as type I collagen, and then be exposed to high local levels of bone-derived growth factors.

Although bone provides a favorable environment for proliferation of metastatic breast cancer cells, its fertility alone cannot account for the special predilection of these cells to thrive in it, since other metastatic cancer cells passing through it are likely to be exposed to the same factors. As mentioned above, breast cancer may possess other intrinsic capacities to survive and proliferate in the presence of a mixture of growth-promoting and -inhibiting factors. The local release of factors, such as PTHrP by breast cancer cells could stimulate osteoclastic resorption resulting in the release of more growth factors from bone (Figure 2), thus establishing a stimulatory cycle between these cell types (see below). It is clear that to understand the mechanisms of preferential metastasis of breast cancer to bone, both the bone microenvironment and breast cancer capacities need to be considered in parallel.

C. Experimental Approach to Study Breast Cancer Metastasis to Bone

Because metastasis is a multistep process involving complex interactions between cancer cells and host cells, animal models are essential to study the metastatic behavior of cancers, the pathophysiology of cancer-associated conditions, and the effects of therapeutic interventions. There are essentially two types of metastasis models: experimental and spontaneous. In the former, cancer cells are injected directly into the blood stream and in the latter, they are inoculated into soft tissue or into the organs from which they arose and from there they disseminate to distant organs. Although spontaneous models more closely mimic the behavior of tumors in cancer patients than experimental ones, they are extremely difficult to establish and thus few are available (Orr et al., 1995). There are no spontaneous and few experimental models of bone metastasis of breast cancer.

The cellular and molecular mechanisms of organ-preference of metastatic tumors can be studied in experimental models without spontaneous spread to bone because the steps involved before cancer cells reach their preferential target organs are likely to be non-specific and independent of organ selectivity. To this end, we recently developed an animal model of experimental bone metastasis of human breast cancer in nude mice (Yoneda et al., 1994) by modifying the model originally described by Arguello et al. (1988). We injected the human breast cancer cell lines MDA-MB-231 (estrogen-independent) and MCF-7 (estrogen-dependent) into the arterial circulation through the left cardiac ventricle of female nude mice. As a unique feature of this model and by as yet unknown mechanisms, these breast cancer cells selectively cause osteolytic bone metastases and rarely form metastases in other organs. The development and growth of osteolytic lesions were monitored by serial radiography (Figure 3) prior to sacrifice of tumor-bearing animals and the size of the lesions was quantitated by computer-assisted image analysis. Animals rarely became hypercalcemic, but frequently developed cachexia (loss of body weight, fat, and muscle) and occasionally paraplegia, a sign of vertebral metastases. Histological analysis of bones from these mice showed replacement of the bone and marrow cavity at the ends of long bones with metastatic breast cancer cells (Figure 4) and resorption of the endosteal bone surface by numerous multinucleated osteoclasts (see Figure 1) (Sasaki et al., 1995). Similar histological observations have been reported using a human melanoma cell line A375 (Nakai et al., 1992; Hiraga et al., 1995).

There are several other animal models of bone metastasis (Orr et al., 1995). Walker 256 carcinosarcoma cells metastasize spontaneously to bone after intramuscular inoculation (Kostenuik et al., 1992). Tail vein injection of the human PC3 prostate cancer into nude mice results in vertebral metastases if the vena cava is compressed beforehand to force the flow of blood

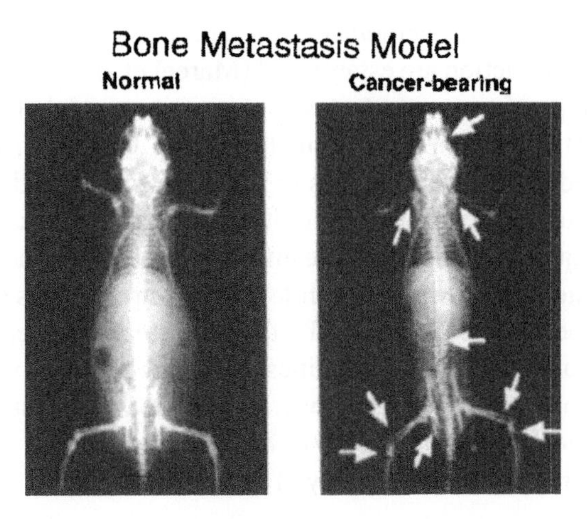

Figure 3. Radiographs of a normal mouse (left) and a tumor-bearing mouse (right). MDA-MB-231 cells (1 x 10^5/mouse) were inoculated into the left cardiac ventricle of a 4-week-old female nude mouse. The radiographs were taken four weeks after cell inoculation. Arrows indicate osteolytic metastases. Note the kyphosis and decreased body size due to cachexia in the cancer-bearing mouse which also developed paraplegia due metastases to vertebrae.

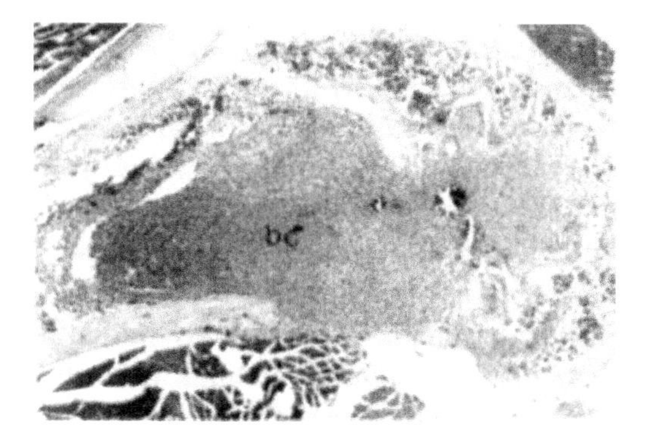

Figure 4. Decalcified section of an osteolytic metastasis from a tumor-bearing nude mouse. Metastatic MDA-MB-231 cells (bc) have caused destruction of most of the cancellous and some of the cortical bone and part of the epiphyseal plate at the lower end of the femur of this nude mouse which was inoculated with tumor cells four weeks previously. Hematoxylin and eosin, magnification x 17.

and the injected tumor cells into the vertebrae via Barton's vertebral venous complex (Shevrin et al., 1988).

Analysis of the mechanisms involved in tumor cell metastasis and growth in bone requires the establishment of reproducible, quantitative, and convenient *in vitro* assays. *In vitro* models for invasion, attachment to extracellular matrix (ECM), chemotactic migration, and matrix metalloproteinase (MMP) production are established (Mareel et al., 1991a) for study of the general steps of cancer metastasis, but *in vitro* models for the specific study of metastasis to bone are difficult to develop. Currently, organ cultures of radiolabeled fetal rat long bones or mouse calvariae and bone marrow cells cultured in the presence of the culture supernatants of cancer cells are used most frequently to examine the effects of cancer products on bone resorption or osteoclast formation, while culture supernatants of bone or osteoblasts can be assessed for their effects on proliferation of cancer cells and their production of proteases, cytokines, and growth factors which promote osteoclastic bone resorption. However, additional *in vitro* assays which can represent a specific part of the *in vivo* processes of bone metastasis are needed to further advance the study of cancer metastasis to bone.

D. Metastatic Cancer Growth and the Bone Microenvironment

To determine whether our concept that the bone microenvironment provides the fertile soil for metastatic breast cancer proliferation is correct, we inoculated breast cancer cells into the left cardiac ventricle of nude mice and modulated their bone turnover by administering the bisphosphonate, risedronate, in three different regimens, viz. either before, simultaneously with, or after breast cancer cell inoculation. In each experiment, risedronate either prevented the development of new osteolytic bone metastases or decreased the progression of those already established. Importantly, histomorphometric analysis revealed that in risedronate-treated mice, the tumor burden in bone was markedly decreased compared with that in untreated mice (Sasaki et al., 1995; Yoneda et al., 1997). These findings and a similar study in rats (Hall and Stoica, 1994) suggest that reduced growth factor release from bone due to the inhibition of osteoclastic bone resorption by risedronate may impair metastatic breast cancer growth in bone.

Cancer cells rarely metastasize spontaneously to calvariae in this model. We speculated that this was due to a lower bone turnover rate in calvariae than in other bones and then examined whether increasing calvarial bone turnover would lead to the development of metastases at this site. To increase bone turnover, human recombinant IL-1α was injected subcutaneously over the calvariae of nude mice for three days according to the methods described by Boyce et al. (1989). Breast cancer cells were inoculated into the left cardiac ventricle the next day, the mice were left untreated and examined for the development of cancer metastases in the calvariae at four weeks after cell inoculation. In IL-1-treated mice, metastatic tumor deposits were clearly visible in the calvariae (Sasaki et al., 1994b) and radiologic examination revealed that these were osteolytic. In contrast, phosphate-buffered saline-treated mice showed no metastatic calvarial tumor deposits. Furthermore, when mice were treated with risedronate prior to IL-1 injections to suppress the IL-1-induced increase in bone turnover, metastatic tumor formation was profoundly diminished (Sasaki et al., 1994b).

To examine in more detail the mechanisms involved in tumor cell growth in bone, the supernatants from resorbing neonatal mouse calvariae were added to cultures of breast cancer cells. These culture supernatants strongly increased the proliferation of breast cancer cells (Yoneda et al., 1995). However, when bone resorption was inhibited *in vitro* by risedronate, there was no stimulation of breast cancer cell proliferation. Furthermore, neutralizing antibodies to IGF-I receptors markedly impaired the growth-stimulating effects of the resorbing bone culture supernatants on the tumor cells (Yoneda et al., 1995). These results show that bone-derived growth factors, including IGFs, that are released from bone during bone resorption are able to promote breast cancer cell proliferation in culture.

Because osteoblasts produce growth factors which are stored in bone, it is conceivable that metastatic cancer cells are also affected by osteoblasts. Recently, it was reported that the culture supernatants of osteoblasts increase chemotactic migration and MMP production by breast cancer cells *in vitro* (Giunciuglio et al., 1995). However, further studies are needed to fully examine the role of osteoblasts in cancer colonization in bone.

E. General Cancer Cell Properties Involved in Bone Metastasis

Cell Adhesion Molecules

CAMs are likely to play a key role in several critical steps involved in cancer cell invasion and metastasis. They mediate cell-to-cell and cell-to-substratum communications by normal cells and their expression by cancer cells may be either decreased or increased depending on the stage of metastasis development and sites of metastasis. They have been shown to regulate tumor cell invasiveness and proliferation by influencing cancer cell adhesion to normal host cells and to ECM (Albelda and Buck, 1990). E-cadherin expression in cancer cells is reversibly modulated according to culture conditions *in vitro* and environmental factors *in vivo* (Mareel et al., 1991b).

Integrins. Integrins, the most abundant CAMs (Haynes, 1992), have been implicated in cancer dissemination (Juliano and Varner, 1993). They mediate cancer cell attachment to vascular endothelial cells and to matrix proteins underlying endothelium, such as laminin and fibronectin (Albelda and Buck, 1990). Human melanoma cells express high levels of the $\alpha_v\beta_3$ integrin (vitronectin receptor) on the cell surface when they bind to and invade the basement membrane matrix, matrigel (Seftor et al., 1992), and neutralizing antibodies to $\alpha_v\beta_3$ integrins inhibit tumor growth and invasion *in vivo* (Brooks et al., 1994). Their role in bone metastasis has not been studied, as yet.

Laminin. Laminin is a major component of basement membrane and has been implicated in cancer metastasis (Nomizu et al., 1995). We found that synthetic antagonists to laminin inhibited the development of osteolytic bone metastases by A375 human melanoma cells which we had inoculated into the left ventricle of nude mice (Nakai et al., 1992).

E-cadherin. E-cadherin (Uvomorulin) is a 120 kDa cell surface glycoprotein involved in calcium-dependent epithelial cell-cell adhesion. It has homophilic properties in cell-cell adhesion and thus may cause homotypic cell aggregation. It appears to play a suppressive role in cancer invasion and metastasis (Takeichi, 1993), and its expression in human tumors has been found to be inversely correlated with breast cancer metastasis (Oka et al., 1993). Its expression is increased in populations of MCF-7 breast cancer cells with reduced invasiveness, but undetectable in the highly invasive MDA-MB-231 breast cancer cells (Sommers et al., 1991).

The results reported to date show only the inhibitory effects of E-cadherin on local cancer invasiveness and very little is known about the possible role of E-cadherin in cancer metastasis to distant organs, including bones. We have found that MDA-MB-231 (low E-cadherin expression) cells much more effectively develop osteolytic bone lesions *in vivo* than MCF-7 (high E-cadherin expression) cells. Furthermore, stable transfection of MDA-MB-231 cells with E-cadherin markedly decreases the number of osteolytic metastases (Mbalaviele et al., 1996), suggesting that increased cell-cell adhesiveness may reduce their capacity to grow in bone.

Organ-Selective Adherence of Cancer Cells

In studies designed to investigate the organ selectivity of metastasizing cancer cells, Kieran and Longnecker (1983) and Netland and Zetter (1984) used ^{57}Cr-labeled cancer cells and showed that they bound selectively to fresh cryostat sections of certain host organs. In similar, but more convincing studies, Nicholson (1988) found that B16 murine melanoma cells which formed brain metastases adhered preferentially to brain-derived endothelial cells compared with B16 melanoma cells which metastasize to the lungs. More recently, Haq et al. (1992) showed that the rat Dunning prostate carcinoma cell line preferentially adheres to cultures of bone marrow stromal cells enriched for endothelial cells.

Matrix Metalloproteinases

Cancer cells which are highly invasive produce large amounts of MMPs, a family of at least eight zinc-dependent endopeptidases which have related structures, but differ in their substrate specificity. Augmented levels of MMPs have been correlated with the development of invasion and metastasis in cancers of human breast, colon, stomach, thyroid, lung, and liver (Seftor et al., 1992; Zucker et al., 1993). It is likely that individual cancer cells utilize several MMPs, as well as other classes of destructive enzymes, to cross the various tissue boundaries they encounter as they invade and metastasize.

We have found that MDA-MB-231 cells cultured on plastic produce 92 kDa (MMP-9) and 72 kDa (MMP-2) MMPs in latent forms. Furthermore, when the cells were cultured on bone ECM laid down by osteoblasts they released active forms of both MMPs, while cells cultured on laminin, fibronectin, type I collagen, matrigel and poly L-lysine released latent forms of MMPs (Sasaki et al., 1994a).

Tissue Inhibitors of Matrix Metalloproteinases (TIMPs)

Cancer invasiveness and metastatic capacity is controlled not only by levels of MMPs, but also by their corresponding inhibitors, TIMPs, at least two of which are ubiquitously distributed (Liotta, 1992). TIMPs function as metastasis-suppressor proteins, and the invasive capacity of cancer cells may depend on the balance between MMP and TIMP production. Overexpression of the TIMP-2 gene resulted in inhibition of invasion and metastasis in animals, and injections of recombinant TIMP-2 have blocked metastasis to other organs (DeClerck et al., 1992). We transfected MDA-MB-231 cells with TIMP-2 cDNA and in a preliminary experiment observed that mice inoculated with MDA-MB-231 cells overexpressing TIMP-2 had fewer osteolytic lesions than mice inoculated with MDA-MB-231 cells transfected with the empty vector (Williams et al., 1995).

F. PTHrP in Metastatic Breast Cancer

PTHrP has been detected by immunohistochemistry in 56% of 155 primary breast tumors from normocalcemic women and its expression appears to correlate with the development of bone metastases (Bundred et al., 1992). For example, a positive signal for PTHrP has been found by immunohistochemistry (Powell et al., 1991; Kohno et al., 1994) and *in situ* hybridization (Vargas et al., 1992) in 80–90% of breast cancer metastases in bone compared with only 17% (Powell et al., 1991) of similar metastases to non-bone sites or with 38% to lung (Kohno et al., 1994). These findings support a major potential role for tumor-produced PTHrP to mediate the increased bone resorption around osteolytic breast metastases. However, there have been no consistent correlations between PTHrP expression in the primary breast tumor and standard prognostic factors, recurrence or survival (Bundred et al., 1992; Liapis et al., 1993).

In view of these clinical observations, we have used the mouse model of bone metastasis to examine further the role of PTHrP in the development of metastases of MDA-MB-231 cells which produce low amounts of PTHrP *in vitro*. Transfection of these cells with the cDNA for human preproPTHrP increased their production of PTHrP *in vitro* and the number of osteolytic metastases *in vivo* (Guise et al., 1994). In contrast, when mice were treated with monoclonal antibodies directed against the 1-34 region of PTHrP prior to inoculation with parental MDA-MB-231 cells, the number and size of observed osteolytic lesions were dramatically less than in similar animals treated with control (Guise et al., 1996). Treatment of mice with established osteolytic metastases with the PTHrP antibody decreased the rate of progression of metastases when compared with mice given a control injection (Yin et al., 1995). Taken together, these data strongly suggest that PTHrP expression by breast cancer cells is important for the development and progression of breast cancer metastases in bone. It stands to reason, then, that production of other osteoclast-stimulating factors could potentiate the development of bone metastases as well.

To determine whether growth factors, such as TGFβ, which is stored in bone (Hauschka et al., 1986), is released during osteoclastic bone resorption (Pfeilschifter and Mundy, 1986) and increases PTHrP expression by MDA-MB-231 cells *in vitro* could potentiate the development of bone metastases, we transfected MDA-MB-231 cells with a cDNA encoding a TGFβ type II receptor lacking a cytoplasmic domain (TβRIIΔcyt) and inoculated the cells into the left ventricle of nude mice. This receptor binds TGFβ, but signal transduction is not initiated so it acts in a dominant-negative fashion to block the biologic effects of TGFβ. Stable clones expressing TβRIIβcyt did not increase PTHrP secretion in response to TGFβ stimulation compared with controls of untransfected MDA-MB-231 cells or those transfected with the empty vector. Mice inoculated with MDA-MB-231 cells expressing TβRIIΔcyt had fewer and smaller osteolytic lesions measured radiographically than control mice given parental cells or cells transfected with the empty vector (Yin et al., 1996). These effects were reversed when the MDA-MB-231 cells expressing TβRIIΔcyt were transfected with a constitutively active type-I TGFβ receptor (Selander et al., 1997). Thus, MDA-MB-231 cells may increase their expression of PTHrP in bone in response to TGFβ and so further stimulate osteoclastic bone resorption locally, causing release of more TGFβ and other growth factors into the bone microenvironment and the establishment of a cycle that promotes more and more tumor growth and bone destruction.

G. Metastatic Prostate Cancer

Bone metastases occur in up to 70% of patients with advanced prostatic cancer (Vest, 1954). Bone is the second most common metastatic site after regional lymph nodes (Galasko, 1981) and most lesions are osteoblastic. They are found most frequently in lumbar vertebrae and pelvic bones following retrograde spread via Barton's vertebral venous plexus, but diffuse skeletal involvement is relatively common. The cancer cells appear to cause osteoblasts on fully calcified, "quiescent" bone surfaces to lay down new matrix on pre-existing bone without preceeding resorption and also to stimulate osteoblast precursors in the bone marrow to proliferate and lay down woven bone matrix between pre-existing bone trabeculae (Valentin Opran et al., 1980). These lesions are typically "hot" on bone scan and, although osteosclerotic, may result in vertebral collapse and paraplegia due to the intrinsically low strength of woven bone and/or concomitant osteolysis.

Like breast cancer, prostatic cancer has a distinct predilection for metastasis to and growth within bone. Normal and malignant prostatic cells express a host of growth factors (see Table 2) and some of their receptors (reviewed by Koutsilieris, 1995), including TGFβ, bone morphogenetic proteins, IGFs, and fibroblast growth factors, all of which can cause osteoblast proliferation, but they can also produce osteoclast-stimulating factors, such as PTHrP (Iwamura et al., 1993), platelet derived growth factor and TGFα. Furthermore, bone marrow stromal cells (Chackal-Roy et al., 1989) and, in particular, cells in the osteoblast lineage (Gleave et al., 1992), produce factors which are mitogenic for prostatic cancer cells, indicating that there may be bidirectional interactions which favor the growth of the tumor cells and osteoblasts in close proximity to one another.

In addition to these growth factors, prostatic cancer cells produce urokinase-type plasminogen activator (uPA) and endothelin-1 (Nelson et al., 1995) which enhance the growth of osteoblasts at the metastatic site. Indeed, plasma levels of endothelin-1 are elevated in patients with metastatic

Table 2. Factors Produced by TumorCells that Stimulate Bone Formation

Transforming growth factor β (TGFβ)
Insulin-like growth factors (IGF-I and II)
Fibroblast growth factors (FGF acidic and basic)
Bone morphogenetic proteins (BMPs)
Platelet-derived growth factor (PDGF)
Bone-derived growth factor (BDGF, B2 microglobulin)
Prostaglandins
Interleukin-1 (IL-1)
Tumor necrosis factor α (TNFα)
Tumor necrosis factor β (lymphotoxin)
Macrophage-derived growth factor (MDGF)
Urokinase (urinary plasminogen activator, uPA)
Endothelin-1

prostatic cancer (Nelson et al., 1995). uPA appears to not only stimulate the proliferation of osteoblasts by hydrolyzing IGF binding proteins and thus activating the growth factors (Koutsilieris et al., 1993), but overexpression of it by transfection of rat prostate cancer cells with full length uPA cDNA promotes the growth of the tumor cells themselves and causes earlier development of metastases after intracardiac injection of the transfected cells compared with controls (Koutsilieris et al., 1993).

The production of osteoclast stimulating factors by prostate cancer cells could account for the osteolysis seen in some metastases, and the level of expression of these relative to that of osteoblast stimulating factors is likely to determine the local effects on bone cells at each metastatic focus. Recent studies of prostate-specific antigen, a serine protease, homologous to the kallikrien family of proteases (Riegman et al., 1989) have shown that it can cleave PTHrP and completely abolish its ability to stimulate cAMP production (Cramer et al., 1996). Thus, although PTHrP is produced by many prostate cancers, its osteoclast stimulating activity may be destroyed by prostate-specific antigen, thus allowing an osteoblast response to predominate in most metastases.

III. DISTANT EFFECTS OF TUMORS ON BONE

A. Humoral Hypercalcemia of Malignancy

Fuller Albright first described the syndrome of ectopic hormone production in 1941, in a patient with malignancy-associated hypercalcemia (Albright, 1941). He postulated that the tumor produced PTH, but when he assayed for parathyroid hormone (PTH), none was detected. Forty-six years later, PTHrP was purified from human lung cancer (Moseley et al., 1987), breast cancer (Burtis et al., 1987), and renal cell carcinoma (Strewler et al., 1987), and was cloned shortly thereafter (Suva et al., 1987). It is now evident that PTHrP, and not PTH, is a major mediator of humoral hypercalcemia of malignancy (Strewler and Nissenson, 1990; Wysolmerski and Broadus, 1994), although four cases of authentic tumor-produced PTH have been reported (Yoshimoto et al., 1989; Nussbaum et al., 1990; Strewler et al., 1993; Rizzoli et al., 1994). In addition to these cancer-related functions, work in the past decade has clearly established that PTHrP has many important functions in normal physiology related to growth and development, reproductive function, and smooth muscle relaxation.

PTHrP has 70% homology to the first 13 amino acids of the N-terminal portion of PTH (Suva et al., 1987), binds to PTH receptors (Abou-Samra et al., 1992) and shares similar biologic activity to PTH (Horiuchi et al., 1987). Specifically, it stimulates adenylate cyclase in renal and bone systems (Burtis et al., 1987; Strewler et al., 1987; Horiuchi et al., 1987; Kemp et al., 1987; Yates et al., 1988), increases renal tubular reabsorption of calcium and osteoclastic bone resorption (Kemp et al., 1987; Yates et al., 1988), decreases renal phosphate uptake (Horiuchi et al., 1987; Kemp et al., 1987; Sartori et al., 1987), and stimulates 1α-hydroxylase (Horiuchi et al., 1987). PTHrP has been found in a variety of tumor types associated with hypercalcemia including squamous, breast, and renal carcinoma (Danks et al., 1989; Asa et al., 1990). Although the majority of squamous cell carcinomas produce PTHrP (Dunne et al., 1993), the capacity to cause hypercalcemia may depend on the level of PTHrP gene expression, which in turn may be determined by differential transcription of the PTHrP gene promoter (Wysolmerski et al., 1996).

In the humoral hypercalcemia of malignancy (HHM), tumor-produced PTHrP interacts with PTH receptors in bone and kidney to cause hypercalcemia, osteoclast-mediated bone resorption, increased nephrogenous cAMP, and phosphate excretion (Stewart et al., 1980). Despite similarities between primary hyperparathyroidism (1°HPT) and HHM and the similar biologic actions of the N-terminal ends of the proteins, unexplained differences between HHM and 1°HPT exist. First, patients with HHM mediated by PTHrP have low serum concentrations of 1,25-dihydroxyvitamin D_3 compared with patients with 1°HPT, despite the fact that both proteins stimulate renal 1α-hydroxylase activity. In clinical studies in which normal humans received short-term infusions of PTHrP-(1-34) (Fraher et al., 1992) or PTHrP-(1-36) (Everhart-Caye et al., 1996) increased serum 1,25-dihydroxyvitamin D_3 concentrations were comparable to those in patients who received similar infusions of PTH-(1-34). Female nude mice infused with synthetic PTHrP-(1-40) for seven days developed hypercalcemia, hypophosphatemia, and increased serum 1,25-dihydroxyvitamin D_3 concentrations (Rosol et al., 1988), while male nude mice bearing Chinese hamster ovary (CHO) cell tumors transfected with the cDNA for human preproPTHrP or preproPTH had equally elevated blood calcium and 1,25-dihydroxyvitamin D_3 concentrations when compared with control animals bearing untransfected CHO tumors (Guise et al., 1992). Additionally, similar increases in blood ionized calcium and 1,25-dihydroxyvitamin D_3 concentrations were observed in nude mice bearing CHO tumors that were engineered to secrete PTHrP mutants truncated at the carboxyl-terminus

(Guise et al., 1993). Second, human studies using either quantitative bone histomorphometry (Stewart et al., 1982; Ralston et al., 1989) or biochemical markers of bone turnover (Nakayama et al., 1996) have demonstrated that, although patients with either HHM or 1°HPT have increased osteoclastic bone resorption, some patients with HHM do not have the coupled increase in osteoblastic bone formation seen typically in those with 1°HPT. In the studies in which normal humans received infusions of PTH or PTHrP (Fraher et al., 1992; Everhart-Caye et al., 1996), bone histomorphometry or biochemical markers of bone turnover were not measured. However, such studies done in animals have found increased bone formation as well as increased osteoclastic bone resorption, as assessed by dynamic bone histomorphometry (Rosol et al., 1989), suggesting that in HHM patients with uncoupled bone turnover factors other than PTHrP are likely to be responsible for the suppressed osteoblast activity.

Although many explanations have been postulated for the differences between HHM and 1°HPT—such as the continuous secretion of PTHrP and the pulsatile secretion of PTH, suppression of bone formation and 1α-hydroxylase activity by other tumor-associated factors or biologically active PTHrP fragments—the reasons for these differences have not been adequately elucidated. Regardless of the explanation for the clinical differences between HHM and 1°HPT, there is clear evidence that other tumor-produced factors can modulate the end-organ effects of PTHrP as well as its secretion from tumors.

Using an *in vivo* model of PTH and PTHrP-mediated hypercalcemia, Uy et al. (1995b) demonstrated that both proteins, delivered in a tumor-produced fashion to mice, caused similar levels of hypercalcemia as well as similar increases in osteoclastic bone resorption, more committed marrow mononuclear osteoclast precursors and mature osteoclasts. No stimulatory effects were seen on the multipotent osteoclast precursors, the granulocyte/macrophage colony-forming unit. In a similar model system, IL-6 potentiated the hypercalcemia and bone resorption mediated by PTHrP *in vivo* by stimulating production of early osteoclast precursors (De La Mata et al., 1995). Likewise, TGFα has been shown to enhance the hypercalcemic effects of PTHrP in an animal model of malignancy-associated hypercalcemia (Guise et al., 1993) and to modulate the renal and bone effects of PTHrP (Pizurki et al., 1990, 1991). Sato et al. (1989) demonstrated that IL-1 α and PTHrP may have synergistic effects *in vivo*.

Tumor-associated factors also appear to be important regulators of PTHrP expression in cancer. For example, epidermal growth factor stimulates PTHrP expression by a keratinocyte (Allinson and Drucker, 1992) and a mammary epithelial cell line (Sebag et al., 1994), while TGFα enhances PTHrP expression in a human squamous cell carcinoma of the lung (Burton et al., 1990). IL-6, tumor necrosis factor, IGF-I, and IGF-II increased the production of PTHrP *in vitro* by a human squamous cell carcinoma (Rizzoli et al., 1994). TGFβ, which is present in high concentrations in bone matrix and is expressed by some breast cancers (Dublin et al., 1993) and cancer-associated stromal cells (van Roozendaal et al., 1995), has been shown to enhance secretion of and stabilize the message for PTHrP in a renal (Zakalik et al., 1992) and squamous cell carcinoma (Kiriyama et al., 1992; Merryman et al., 1994). Recent data demonstrate that this relationship also exists in the human breast carcinoma cell line, MDA-MB-231 (Firek et al., 1994; Guise et al., 1994).

B. Oncogenic Osteomalacia

This is an uncommon condition which although typically associated with benign mesenchymal tumors has been reported in patients with malignant tumors, particularly of soft tissues and bone, but also of prostate (see review by Nelson et al., 1997). In most cases, the basis of the mineralization defect appears to be related to hypophosphatemia due to renal phosphate leakage, and excision of the tumor usually is followed by normalization of blood and urine phosphate concentrations and eventual healing of the osteomalacia. Thus, tumor production of a phosphaturic hormone has been postulated (Salassa et al., 1970), but its identity remains unknown. Normal PTH plasma concentrations in most patients have ruled out PTH as the culprit (Siris et al., 1987). Plasma concentrations of 25(OH) vitamin D_3 are normal and those of $1,25(OH)_2$ vitamin D_3 are low in many patients and these typically return to normal rapidly after tumor excision (Siris et al., 1987), suggesting the production of an inhibitor of 1-α-hydroxylase activity (Fukumoto et al., 1989) whose identity also remains unknown.

IV. CONCLUSIONS

Bone metastases and hypercalcemia are common in patients with advanced cancer and typically are sinister prognostic signs. Study of the local and systemic effects of malignant cells on bone has greatly increased our understanding of the close interaction that takes place between cancer cells and bone cells and has lead to the discovery and characterization of novel hormones, such as PTHrP. Despite these advances, the specific molecular mechanisms by which cancer cells spread to and destroy bone remain poorly understood, and pharmacologic agents with proven therapeutic efficacy to prevent or reverse metastatic bone disease have yet to be developed. Since many of the factors produced in excess by tumor cells are likely to be released locally in much lower concentrations by normal cells within bone and to be involved in the regulation of bone remodeling, continued study of the effects of cancer cells on bone will not only benefit patients with cancer, but also lead to better understanding of bone turnover in normal and disease states.

ACKNOWLEDGMENTS

We thank Beryl Story, Arlene Farias, Paul Williams, and Suzanne D. Taylor for technical assistance. The work was supported by NIH grants P01 CA40035, R01 CA63628, ARO 1899, and DAMD 17-94-J-4213.

REFERENCES

Abou-Samra, A., Juppner, H., Force, T., Freeman, M.W., Kong, X., Schipani, E., Urena, P., Richards, J., Bonventre, J.V., Potts, J.T., Kronenberg, H.M., and Segre, G.V. (1992). Expression cloning of a common receptor for parathyroid hormone and parathyroid hormone-related peptide from rat osteoblastlike cells: A single receptor stimulates intracellular accumulation of both cAMP and inositol triphosphates and increases intracellular free calcium. Proc. Natl. Acad. Sci. U.S.A. 89, 2732-2736.

Albelda, S.M. and Buck, C.A. (1990). Integrins and other cell adhesion molecules. FASEB J. 4, 2868-2880.

Albright, F. (1941). Case Records of the Massachusetts General Hospital (case 27461). N. Engl. J. Med. 225, 789-791.

Allinson, E.T. and Drucker, D.J. (1992). Parathyroid hormonelike peptide shares features with members of early response gene family: Rapid induction by serum growth factors and cycloheximide. Cancer Res. 52, 3103-3109.

Arguello, F., Baggs, R.B., and Frantz, C.N. (1988). A murine model of experimental metastasis to bone and bone marrow. Cancer Res. 48, 6876-6881.

Asa, S.L., Henderson, J., Goltzman, D., and Drucker, D.J. (1990). Parathyroid hormonelike peptide in normal and neoplastic human endocrine tissues. J. Clin. Endocrinol. Metab. 71, 1112-1118.

Auerbach, R., Lu, W.C., Pardon, E., Gumkowski, F., Kaminska, G., and Kaminski, M. (1987). Specificity of adhesion between murine tumor cells and capillary endothelium: An in vitro correlate of preferential metastasis in vivo. Cancer Res. 47, 1492-1496.

Boyce, B.F., Aufdemorte, T.B., Garrett, I.R., Yates, A.J.P., and Mundy, G.R. (1989). Effects of interleukin-1 on bone turnover in normal mice. Endocrinol. 125, 1142-1150.

Brooks, P.C., Montgomery, A.M.P., Rosenfeld, M., Reisfeld, R.A., Hu, T., Klier, G., and Cheresh, D.A. (1994). Integrin $\alpha_v\beta_3$ antagonists promote tumor regression by inducing apoptosis of angiogenic blood vessels. Cell 79, 1157-1164.

Bundred, N.J., Walker, R.A., Ratcliffe, W.A., Warwick, J., Morrison, J.M., and Ratcliffe, J.G. (1992). Parathyroid hormone related protein and skeletal morbidity in breast cancer. Euro. J. Cancer 28, 690-692.

Burtis, W.J., Wu, T., Bunch, C., Wysolmerski, J., Insogna, K., Weir, E., Broadus, A.E., and Stewart, A.F. (1987). Identification of a novel 17,000-dalton parathyroid hormonelike adenylate cyclase-stimulating protein from a tumor associated with humoral hypercalcemia of malignancy. J. Biol. Chem. 262 (15), 7151-7156.

Burton, P.B.J., Moniz, C., and Knight, D. (1990). Parathyroid hormone-related peptide can function as an autocrine growth factor in human renal cell carcinoma. Biochem. Biophys. Res. Commun. 167, 1134-1138.

Chackal-Roy, M., Niemeyer, C., Moore, M., and Zetter, B.R. (1989). Stimulation of human prostatic carcinoma cell growth by factors present in human bone marrow. J. Clin. Invest. 84, 43-50.

Cifuentes, N. and Pickren, J.W. (1979). Metastases from carcinoma of mammary gland: An autopsy study. J. Surg. Oncol. 11, 193-205.

Coleman, R.E. and Rubens, R.D. (1987). The clinical course of bone metastases from breast cancer. Br. J. Cancer 55, 61-66.

Cramer, S.D., Chen, Z., and Peehl, D.M. (1996). Prostate specific antigen cleaves parathyroid hormone-related protein in the PTH-like domain: Inactivation of PTHrP-stimulated cAMP accumulation in mouse osteoblasts. J. Urology. 156, 526-531.

Danks, J.A., Ebeling, P.R., Hayman, J., Chou, S.T., Moseley, J.M., Dunlop, J., Kemp, B.E., and Martin, T.J. (1989). Parathyroid hormone-related protein: Immunohistochemical localization in cancers and in normal skin. J. Bone Min. Res. 4, 273-278.

DeBruyn, P.P.H. (1981). Structural substrates of bone marrow function. Semin. Hematol. 18, 179-193.

DeClerck, Y.A., Perez, N., Shimada, H., Boone, T.C., Langley, K.E., and Taylor, S.M. (1992). Inhibition of invasion and metastasis in cells transfected with an inhibitor of metalloproteinases. Cancer Res. 52, 701-708.

De La Mata J., Uy H.L., Guise T.A., Story B., Boyce B.F., Mundy G.R., and Roodman, G.D. (1995). IL-6 enhances hypercalcemia and bone resorption mediated by PTHrP in vivo. J. Clin. Invest. 95, 2846-2852.

Dong, J.T., Lamb, P.W., Rinker-Schaeffer, C.W., Vukanovic, J., Ichikawa, T., Issacs, J.T., and Barrett, J.C. (1995). *KA11*, a metastasis suppressor gene for prostate cancer on human chromosome 11p11.2. Science 268, 884-886.

Dublin, E.A., Barnes, D.M., Wang, D.Y., King, J.R., and Levison, D.A. (1993). TGFα and TGFβ expression in mammary carcinoma. J. Pathol. 170, 15-22.

Dunne, F.P., Lee, S., Ratcliffe, W.A., Hutchesson, A.C., Bundred, N.J., and Heath, D. (1993). Parathyroid hormone-related protein (PTHrP) gene expression in solid tumors associated with normocalcemia and hypercalcemia. J. Pathol. 172, 215-221.

Eilon, G. and Mundy, G.R. (1978). Direct resorption of bone by human breast cancer cells in vitro. Nature 276, 726-728.

Everhart-Caye, M., Inzucchi, S.E., Guinness-Henry, J., Mitnick, M.A., and Stewart, A.F. (1996). Parathyroid hormone (PTH)-related protein (1-36) is equipotent to PTH (1-34) in humans. J. Clin. Endocrinol. Metab. 81, 199-208.

Fidler, I.J. (1990). Critical factors in the biology of human cancer metastasis. Cancer Res. 50, 6130-6138.

Fidler, I.J. (1995). Modulation of the organ microenvironment for treatment of cancer metastasis. J. Natl. Cancer Inst. 87, 1588-1592.

Firek, A., Jennings, J., Tabuenca, A., Caulfield, M., Garberoglio, C., and Linkhart, T. (1994). TGFβ stimulates PTHrP release from human mammary cancer. Program and Abstracts Endocrine Society Meeting 769A, 395.

Fraher, L.J., Hodsman, A.B., Jonas, K., Saunders, D., Rose, C.L., Henderson, J.E., Hendy, G.N., and Goltzman, D. (1992). A comparison of the in vivo biochemical responses to exogenous parathyroid hormone-(1-34) [PTH-(1-34)] and PTH-related peptide-(l-34) in man. J. Clin. Endocrinol. Metab. 75, 417-423.

Francini, G., Petrioloi, R., Maioli, E., Gonnelli, S., Marsili, S., Aquino, A., and Bruni, S. (1993). Hypercalcemia in breast cancer. Clin. Exp. Metastasis 11, 359-367.

Fukumoto, Y., Matsumoto, T., Yamamoto, H., Kawashima, H., Euyama, Y., Tamaoki, N., and Ogata, E. (1989). Suppression of serum 1,25-dihydroxyvitamin D in humoral hypercalcemia of malignancy is caused by elaboration of a factor that inhibits renal 1,25 hydroxyvitamin D production. Endocrinol. 124, 2057-2062.

Galasko, C.S.B. (1981). The Anatomy and Pathways of Skeletal Metastasis. In: Bone Metastasis. (Weiss and Gilbert, Eds.), pp. 49-63. Hall, Boston.

Giunciuglio, D., Cai, T., Filanti, C., Manduca, P., and Albini, A. (1995). Effect of osteoblast supernatants on cancer cell migration and invasion. Cancer Lett. 97, 69-74.

Gleave, M., Hsieh, J.T., Gao, C.A., von Eschenbach, A.C., and Chung, L.W. (1991). Acclerationof human prostate cancer growth in vivo by factors produced by prostate and bone fibroblasts. Cancer Res. 51, 3753-3761.

Guise, T.A., Chirgwin, J.M., Favarato, G., Boyce, B.F., and Mundy, G.R. (1992). Chinese hamster ovarian cells transfected with human parathyroid hormone-related protein cDNA cause hypercalcemia in nude mice. Lab. Invest. 67, 477-485.

Guise, T.A., Yoneda, T., Yates, A.J.P., and Mundy, G.R. (1993b). The combined effect of tumor produced parathyroid hormone-related protein and transforming growth factor α enhance hypercalcemia in vivo and bone resorption in vitro. J. Clin. Endocrinol. Metab. 77, 40-45.

Guise, T.A., Chirgwin, J.M., Taylor, S., Boyce, B.F., Dunstan, C.R., and Mundy, G.R. (1993ab). Deletions of C-terminal end of parathyroid hormone-related protein (PTHrP) have equivalent effects on bone and calcium homeostasis in vivo. J. Bone Miner. Res. 8, S174.

Guise, T.A., Taylor, S.D., Yoneda, T., Sasaki, A., Wright, K., Boyce, B.F., Chirgwin, J.M., and Mundy, G.R. (1994). PTHrP expression by breast cancer cells enhances osteolytic bone metastases in vivo. J. Bone Miner. Res. 9, S128.

Guise, T.A., Yin, J.J., Taylor, S.D., Kumagai, Y., Dallas, M., Boyce B.F., Yoneda, T., Mundy, G.R. (1996). Evidence for a causal role of parathyroid hormone-related protein in breast cancer–mediated osteolysis. J. Clin. Invest. 98, 1544-1548.

Hall, D.G. and Stoica, G. (1994). Effect of the bisphosphonate risedronate on bone metastases in a rat mammary adenocarcinoma model system. J. Bone Miner. Res. 9, 221-230.

Haq, M., Goltzman, D., Tremblay, G., and Brodt, P. (1992). Rat prostate adenocarcinoma cells disseminate to bone and adhere preferentially to bone marrow-derived endothelial cells. Cancer Res. 52, 4613-4619.

Hart, I.R. and Fidler, I.J. (1981). Role of organ selectivity in the determination of metastatic patterns of B16 melanoma. Cancer Res. 41, 1281-1287.

Hauschka, P.V., Mavrakos, A.E., Iafrati, M.D., Doleman, S.E., and Klagsbrun, M. (1986). Growth factors in bone matrix. Isolation of multiple types by affinity chromatography on heparin-sepharose. J. Biol. Chem. 261, 12665-12674.

Haynes, R.O. (1992). Integrins: versatility, modulation, and signaling in cell adhesion. Cell 69, 11-25.

Hennessy, C., Henry, J.A., May, F.E.B., Westley, B.R., Angus, B., and Lennard, T.W.J. (1991). Expression of the antimetastatic gene *nm23* in human breast cancer: An association with good prognosis. J. Natl. Cancer Inst. 83, 281-285.

Hiraga, T., Nakajima, T., and Ozawa, H. (1995). Bone resorption induced by a metastatic human melanoma cell line. Bone 16, 349-356.

Hirayama, R., Sarvai, S., Takagi, Y., Mishima, Y., Kimura, N., Shimada, N., Esaki, Y., Kurashima, C., Utsuyama, M., and Hirokawa. K. (1991). Positive relationship between expression of antimetastatic factor (*nm23* gene product or nucleoside diphosphate kinase) and good prognosis in human breast cancer. J. Natl. Cancer Inst. 83, 1249-1250.

Horiuchi, N., Caulfield, M., Fisher, J.E., Goldman, M., Mckee, R., Reagan, J., Levy, J., Nutt, R., Rodan, S., Schoefield, T., Clemens, T., and Rosenblatt, M. (1987). Similarity of synthetic peptide from human tumor to parathyroid hormone in vivo and in vitro. Science 238, 1566-1568.

Hughes, D.E., Wright, K.R., Mundy, G.R., and Boyce, B.F. (1994). TGFβ_1 induces osteoclast apoptosis in vitro. J. Bone Miner. Res. 9, S138.

Iwamura, M., di Sant'Agnese, P.A., Wu, G., Benning, C.M., Cockett, A.T., Deftos, L.J., and Abrahamsson, P.A. (1993). Immunohistochemical localization of parathyroid hormone–related protein in human prostate. Cancer Res. 53, 1724-1726.

Juliano, R.L. and Varner, J.A. (1993). Adhesion molecules in cancer: The role of integrins. Curr. Opin. Cell. Biol. 5, 812-818.

Kemp, B.E., Moseley, J.M., Rodda, C.P., Ebeling, P.R., Wettenhall, R.E.H., Stapleton, D., Dieffenbach-Jagger, H., Ure, F., Michelangeli, B.P., Simmons, H.A., Raisz, L.G., and Martin, T.J. (1987). Parathyroid hormone-related protein of malignancy: Active synthetic fragments. Science 238, 1568-1570.

Kieran, M.W. and Longenecker, B.M. (1983). Organ-specific metastasis with special reference to avian systems. Cancer Metast. Rev. 2, 165-182.

Kiriyama T., Gillespie M.T., Glatz, J.A., Fukumoto S., Moseley J.M., and Martin T.J. (1992). Transforming growth factor β stimulation of parathyroid hormone-related protein (PTHrP): A paracrine regulator? Mol. Cell Endocrinol. 92, 55-62.

Kohno, N., Kitazawa, S., Fukase, M., Sakoda, Y., Kanabara, Y., Furuya, Y., Ohashi, O., Ishikawa, Y., and Saitoh, Y. (1994). The expression of parathyroid hormone–related protein in human breast cancer with skeletal metastases. Surgery Today 24, 215-220.

Kostenuik, P.J., Sigh, G., Suyama, K.L., and Orr, F.W. (1992). A quantitative model for spontaneous bone metastasis: Evidence for a mitogenic effect of bone on Walker 256 cancer cells. Clin. Exp. Metast. 10, 403-410.

Koutsilieris, M. (1995). Skeletal metastases in advanced prostate cancer: cell biology and therapy. Crit. Rev. Oncol. Hemato. 18, 51-64.

Koutsilieris, M., Frenette, G., Lazure, C., Lehoux, J.G., Govindan, M.V., and Polychronakos, C. (1993). Urokinase-type plasminogen activator: A paracrine factor regulating the bioavailabilty of IGFs in PA-III cell-induced osteoblastic metastases. Anticancer Res. 13, 481-486.

Liapis, H., Crouch, E.C., Grosso, L.E., Kitazawa, S., and Wick, M.R. (1993). Expression of parathyroidlike protein in normal, proliferative, and neoplastic human breast tissue. Am. J. Pathol. 143, 1169-1178.

Liotta, L.A. (1992). Cancer cell invasion and metastasis. Sci. Amer. 266 (2), 54-63.

Mareel, M.M., De Baetselier, P., and Van Roy, F.M. (1991a). Bioassays for invasion and metastasis. In: Mechanisms of Invasion and Metastasis. 41-53. CRC Press Inc, Boca Raton, FL.

Mareel, M.M., Behrens, J., Birchmeier, W., De Bruyne, G.K., Vleminchex, K., Hoogewijs, A., Fiers, W.C., and Van Roy, F.M. (1991b). Downregulation of E-cadherin expression in Madin-Darby canine kidney (MDCK) cells inside tumors of nude mice. Int. J. Cancer 47, 922-928.

Mbalaviele, G., Dunstan, C.R., Sasaki, A., Williams, P.J., Mundy, G.R., and Yoneda, T. (1996). E-cadherin expression in human breast cancer cells suppresses the development of osteolytic bone metastases in an experimental metastasis model. Cancer Res. 56, 4063-4070.

Merryman, J.I., DeWille, J.W., Werkmeister, J.R., Capen, C.C., and Rosol, T.J. (1994). Effects of transforming growth factors on parathyroid hormone–related protein production and ribonucleic acid expression by a squamous carcinoma cell line in vitro. Endocrinol. 134, 2424-2240.

Morikawa, K., Walker, S.M., Nakajima, M., Pathak, S., Jessup, J.M., and Fidler, I.J. (1988). Influence of organ environment on the growth, selection and metastasis of human colon carcinoma cells in nude mice. Cancer Res. 48, 6863-6871.

Moseley, J.M., Kubota, M., Diefenbach-Jagger, H., Wettenhall, R.E.H., Kemp, B.E., Suva, L.J., Rodda, C.P., Ebeling, P.R., Hudson, P.J., Zajac, J.D., and Martin, T.J. (1987). Parathyroid hormone-related protein purified from a human lung cancer cell line. Proc. Natl. Acad. Sci. U.S.A. 84, 5048-5052.

Mundy, G.R., DeMartino, S., and Rowe, D.W. (1981). Collagen and collagen fragments are chemotactic for tumor cells. J. Clin. Invest. 68, 1102-1105.

Mundy, G.R. and Martin, T.J. (1993). Pathophysiology of skeletal complications of cancer. In: Physiology and Pharmacology of Bone. Handbook of Experimental Pharmacology. (Mundy, G.R and Martin, T.J., Eds.), pp. 641-671, Springer, Berlin.

Mundy, G.R. (1995). Myeloma bone disease. In: Bone Remodeling and Its Disorders. (Mundy, G.R., Ed.), pp. 123-136, Martin Dunitz, London.

Mundy, G.R. and Yoneda, T. (1995). Facilitation and suppression of bone metastasis. Clin. Orthop. Relat. Res. 312, 34-44.

Nakai, M., Mundy, G.R., Williams, P.J., Boyce, B., and Yoneda, T. (1992). A synthetic antagonist to laminin inhibits the formation of osteolytic metastases by human melanoma cells in nude mice. Cancer Res. 52, 5395-5399.

Nakajima, M., Morikawa, K., Fabra, A., Bucana, C., and Fidler, I.J. (1990). Influence of organ environment on extracellular matrix degradative activity and metastasis of human colon carcinoma cells. J. Natl. Cancer Inst. 82, 1890-1898.

Nakayama, K., Fukumoto, S., Takeda, S., Takeuchi, Y., Ishikawa, T., Miura, M., Hata, K., Hane, M., Tamura, Y., Tanaka, Y., Kitaoka, M., Obara, T., Ogata, E., and Matsumoto, T. (1996). Differences in bone and vitamin D metabolism between primary hyperparathyroidism and malignancy-associated hypercalcemia. J. Clin. Endocrinol. Metab. 81, 607-611.

Nelson, A.E., Robinson, B.G., and Mason, R.S. (1997). Oncogenic osteomalacia: is there a new phosphate regulating hormone? Clin. Endocrin. 47, 635-642.

Nelson, J.B., Hedican, S.P., George, D.J., Reddi, A.H., Piantadosi, S., Eisenberger, M.A., Simons, J.W. (1995). Identification of endothelin-1 in the pathophysiology of metastatic adenocarcinoma of the prostate. Nat. Med. 1, 944-949.

Netland, P.A. and Zetter, B.R. (1984). Organ-specific adhesion of metastatic tumor cells in vitro. Science 224, 1113-1115.

Nicholson, G.L. (1988). Organ specificty of tumor metastasis: Role of preferential adhesion, invasion and growth of malignant cells at specific secondary sites. Cancer Metast. Rev. 7, 143-188.

Nijweide, P.J., Burger, E.H., and Feyen, J.H. (1986). Cells of bone: Proliferation, Differentiation, and hormonal regulation. Physiol. Rev. 66, 855-886.

Nomizu, M., Kim, W.H., Yamamura, K., Utani, A., Song, S-Y., Otaka, A., Roller, P.P., leinman, H.K., and Yamada, Y. (1995). Identification of cell binding sites in the laminin α_1-chain carboxyl-terminal globular domain by systematic screening of synthetic peptides. J. Biol. Chem. 270, 20583-20590.

Nussbaum, S., Gaz, R., and Arnold, A. (1990). Hypercalcemia and ectopic secretion of parathyroid hormone by an ovarian carcinoma with rearrangement of the gene for parathyroid hormone. N. Engl. J. Med. 323, 1324-1326.

Oka, H., Shiozaki, H., Kobayashi, K., Inoue, M., Tahara, H., Kobayashi, T., Takatsuka, Y., Matsuyoohi, N., Hirano, S., Takeichi, M., and Mori, T. (1993). Expression of E-cadherin cell adhesion molecules in human breast cancer tissues and its relationship to metastasis. Cancer Res. 53, 1696-1701.

Orr, F.W., Varani, J., Gondek, M.D., Ward, P.A., and Mundy, G.R. (1979). Chemotactic response of tumor cells to products of resorbing bone. Science 203, 176-179.

Orr, F.W., Sanchez-Sweatman, O.H., Kostenuik, P., and Singh, G. (1995). Tumor-bone interactions in skeletal metastasis. Clin. Orthop. Relat. Res. 312, 19-33.

Paget, S. (1889). The distribution of secondary growths in cancer of the breast. Lancet 1, 571-573.

Pfeilschifter, J. and Mundy, G.R. (1987). Modulation of transforming growth factor β activity in bone cultures by osteotropic hormones. Proc. Natl. Acad. Sci. U.S.A. 84, 2024-2028.

Pizurki, L., Rizzoli, R., Caverzasio, J., and Bonjour, J. (1990). Effect of transforming growth factor α and parathyroid hormone–related protein on phosphate transport in renal cells. Am. J. Physiol. 259, F929-F935.

Pizurki, L., Rizzoli, R., Caverzasio, J., and Bonjour, J. (1991). Stimulation by parathyroid hormone–related protein and transforming growth factor α of phosphate transport in osteoblastlike cells. J. Bone Miner. Res. 6, 1235-1241.

Powell, G.J., Southby, J., Danks, J.A., Stillwell, R.G., Hayman, J.A., Henderson, M.A., Bennett, R.C., and Martin, T.J. (1991). Localization of parathyroid hormone–related protein in breast cancer metastases: Increased incidence in bone compared with other sites. Cancer Res. 51, 3059-3061.

Ralston, S.H., Boyce, B.F., Cowan, R.A., Gardner, M.D., Fraser, W.D., and Boyle, I.T. (1989). Contrasting mechanisms of hypercalcemia of malignancy. J. Bone Miner Res. 4, 103-111.

Rasmussen, H. and Bordier, P.J. (1974). *The Physiological and Cellular Basis of Metabolic Bone Disease*. Williams, and Wilkins, Baltimore, Maryland.

Riegman, P.H., Vliestra, R.J., Klaassen, P., van der Korput, J.A., Geurts van Kessel, A., Romijn, J.C., and Trapman, J. (1989). The prostate-specific antigen gene and the human glandular kallikrein-1 gene are tandemly located on chromosome 19. FEBS Lett. 247, 123-126.

Rizzoli, R., Feyen, J.H.M., Grau, G., Wohlwend, A., Sappino, A.P., and Bonjour, J-P. (1994). Regulation of parathyroid hormone–related protein production in a human lung squamous cell carcinoma line. J. Endocrinol. 143, 333-341.

Rizzoli, R., Pache, J.C., Didierjean, L., Burgee, A., Bonjour, J.P. (1994b). A thymoma as a cause of true ectopic hyperparathyroidism. J. Clin. Endocrinol. Metab. 79, 912-915.

Rosol, T.J., Capen, C.C., and Horst, R.L. (1988). Effects of infusion of human parathyroid hormone–related protein-(1B40) in nude mice: Histomorphometric and biochemical investigations. J. Bone Miner. Res. 3, 699-706.

Rusciano, D. and Burger, M.M. (1992). Why do cancer cells metastasize into particular organs? Bio. Essays 14, 185-193.

Salassa, R.M., Jowsey, J., and Arnaud, C.D. (1970). Hypophosphatemic osteomalacia associated with "nonendocrine" tumors. N. Eng. J. Med. 283, 65-70.

Sargent, N.S.E, Oestreicher, M., Haidvogl, H., Madnick, H.M., and Burger, M.M. (1988). Growth regulation of cancer metastases by their host organ. Proc. Natl. Acad. Sci. U.S.A. 85, 7251-7255.

Sartori, L., Weir, E.C., Stewart, A.F., Broadus, A.E., Mangin, M., Barrett, P.Q, and Insogna, K.L. (1987). Synthetic and partially-purified adenylate cyclase-stimulating proteins from tumors associated with humoral hypercalcemia of malignancy inhibit phosphate transport in a PTH-responsive renal cell line. J. Clin. Endocrinol. Metab. 66, 459-461.

Sasaki, A., Niewolna, M., Mundy, G.R., and Yoneda, T. (1994a). Secretion of matrix metalloproteinases (MMPs) by human breast cancer is stimulated by bone extracellular matrix (BECM). J. Bone Miner. Res. 9, S150.

Sasaki, A., Williams, P., Mundy, G.R., and Yoneda, T. (1994b). Osteolysis and tumor growth are enhanced in sites of increased bone turnover in vivo. J. Bone Miner. Res. 9, S294.

Sasaki, A., Boyce, B.F., Story, B., Wright, K.R., Chapman, M., Boyce, R., Mundy, G.R., and Yoneda, T. (1995). Bisphosphonate risedronate reduces metastatic human breast cancer burden in bone in nude mice. Cancer Res. 55, 3551-3557.

Sato, K., Fujii, Y., Kasono, K., Ozawa, M., Imamura, H., Kanaji, Y., Kurosawa, H., Tsushima, T., and Shizume, K. (1989). Parathyroid hormone–related protein and interleukin-1α synergistically stimulate bone resorption in vitro and increase the serum calcium concentration in mice in vivo. Endocrinol. 124, 2172-2178.

Sebag M., Henderson J., Goltzman, D., and Kremer, R. (1994). Regulation of parathyroid hormone–related peptide production in normal human mammary epithelial cells in vitro. Am. J. Physiol. 267, C723-C730.

Seftor, R.E.B., Seftor, E.A., Gehlsen, K.R., Stetler-Stevenson, W.G., Brown, P.D., Ruoslahti, E., and Hendrix, M.J.C. (1992). Role of the $\alpha_v\beta_3$ integrin in human melanoma cell invasion. Proc. Natl. Acad. Sci. U.S.A. 89, 1557-1561.

Selander, K.S., Chirgwin, J.M., Yin, J.J., Dallas, M. Grubbs, B.G., Wieser, R., Massagué, J., Mundy, G.R., and Guise, T.A. (1997). A constitutionally active transforming growth factor-β (TGFβ) type-I receptor increases parathyroid hormone–related protein (PTHrP) production by breast cancer cells in vitro and bone metastes in vivo. J. Bone Miner. Res. 12(Suppl.), 55A, S116.

Shevrin, D.H., Kukreja, S.C., Ghosh, L., and Lad, T.E. (1988). Development of skeletal metastasis by human prostate cancer in athymic nude mice. Clin. Exp. Met. 6, 401-409.

Siris, E.S., Clemens, T.L., Demster, D.W., Shane, E., Segre, G.V., Lindsay, R., Bilezikian, J.P. (1987). Tumor-induced osteomalacia. Kinetics of calcium, phosphorus and vitamin D metabolism and characteristics of bone histomorphometry. Am. J. Med. 82, 307-312.

Sommers, C.L., Thompson, E.W., Torri, J.A., Kemler, R., Gelmann, E.P., and Byers, S.W. (1991). Cell adhesion molecule uvomorulin expression in human breast cancer cell lines: Relationship to morphology and invasive capacities. Cell Growth Diff. 2, 365-371.

Steeg, P.S., Bevilacqua, G., Kopper, L., Thorgeirsson, U.P., Talmadge, J.E., Liotta, L.A., and Sobel, M.E. (1988). Evidence for a novel gene associated with low tumor metastatic potential. J. Natl. Cancer Inst. 80, 200-204.

Stewart, A.F., Vignery, A., and Silvergate, A. (1982). Quantitative bone histomorphometry in humoral hypercalcemia of malignancy: uncoupling of bone cell activity. J. Clin. Endocrinol. Metab. 55, 219-227.

Stewart, A.F., Horst, R., Deftos, L.J., Cadman, E.C., Lang, R., and Broadus, A.E. (1980). Biochemical evaluation of patients with cancer-associated hypercalcemia: Evidence for humoral and nonhumoral groups. N. Eng. J. Med. 303, 1377-1383.

Strewler, G.J., Stern, P., Jacobs, J., Eveloff, J., Klein, R.F., Leung, S.C., Rosenblatt, M., and Nissenson, R. (1987). Parathyroid hormone–like protein from human renal carcinoma cells. Structural and functional homology with parathyroid hormone. J. Clin. Invest. 80, 1803-1807.

Strewler, G.J. and Nissenson, R.A. (1990). Hypercalcemia in malignancy. West. J. Med. 153, 635-640.

Strewler, G.J., Budayr, A.A., Clark, O.H., and Nissenson, R.A. (1993). Production of parathyroid hormone by a malignant nonparathyroid tumor in a hypercalcemic patient. J. Clin. Endocrinol. Metab. 76, 1373-1375.

Suva, L.J., Winslow, G.A., Wettenhall, R.E.H., Hammonds, R.G., Moseley, J.M., Dieffenbach-Jagger, Rodda, C., Kemp, B.E., Rodriguez, H., Chen, E., Hudson, P.J., Martin, T.J., and Wood, W.I. (1987). A parathyroid hormone–related protein implicated in malignant hypercalcemia: Cloning and expression. Science 237, 893-896.

Takeichi, M. (1995). Morphogenetic roles of classic cadherins. Curr. Opin. Cell. Biol. 7, 619-627.

Takeichi, M. (1993). Cadherins in cancer: Implications for invasion and metastasis. Curr. Opin. Cell. Biol. 5, 806-811.

Tao, T., Matter, A., Vogel, K., and Burger, M.M. (1979). Liver-colonizing melanoma cells selected from B16 melanoma. Int. J. Cancer 23, 854-857.

Taube, T., Elomaa, I., Blomqvist, C., Beneton, M.N.C., and Kanis, J.A. (1994). Histomorphometric evidence for osteoclast-mediated bone resorption in metastatic breast cancer. Bone 15, 161-166.

Thomson, B.M., Mundy, G.R., and Chambers, T.J. (1987). Tumor Necrosis factors α and β induce osteoblastic cells to stimulate osteoclastic bone resorption. Immunol 138, 775-779.

Thomson, B.M., Saklatvala, J., and Chambers, T.J. (1986). Osteoblasts mediate interleukin 1 stimulation of bone resorption by rat osteoclasts. Exp. Med. 164, 104-112.

Toh, Y., Pencil, S.D., and Nicolson, G.L. (1994). A novel candidate metastasis-associated gene, *mta2*, differentially expressed in highly metastatic mammary adenocarcinoma cell lines. cDNA cloning, expression, and protein analyses. J. Biol. Chem. 269, 22958-22963.

Uy, H.L., Dallas, M., Calland, J.W., Boyce, B.F., Mundy, G.R., and Roodman, G.D. (1995a). Use of an in vivo model to determine the effects of the Interleukin-1 on cells at different stages in the osteoclast lineage. J. Bone Miner. Res. 10, 295-301.

Uy, H.L., Guise, T.A., De La Mata, J., Taylor, S.D., Story, B.M., Dallas, M.R., Boyce, B.F., Mundy, G.R., and Roodman, G.D. (1995b). Effects of parathyroid hormone (PTH)–related protein and PTH on osteoclasts and osteoclast precursors in vivo. Endocrinol. 136, 3207-3212.

Valentin-Opran, A., Edouard, C., Charhon, S., and Meunier, P.J. (1980). Histomorphometic analysis of iliac bone metastases of prostatic origin. In: *Bone and Tumors.* (Donath, A. and Huber, H., Eds.) pp. 24-28. Médicine et hygiène, Genève.

van Roozendaal, C.E.P., Klijn, J.G.M., van Ooijen, B., Claassen, C., Eggermont, A.M.M., Henzen-Logmans, S.C., and Foekens, J.A. (1995). Transforming growth factor β secretion from primary breast cancer fibroblasts. Mol. Cell. Endocrinol. 111, 1-6.

Vargas, S.J., Gillespie, M.T., Powell, G.J., Southby, J., Danks, J.A., Moseley, J.M., and Martin, T.J. (1992). Localization of parathyroid hormone–related protein mRNA expression in metastatic lesions by in situ hybridization. J. Bone Miner. Res. 7, 971-980.

Walther, H.E. (1948). Krebsmetastasen. Bens Schwabe Verlag, Basel, Switzerland.

Weiss, L. (1992). Comments on hematogenous metastatic patterns in humans as revealed by autopsy. Clin. Exp. Met. 10, 191-199.

Williams, P., Mbalaviele, G., Sasaki, A., Dunstan, C., Adams, R., Bauss, F., Mundy, G.R., and Yoneda, T. (1995). Multistep inhibition of breast cancer metastasis to bone. J. Bone Miner. Res. 10, S169.

Wysolmerski, J.J. and Broadus, A.E. (1994). Hypercalcemia of malignancy: the central role of parathyroid hormone–related protein. Ann. Rev. Med. 45, 189-200.

Wysolmerski, J.J., Vasavada, R., Foley, J., Weir, E.C., Burtis, W.J., Kukreja, S.C., Guise, T.A., Broadus, A.E., and Philbrick, W.M. (1996). Transactivation of the PTHrP gene in squamous carcinomas predicts the occurrence of hypercalcemia in athymic mice. Cancer Res. 56, 1043-1049.

Yates, A.J.P., Gutierrez, G.E., Smolens, P., Travis, P.S., Katz, M.S., Auftemorte, T.B., Boyce, B.F., Hymer, T.K., Poser, J.W., and Mundy, G.R. (1988). Effects of a synthetic peptide of a parathyroid hormone–related protein on calcium homeostasis, renal tubular calcium reabsorption, and bone metabolism in vivo and in vitro in rodents. J. Clin. Invest. 81, 932-938.

Yin, J.J., Taylor, S.D., Yoneda, T., Dallas, M., Boyce, B.F., Kumagai, Y., Mundy, G.R., and Guise, T.A. (1995). Evidence that parathyroid hormone–related protein (PTHrP) causes osteolytic metastases without hypercalcemia. J. Bone Miner. Res. 10, S169.

Yin, J.J., Taylor, S.D., Dallas, M., Massagué, J., Mundy, G.R., and Guise, T.A. (1996). Dominant negative blockade of the transforming growth factor βtype-II receptor decreases breast cancer-mediated osteolysis. J. Bone Miner. Res. (In press.)

Yoneda, T., Sasaki, A., and Mundy, G.R. (1994). Osteolytic bone disease in breast cancer. Breast Cancer Res. Treat. 32, 73-84.

Yoneda, T., Williams, P., Dunstan, C., Chavez, J., Niewolna, M., and Mundy, G.R. (1995). Growth of metastatic cancer cells in bone is enhanced by bone-derived insulinlike growth factors (IGFs). J. Bone Miner. Res. 10, S269.

Yoneda, T., Sasaki, A., Dunstan, C., Williams, P.J., Bauss, F., DeClerck, Y.A., and Mundy, G.R. (1997). Inhibition of osteolytic bone metastases of breast cancer by combined treatment with the bisphosphonate ibandronate and tissue inhibitor of the matrix metalloproteinase-2. J. Clin. Invest. 99, 2509-2517.

Yoshimoto, K., Yamasaki, R., Sakai, H., Tezuka, U., Takahashi, M., Iizuka, M., Sekiya, T., and Saito, S. (1989). Ectopic production of parathyroid hormone by small cell lung cancer in a patient with hypercalcemia. J. Clin. Endocrinol. Metab. 68, 976-981.

Zakalik, D., Diep, D., Hooks, M.A., Nissenson, R.A., and Strewler, G.J. (1992). Transforming growth factor β increases stability of parathyroid hormone–related protein messenger RNA. J. Bone Miner. Res. 7, S118.

Zucker, S., Lysik, R.M., Zarrabi, M.H., and Moll, U. (1993). M_r 92,000 Type-IV collagenase is increased in plasma of patients with colon cancer and breast cancer. Cancer Res. 53, 140-146.

MECHANISMS OF IMMUNOSUPRESSANT-INDUCED BONE DISEASE

Grant R. Goodman and Solomon Epstein

Advances in Organ Biology
Volume 5C, pages 739-763.

ISBN: 0-7623-0390-5

I. INTRODUCTION

Immunosuppressants are a group of drugs used in the treatment of immunologic disorders such as psoriasis, rheumatoid arthritis, nephrotic syndrome, and inflammatory bowel disease (Russell et al., 1992b). They are also used extensively, in various combinations, in the prevention of organ rejection in transplant patients and have significantly altered the outcome of these procedures (Kahan, 1989, 1992; Graziani et al., 1991; Stepkowski and Kahan, 1993). They have numerous side effects that vary from organ to organ. This chapter will concentrate on skeletal adverse effects, viz. osteoporosis with an increased risk of fractures.

Immunosuppressants used in either clinical practice or clinical trials include glucocorticoids, cyclosporine, tacrolimus, methotrexate, azathioprine, rapamycin, mycophenolate mofetil, deoxyspergualin, brequinar sodium, mizoribine, leflunomide, and azasperane. The mechanisms of action, with respect to skeletal side effects, have not been fully elucidated. Glucocorticoids and cyclosporine are two of the most commonly used immunosuppressants and their effects on the skeletal system have been extensively studied.

II. GLUCOCORTICOIDS

Glucocorticoids have significant anti-inflammatory, immunosuppressive, and antineoplastic effects. They are used in organ transplantation and in the treatment of diseases such as temporal arteritis, polymyalgia rheumatica, chronic obstructive pulmonary disease, rheumatoid arthritis, and multiple myeloma.

Glucocorticoids are well known to cause osteoporosis, which affects about 50% of patients receiving therapy (Libanati and Baylink, 1992). The reported prevalence of fractures due to excess glucocorticoids varies from 30% to 50% (Gennari, 1994; Shane and Epstein, 1994). Trabecular (cancellous) bone is more commonly affected than cortical bone (Gennari, 1994), thus sites such as the ribs, vertebrae and the distal ends of long bones are high risk sites for fractures. The incidence and severity of the osteoporosis caused by glucocorticoids is a direct function of the dosage used and the length of treatment (Wolinsky-Friedland, 1995). While there is progressive demineralization (Bockman and Weinerman, 1990), the rate of bone loss is not constant. The greatest rate of demineralization occurs during the first 12 to 15 months of treatment. Systemic daily doses of prednisone of 7.5 mg to

10 mg, or more, may result in osteopenia (Bohannon and Lyles, 1994; Shane and Epstein, 1994). Recent work has shown that lower doses, particularly of inhaled steroids, may also cause bone loss (Boe and Skoogh, 1992; Ip et al., 1994). In addition, it has been found that glucocorticoid-induced osteoporosis is partially reversible when therapy is stopped (Gennari, 1994).

Glucocorticoids produce low turnover osteoporosis that is evidenced by reduced serum levels of osteocalcin (bone gla protein, BGP), a noncollagenous vitamin K dependent protein of the bone matrix synthesized by osteoblasts (Dempster, 1989; Peretz et al., 1989). Histology of patients exposed to excess glucocorticoids shows uncoupling of the bone remodeling cycle (i.e., an increased rate of bone resorption with a decreased rate of bone formation) (Dempster, 1989). This results in a reduced trabecular volume. Static bone parameters in glucocorticoid excess show increased trabecular resorption surfaces, an increase in osteoclast number, and an increase in the activity of, and the number of, active resorption surfaces. There is also a relative increase in osteoid volume, a normal or increased percentage of osteoid surface, and a normal or decreased osteoid seam thickness. Dynamic parameters in glucocorticoid excess show a normal or decreased osteoblast apposition rate, a decreased mineral apposition rate and a decreased adjusted osteoblast apposition rate (Dempster, 1989). There is a decrease in the total amount of bone replaced in each remodeling cycle, as determined by the mean completed wall thickness (Dempster, 1989).

At the molecular level, glucocorticoids diffuse passively through cell membranes, and bind to specific receptors in the cytoplasm and nucleus. This produces an activated receptor complex which binds to specific nuclear DNA/membrane components within the nucleus. This alters the expression of adjacent genes which induces DNA transcription and hence new protein synthesis which trigger a cascade of events that change cell function. The glucocorticoid-receptor complex then dissociates, releasing a steroid molecule that is usually inactive and a receptor molecule that is recycled (Bockman and Weinerman, 1990). The immunosuppressive action of the glucocorticoids is mediated by T cells through the inhibition of expression of a number of cytokines including interleukin-1 (IL-1), interleukin-2 (IL-2), interleukin-6 (IL-6), interferon gamma (IFN-γ), and tumor necrosis factor α (TNF-α) (Suthanthiran and Strom, 1994).

Glucocorticoid excess results in rapid loss of bone, the mechanism of which is multifactorial, and is due to both pathways of bone regulation being negatively affected, viz. bone resorption is increased and bone formation is decreased (Libanati and Baylink, 1992). Mechanisms of glucocorticoid-induced osteoporosis include systemic effects, and direct effects on the osteoclast, osteoblast, and musculature.

Systemic mechanisms of glucocorticoid-induced osteoporosis include effects on calcium, parathyroid hormone (PTH), calcitonin, and sex steroids. Glucocorticoid excess causes a decrease in gastrointestinal calcium absorption, in a dose dependent manner, and an increase in renal excretion of calcium. The decrease in gastrointestinal calcium absorption is probably due to impairment in the intestinal cell calcium transport process, including a decrease in the synthesis of calcium-binding proteins (Feher and Wasserman, 1979). The increase in renal calcium excretion is initially due to a rapid decrease in bone formation, but in patients on long-term glucocorticoid therapy, the fasting hypercalciuria is due to increased mobilization of calcium from the skeleton and a reduction in renal tubular reabsorption of calcium (Caniggia et al., 1981; Reid and Ibbertson, 1987). Renal loss of calcium is exacerbated by a high sodium intake and is decreased by sodium restriction and thiazide diuretics (Lukert, 1992). The alterations in calcium metabolism lead to secondary hyperparathyroidism (Libanati and Baylink, 1992), although data on this is not conclusive (Marcus, 1992). In addition, glucocorticoids may also directly stimulate PTH release (Burckhardt, 1984; Libanati and Baylink, 1992). Glucocorticoid-induced phosphaturia results from the secondary hyperparathyroidism (Lukert and Raisz, 1990).

Calcitonin, a powerful antiresorptive agent, is a 32 amino acid polypeptide secreted by the parafollicular cells in the thyroid gland. Glucocorticoids impair calcitonin secretion, thus allowing increased bone resorption (Jenkinson and Bhalla, 1993).

Glucocorticoids decrease gonadal hormone secretion from the ovaries, testicles, and adrenals. Testosterone is known to be a potent bone anabolic agent in both males and females (Libanati and Baylink, 1992; Orwoll, 1996). Glucocorticoids decrease the production of estrogen and testosterone by direct means and by inhibiting pituitary gonadotropin (luteinizing hormone and follicle-stimulating hormone) secretion (Doerr and Pirke, 1976; Hsueh and Erickson, 1978). Adrenocorticotropic hormone (ACTH) suppression and adrenal atrophy, due to exogenous glucocorticoid therapy, result in low levels of adrenal androgens (Libanati and Baylink, 1992).

Glucocorticoids may directly stimulate osteoclastic activity. An *in vitro* experiment using fetal rat parietal bones has shown that glucocorticoids can cause a transient increase in bone resorption by increasing osteoclast number and activity (Gronowicz et al., 1990). However, another *in vitro* experiment has shown that glucocorticoids can inhibit bone resorption (Raisz et al., 1972); this may be due to the *in vitro* toxic effects of high-dose glucocorticoid on osteoclasts. *In vitro* experiments using mouse and rat tissues have shown that glucocorticoids increase the number of 1,25-dihydroxyvitamin D (1,25-$(OH)_2D$) receptors per osteoclast. This is thought to be due to decreased receptor degradation (Manolagas et al., 1979). Thus the stimulatory effect of 1,25-$(OH)_2D$ on the osteoclast could be enhanced without altering serum levels.

Glucocorticoids directly inhibit osteoblast function, as is evidenced by decreased replication, differentiation, and life span (Lukert and Raisz, 1990). Glucocorticoids decrease synthesis of collagen by osteoblasts (Libanati and Baylink, 1992; Gennari, 1994). This could be due to decreased production by bone cells of transforming growth factor β (TGFβ), prostaglandin E_2 (PGE_2) and insulin-like growth factors (IGFs), which promote collagen synthesis (McCarthy et al., 1990; Lukert, 1992). *In vitro* work on osteoblasts have shown that cortisol decreases transcription of the rat α1(I) procollagen gene (Delany et al., 1995). In addition, glucocorticoids have been shown to decrease osteoblastic synthesis of osteocalcin by modulating the osteocalcin gene promoter (Aslam et al., 1995).

PGE_2, in the presence of physiological levels of cortisol, has anabolic and antiresorptive effects (Jee et al., 1990; Raisz and Fall, 1990). Cortisol *in vitro* has been found to interfere with arachidonic acid release from membrane phospholipids and to inhibit the expression of prostaglandin G/H synthase-2 (PGHS-2) in cultured mouse calvariae (Kawaguchi et al., 1994), thereby decreasing the production of PGE_2, and hence its antiresorptive and anabolic effects.

Growth hormone, through stimulation of IGF-I, induces bone growth (McCarthy et al., 1990). Prednisone inhibits pituitary secretion of growth hormone in response to growth hormone-releasing hormone in normal men (Kaufmann et al., 1988). It is interesting to note, however, that serum levels of growth hormone and IGF-I have been found to be normal in patients receiving glucocorticoids. The effect of IGF-I is also modulated by an IGF-I inhibitor that has been found in the serum of children treated with glucocorticoids (Unterman and Phillips, 1985). In osteoblast cultures, it has been found that glucocorticoids decrease the expression of IGF-I by repression of IGF-I transcripts (McCarthy et al., 1990; Delany and Canalis, 1995). Glucocorticoids also modulate the expression of the IGF binding protein family (IGFBP-1 through -6) in a differential manner, increasing the expression of IGFBP-1 and IGFBP-6, and decreasing the expression of IGFBP-3, IGFBP-4, and IGFBP-5 (Okazaki et al., 1994; Gabbitas and Canalis, 1996). The effect of this modulation of the IGFBP family is uncertain as various studies have shown that the same IGFBP can both inhibit or enhance IGF-I action (Lowe, 1994; Okazaki et al., 1994).

In vitro work on cultured fetal rat parietal bones has shown that glucocorticoids inhibit the synthesis of fibronectin (Gronowicz et al., 1991), an important molecule involved in cell adhesion. Glucocorticoids have also been found to decrease the expression of β_1 integrins. IGF-I has been found to increase the expression of β_1 integrins in fetal rat parietal bone culture (Doherty et al., 1995; Gohel et al., 1995). β_1 integrins belong to a family of transmembrane receptors that mediate cell adhesion to extracellular matrix macromolecules (such as fibronectin and type-I collagen), and affect cell migration and differentiation. It appears that the loss of these cell-attachment receptors may be involved in the production of osteoporosis (Doherty et al., 1995), thus a combined decrease of fibronectin and β_1 integrin synthesis may contribute to the pathogenesis of glucocorticoid-induced osteoporosis.

Glucocorticoids increase the sensitivity of osteoblasts to PTH and 1,25-$(OH)_2D$, both of which are known to inhibit collagen synthesis (Lukert, 1992). Increased sensitivity of osteoblast to PTH is evidenced by increased cyclic adenosine monophosphate (cAMP) production (Reid, 1989). Glucocorticoids appear to increase the amount of 1,25-$(OH)_2D$ receptors expressed by osteoblasts, although this may be cell-cycle dependent (Reid, 1989; Bockman and Weinerman, 1990). Glucocorticoids can affect the musculature, causing wasting and decreased muscle strength (so-called glucocorticoid-induced myopathy). This leads to decreased bone formation (Askari et al., 1976).

III. CYCLOSPORINE

Cyclosporine A (CsA) is a cyclic endecapeptide, initially isolated from the fungus *Tolypocladium inflatum Gams* (Keown and Stiller, 1987) and is used in the treatment of rheumatoid arthritis (del Pozo et al., 1990), psoriasis, nephrotic syndrome, inflammatory bowl disease (Russell et al., 1992a), and in the prevention of organ rejection in transplantation (Loertscher et al., 1983; Graziani et al., 1991; Rich et al., 1992; Shane et al., 1993; Grotz et al., 1994; Sambrook et al., 1994). Debate exists as to CsA's effect on bone mineral metabolism in the human. Some investigators have found that CsA does not affect human bone mineral metabolism (Grotz et al., 1994; Briner et al., 1995). However, an increasingly large number of investigators believe that CsA causes bone loss in the human (Loertscher et al., 1983; Aubia et al.,

1988; Graziani et al., 1991; Rich et al., 1992; Shane et al., 1993; Sambrook et al., 1994; Thiebaud et al., 1996). Elucidating CsA's effect on bone mineral metabolism in the human is made difficult by confounding factors such as the bone effects of the underlying disease and other drugs used in combination with CsA (e.g., glucocorticoids) (Katz and Epstein, 1992; Epstein et al., 1995).

Although it is generally accepted that CsA causes bone loss, some investigators have found the opposite. This is probably due to the inadequacy of the model used, for example, *in vitro* systems are isolated systems that do not adequately reflect the human situation. *In vitro* experiments using fetal rat bones and murine calvariae have shown that CsA inhibits calcemic hormone-induced bone resorption (Stewart et al., 1986; Stewart and Stern, 1988, 1989; Tullberg-Reinert and Hefti, 1991). CsA given to weanling rats produced increased bone formation with decreased bone resorption in one study (Orcel et al., 1989), and had no effect on the absolute rate of cortical bone resorption in another weanling rat study (Klein et al., 1994). Numerous *in vivo* experiments in young and older rats have shown that CsA induces high turnover osteoporosis (Movsowitz et al., 1988, 1989, 1990a,b; Schlosberg et al., 1989; Katz et al., 1991; Stein et al., 1991b; Cvetkovic et al., 1994; Epstein et al., 1994; Mann et al., 1994). The differences found in these *in vivo* rat studies may be explained by the different anatomic sites studied (Katz et al., 1994) or by the age of the rat used, as the weanling rat is a model of excess bone formation where resorption maybe missed (Epstein, 1996). Studies investigating the effect of CsA on adjuvant-arthritis in the rat (del Pozo et al., 1990, 1992) have shown that doses of CsA between 5 mg/kg and 20 mg/kg are effective in treating arthritis and reducing bone loss. This is probably due to inhibition of T cell mediated inflammatory tissue destruction (Russell et al., 1992b). However, the investigators in one of the experiments found that giving CsA at 30 mg/kg did effectively treat the arthritis, but there was also trabecular bone loss. This was thought to be due to the direct effect of CsA on bone mineral metabolism (del Pozo et al., 1992).

In vivo experiments in the non-weanling, non-arthritic rat have shown that CsA produces severe osteoporosis that is dose and duration dependent (Movsowitz et al., 1988). Bone histomorphometry showed decreased percent trabecular bone volume, increased osteoclast number (increased resorption), and increased parameters of bone formation. Serum osteocalcin was also raised, reflecting the increased histomorphometric parameters of bone formation (Movsowitz et al., 1988). CsA did not affect PTH levels in this study. Normocalcemia has been seen in patients receiving CsA and was observed in this study (Movsowitz et al., 1988). An *in vivo* experiment to assess the effect of CsA withdrawal on bone mineral metabolism showed a return of all the histomorphometric parameters, except for bone volume, towards control values within a period of two weeks. There was limited restoration of bone volume by five weeks (Schlosberg et al., 1989). In rat and mouse *in vivo* studies it has been found that CsA increases serum 1,25-$(OH)_2D$ levels by directly increasing 1α-hydroxylase activity (Stein et al., 1991a).

It has been found in various *in vivo* rat studies that CsA-induced osteoporosis is ameliorated, or even reversed, by the administration of PGE_2, 1,25-$(OH)_2D_3$, salmon calcitonin, 2-(2-pyridinyl) ethylidene bisphosphonate (2-PEBP), alendronate, 17β estradiol in estrogen deficiency, and raloxifene (Epstein et al., 1990; Movsowitz et al., 1990a; Stein et al., 1991b; Katz et al., 1992; Bowman et al., 1995; Epstein, 1996). Other *in vivo* rat experiments have shown that IFN-γ does not ameliorate CsA-induced osteoporosis (Mann et al., 1994) and that CsA enhances estrogen deficiency-induced bone resorption (Movsowitz et al., 1989).

At the cellular level, since CsA is lipophilic, it gains access to the cell cytoplasm by diffusion (Russell et al., 1992b). Within cells CsA binds to its intracellular protein targets: the cyclophilin (CyP) family of proteins (Handschumacher et al., 1984; Walsh et al., 1992). In humans, four cDNAs encoding cyclophilin-like proteins have been found, viz. cyclophilin-A through -D (Walsh et al., 1992). CsA has been found to bind competitively to the N-terminal portion of a porcine *cis-trans*-peptidylprolyl isomerase (PPIase, rotamase) which is identical to bovine cyclophilin, thereby inhibiting the action of the PPIase and in doing so blocking the folding of certain proteins. It has been postulated that this may be an important step in the immunosuppressant action of CsA (Siekierka and Sigal, 1992; Walsh et al., 1992). In addition, this CsA-CyP complex inhibits calcineurin activity. Calcineurin (protein phosphatase 2B) is a calcium/calmodulin-dependent protein phosphatase (Siekierka and Sigal, 1992) and is found in neural tissues and in T and B cells. Its inhibition is believed to be involved in the immunosuppressive action of CsA (Siekierka and Sigal, 1992; Walsh et al., 1992). It is has been postulated that the CsA-CyP complex may alter certain nuclear proteins such as NF-AT, AP-3 and NF-κB, thereby regulating gene transcription (Emmel et al., 1989; Shaw et al., 1995).

The interaction of the various components of the immune system is complex and CsA exerts its effect on the immune system in a selective manner. Typically, an antigen is processed by antigen presenting cells such as monocytes, dendritic cells, and B cells. This initial processing appears to be unaffected by CsA (Russell et al., 1992b). During antigen processing, macrophages produce IL-1 which cause T cells to produce IL-2. While CsA does not prevent the production of IL-1 from macrophages, it does however inhibit the *de novo* synthesis of IL-2 by the T cell (Kahan, 1989). CsA also indirectly inhibits the action of IL-2 by increasing the expression of TGFβ, a known inhibitor of IL-2-stimulated T cell proliferation (Suthanthiran and Strom, 1994). In addition, CsA causes partial inhibition of IL-2 cell surface receptor expression, thus affecting the autocrine and paracrine effects of IL-2 on the T cell (Walsh et al., 1992). CsA also inhibits the production of mRNA for some early activation genes, such as c-myc, thus inhibiting the priming of T cells (Kahan, 1989). CsA inhibits the synthesis of IL-3, IL-4, and IFN-γ (Kahan, 1989). CsA does not affect the synthesis of granulocyte macrophage colony stimulating factor (GM-CSF) in T cells (Russell et al., 1992b). CsA inhibits the induction of the major histocompatibility complex class II determinants which are necessary for antigen presentation by macrophages (Kahan, 1989). This maybe due to CsA's inhibition of T cell production of IL-4 and IFN-γ (Russell et al., 1992b).

It is thought that CsA-induced osteoporosis is mediated by the presence of T lymphocytes, as CsA does not produce osteopenia in the T cell depleted, nude rat (Buchinsky et al., 1994). In rat bone, CsA has been shown to increase mRNA expression of IL-1, a cyctokine that is known to be involved in bone resorption (Marshall et al., 1995). The *in vivo* effects of CsA administration on the expression of other cytokines have yet to be elucidated.

Cyclosporine G (CsG) is a natural equipotent immunosuppressive analogue of CsA produced by the fungus *Tolypocladium inflatum Gams*. CsG differs from CsA in that the α-amino butyric acid residue in position 2 is replaced by L-nor-valine (Epstein, 1996). Compared to CsA, CsG has been found to be less nephrotoxic. However, despite this, CsG has similar deleterious effects on rat bone metabolism, as compared to CsA (Cvetkovic et al., 1994).

IV. TACROLIMUS

Tacrolimus (FK-506, fujimycin) is a powerful lipophilic macrolide immunosuppressant produced by *Streptomyces tsukubaensis* (Starzl et al., 1990; Venkataramanan et al., 1990). Tacrolimus has been found to be a more potent immunosuppressive than CsA and has been used as an effective immunosuppressant in organ transplantation in both animal and human studies (Morris et al., 1990b; Starzl et al., 1990; European FK506 Multicentre Liver Study Group, 1994; The U.S. Multicenter FK506 Liver Study Group,

1994). While serious adverse effects such as disturbances of glucose metabolism, renal impairment, and neurological sequelae were more commonly induced by tacrolimus, they were probably less severe than those induced by CsA (Starzl et al., 1990; European FK506 Multicentre Liver Study Group, 1994). Tacrolimus has also been shown to cause trabecular bone loss in cardiac transplant patients (Stempfle et al., 1996). *In vivo* experiments in rats have shown that tacrolimus causes a reduction in percent trabecular area with increased measures of both bone formation and bone resorption (Cvetkovic et al., 1994). The bone losing effect of tacrolimus was greater than that of CsA and CsG (Cvetkovic et al., 1994).

While tacrolimus and CsA are chemically unrelated and have different cytosolic binding sites (Siekierka et al., 1989), they have similar effects on the immune system (Starzl et al., 1990). Tacrolimus binds to it's own class of immunophilins (FK Binding Protein; FKBP). Various forms of FKBP have been isolated, viz. FKBP12, FKBP13, FKBP30, and FKBP60 (Walsh et al., 1992). FKBP has PPIase activity which is inhibited by tacrolimus (Walsh et al., 1992). Like the CsA-CyP complex, the tacrolimus-FKBP complex inhibits calcineurin, thus affecting a common pathway of inhibiting T cell activation (Liu et al., 1991; Fruman et al., 1992; Walsh et al., 1992). Tacrolimus mediates it's immunosuppressive effects by suppressing antigen-stimulated T cell proliferation (Morris et al., 1990b; Macleod and Thomson, 1991). *In vitro* studies have shown that tacrolimus inhibits the release of IL-2, IL-3, IL-4, and IFN-γ from $CD4^+$ T cells and may impair the expression of IL-2 receptors on stimulated T cells (Macleod and Thomson, 1991; Bundick et al., 1992).

While the mechanisms of action of CsA and tacrolimus seem to be similar, the changes in mineral metabolism by the two drugs are not identical. This is evidenced by decreased serum calcium and a resultant increase in serum PTH in tacrolimus treated rats, but not in CsA treated rats; and serum levels of BGP were increased by CsA but not by tacrolimus (Jacobs et al., 1991; Cvetkovic et al., 1994). To date, no further data is available on the bone effects of tacrolimus, particularly at the molecular level.

V. METHOTREXATE

Methotrexate is a folic acid analogue which has been useful in the treatment of a wide variety of conditions, including childhood lymphocytic leukemia, choriocarcinoma, lymphoma, osteogenic sarcoma, carcinoma of the bladder, psoriasis, rheumatoid arthritis, polymyositis, Wegener's granulomatosis, systemic lupus erythematosus, and polyarteritis nodosa. Methotrexate has been found to cause osteoporosis in humans (Bohannon and Lyles, 1994; Wolinsky-Friedland, 1995). Methotrexate osteopathy is an uncommon clinical condition that arises from long-term oral maintenance therapy of methotrexate for childhood neoplasms such as acute lymphocytic leukemia (ALL) (Nesbit et al., 1976; Schwartz and Leonidas, 1984). Typically the patient complains of severe lower extremity pain. Investigations reveal osteoporosis, particularly of the lower extremities, and thick, dense provisional zones of calcification and growth arrest lines which resemble those found in scurvy. Fractures are a risk in this condition (Schwartz and Leonidas, 1984). Current thought is that methotrexate osteoporosis is due to increased bone resorption rather than decreased bone formation, as children who are given methotrexate continue to grow normally (May et al., 1994). However, in a recent study, published in abstract form, it was found that osteocalcin, a marker of bone formation, was increased in a group of postmenopausal women suffering from primary biliary cirrhosis treated with methotrexate (Blum et al., 1996).

Various studies have been done to investigate the effect of methotrexate on bone metabolism using the rodent model. Methotrexate has short- and long-term effects on the rat skeleton. A study done to investigate the short-term effects of methotrexate showed a reduction in net trabecular bone volume of the rat skeleton (Friedlaender et al., 1984). A recent rat experiment done to investigate both the short- and long-term effects of methotrexate (Wheeler et al., 1995) showed that it decreased bone volume, bone formation, and osteoblast activity; there was also increased osteoclast activity. They found that methotrexate had negative effects on both cortical and cancellous bone and that these effects continued long after the initial methotrexate insult (Wheeler et al., 1995). Osteoporosis has been found in rats subjected to prolonged low dose methotrexate (May et al., 1994) and mice subjected to chronic high doses of methotrexate (Freeman-Narrod and Narrod, 1977). An *in vitro* study using human osteoblasts has shown that methotrexate is a potent inhibitor of osteoblast proliferation and may affect bone metabolism and remodeling by altering bone cell turnover (Scheven et al., 1995).

Methotrexate's main mechanism of action is competitive inhibition of folic acid reductase. This inhibits the conversion of dihydrofolate to tetrahydrofolate decreasing the deoxyribonucleotide (dNTP) pool. It also decreases thymidine kinase activity, decreasing the production of thymidylate by the thymidine salvage pathway; thus tissue-cell (including bone marrow) reproduction and repair is impaired (Mazanec and Grisanti, 1989; Abonyi et al., 1992; Kasahara et al., 1993). Methotrexate's effect on tissue-cell reproduction and repair is cumulative, increasing with multiple treatments (Kasahara et al., 1992).

VI. AZATHIOPRINE

Azathioprine is a purine analogue that interferes with DNA synthesis and produces its cytostatic effect on lymphoid cells (Bryer et al., 1995). Azathioprine was the initial drug of choice for renal transplantation, but with the introduction of cyclosporine it became apparent that graft survival was better with cyclosporine (Hollander et al., 1995). In Europe, the combination of azathioprine, cyclosporine, and steroids is currently used in over 80% of renal transplant patients (Moore et al., 1995). However, due to irreversible cyclosporine-induced nephrotoxicity and due to the high cost of cyclosporine, alternative immunosuppressive regimens have been sought. One such regimen is the conversion of cyclosporine to azathioprine three months after transplantation. Prednisone is continued and the dosage is temporarily increased coincident with the conversion. This regimen has been found to be safe and cost-effective in the long-term (Hollander et al., 1995). Azathioprine has also been used in the treatment of severe rheumatoid arthritis (Bryer et al., 1995).

To date there is sparse literature detailing the effect of azathioprine on bone. A prospective clinical study in patients post-renal transplant showed bone remodeling parameters that were significantly worse in the group of patients who were converted to a regimen of cyclosporine/corticosteroid as compared to the patients who remained on a azathioprine/corticosteroid regimen (Aubia et al., 1988). In contrast, another clinical study showed that patients receiving azathioprine (as compared to cyclosporine) were at a higher risk of developing avascular bone necrosis (Wilmink et al., 1989). An *in vivo* study designed to compare the long-term effects of azathioprine and cyclosporine on bone metabolism was performed on the Sprague-Dawley rat. This experiment showed that azathioprine, used alone, seems to be relatively bone sparing. In combination with CsA, azathioprine did not alter the CsA-induced osteoporosis (Bryer et al., 1995). Azathioprine did not affect blood ionized calcium, serum 1,25-$(OH)_2D$ or serum PTH levels. Azathioprine did, however, lower serum osteocalcin levels (i.e., it suppressed osteoblast activity). This effect of azathioprine on osteocalcin could be due to its *in vivo* conversion to 6-mercaptopurine, which is known to suppress bone marrow function. Since osteoblast cells are derived from a pluripotent stromal mesenchymal precursor cell (bone marrow stromal stem cell or connective tissue mesenchymal stem cell) it may be hypothesized that azathioprine can inhibit either the production, differentiation, or activity of mature osteoblasts (Bryer et al., 1995), although longer term studies would be required to assess this.

VII. RAPAMYCIN

Rapamycin (sirolimus) is a recently developed immunosuppressant. Using the rat model, it seems as if rapamycin is bone sparing when given over a 14 day period (Joffe et al., 1993). When given to the rat over a 28 day period, it appeared that rapamycin was relatively bone sparing, retaining cancellous bone volume, despite concomitant accelerated bone modeling and remodeling that was unassociated with changes in serum levels of 1,25-$(OH)_2D$ or osteocalcin (Romero et al., 1995). In the 28 day study there was, however, a decrease in the longitudinal growth rate (Romero et al., 1995). This implies that rapamycin might be problematic to use in children whose bone structure is still growing (Romero et al., 1995). However, rapamycin was found to cause cortical bone loss, while producing no change in trabecular thickness, in an *in vivo* experiment using the rapidly growing rat and very high doses of rapamycin. This was accompanied by a marked increase in the physis thickness, mainly due to an increase in hypertrophic chondrocytes (Bertolini et al., 1993).

Rapamycin exerts it immunosuppressive effects by blocking the messenger system distal to the interaction between IL-2 and it's receptor, thereby suppressing T cell mediated immunity (Bierer et al., 1991; Romero et al., 1995). No further results are available on rapamycin's effects or mechanisms of action on bone.

VIII. MYCOPHENOLATE MOFETIL

Mycophenolate mofetil (mycophenolic acid morpholinoethylester, RS-61443) has been licensed as an immunosuppressant for use in organ transplantation in a number countries including the U.S.A. and U.K. (Gray, 1995). Clinical trials have shown mycophenolate mofetil, in combination with cyclosporine and steroids, to be a well-tolerated and effective immunosuppressant in the treatment of acute rejection in heart (Ensley et al., 1993) and renal transplant patients (European Mycophenolate Mofetil Cooperative Study Group, 1995; RS-61443 Investigation Committee–Japan, 1995). In addition, animal trials (rat and mouse) have shown that mycophenolate mofetil is an effective immunosuppressant (Morris et al., 1990a).

Results from our laboratory show that mycophenolate mofetil does not cause bone loss, as evidenced by histomorphometric examination of male Sprague-Dawley rats treated with oral mycophenolate mofetil for 28 days (Dissanayake et al., 1998). *In vivo*, mycophenolate mofetil is de-esterified to mycophenolic acid (the active form) (European Mycophenolate Mofetil Cooperative Study Group, 1995) and acts by blocking purine metabolism in a manner similar to azathioprine. However, it is thought that mycophenolate mofetil should have greater immunosuppressive activity than azathioprine. This is because it inhibits the *de novo* pathway of purine synthesis used preferentially by lymphocytes, thereby inhibiting the proliferation of T and B cells (European Mycophenolate Mofetil Cooperative Study Group, 1995; Gray, 1995). It has less myelosuppressive effects than therapeutically equivalent doses of azathioprine (Ensley et al., 1993). At present, no further studies on the bone effects of mycophenolate mofetil are available.

IX. DEOXYSPERGUALIN

15-Deoxyspergualin (heptanamide), a derivative of spergualin and derived from *Bacillus laterosporus*, a strain of *Bacillus subtilis* (Takahashi et al., 1990; Tepper et al., 1991), was initially developed for its antibiotic and antitumor activity. Clinical trials have also shown that 15-deoxyspergualin is effective in treating acute rejection in patients with renal transplants (Suzuki et al., 1990; Takahashi et al., 1990). It has also proven to be an effective and potent immunosuppressive in animal experiments (Yuh and Morris, 1991). In an *in vivo* experiment using a murine model, Yuh and Morris (1991) showed that 15-deoxyspergualin is a more potent and effective immunosuppressant than cyclosporine, and that one of the desirable effects was its ability to arrest ongoing acute graft rejection. An *in vitro* experiment has shown that a methylated form of deoxyspergualin (deoxymethylspergualin) is also an effective immunosuppressant.

Possible immunosuppressive mechanisms of 15-deoxyspergualin include the inhibition of formation and differentiation of cytotoxic T cells, the binding to the family of heat shock proteins and the induction of increased mRNA expression of various cytokines, including IL-2, IL-3, IL-4, IL-10, and IFN-γ (Zhu et al., 1994).

In an experiment on bone marrow cells derived from BALB/c → C3H/He bone marrow chimeras treated with 15-deoxyspergualin, the investigators found that the bone marrow cells had increased mRNA expression of IL-1β, IL-2, IL-4, IL-5, IL-6, IL-10, TNF-α, IFN-γ, TGFβ, granulocyte colony stimulating factor (G-CSF), GM-CSF, and leukemia inhibitory factor (LIF) (Zhu et al., 1994). Many of these cytokines modulate bone mineral metabolism, and since bone marrow cells form part of the bone microenvironment, this may be a possible mechanism of modulation of bone mineral metabolism by 15-deoxyspergualin. However, to date, information on the effect of deoxyspergualin and deoxymethylspergualin on bone remains nonexistent.

X. BREQUINAR SODIUM

Brequinar sodium (DUP-785, NSC-368390) was initially developed as an antimetabolite for the clinical treatment of cancer, psoriasis, and rheumatoid arthritis, and is now being evaluated as an immunosuppressant to be used in organ transplantation (Makowka et al., 1993; Yasunaga et al., 1993; Kahan, 1995). Toxic side effects of brequinar sodium include gastrointestinal effects such as diarrhea, bone marrow hypoplasia, and lymphocyte depletion (Barnes et al., 1993; Makowka et al., 1993).

Brequinar sodium acts by blocking *de novo* pyrimidine metabolism via noncompetitive inhibition of dihydroorotate dehydrogenase and is selective for T and B cells since they rely on the *de novo* pathway for pyrimidine synthesis (Jaffee et al., 1993; Makowka et al., 1993). A literature review has revealed that, to date, no data are available on the effect of brequinar sodium on bone.

XI. OTHER IMMUNOSUPRESSANTS

The specific bone effects of mizoribine, leflunomide, and azasperane have not been elucidated.

XII. CONCLUSIONS

There appears to be a definite paucity of data on the cellular and molecular effects of immunosuppressants on bone. Most of the studies are related to *in vitro*, animal or human clinical studies, which have demonstrated various adverse effects. However, mechanisms of action as regards specific pathways in bone cells still need to be clarified.

XIII. SUMMARY

Immunosuppressants are a group of drugs used in the treatment of immunologic disorders. They are also used extensively, in various combinations, in the prevention of organ rejection in transplant patients and have significantly altered the outcome of these procedures. They have numerous side effects that vary from organ to organ. This chapter concentrated on skeletal adverse effects, viz. osteoporosis with an increased risk of fractures.

Glucocorticoids are well-known to cause osteoporosis; the mechanism of which is multifactorial, and is due to both arms of bone regulation being negatively affected, viz. bone resorption is increased and bone formation is decreased. Mechanisms of glucocorticoid-induced osteoporosis include systemic effects such as alteration of calcium metabolism leading to secondary hyperparathyroidism, suppression of calcitonin secretion, and decrease in gonadal hormone secretion. Glucocorticoids stimulate osteoclastic activity and inhibit osteoblastic function. The effect on the osteoclast is not well-clarified in terms of increasing resorption, and may be secondary to the hyperparathyroidism induced by decreasing calcium absorption. Glucocorticoids decrease collagen synthesis from osteoblasts by decreasing TGFβ, PGE_2, and insulin-like growth factors. Glucocorticoids also increase the sensitivity of osteoblasts to PTH and 1,25-$(OH)_2D$, both of which are known to inhibit collagen synthesis. In addition to the above, glucocorticoid-induced myopathy leads to decreased bone formation.

It is generally accepted that CsA causes bone loss. This is supported by *in vivo* rat experiments which have shown that CsA produces severe osteoporosis that is dose and duration dependent. At the cellular level, CsA binds to its intracellular protein target—the CyP family of proteins. This CsA-CyP complex inhibits calcineurin activity thereby producing its immunosuppressive effects, especially through the inhibition of synthesis and action of IL-2. It is also thought that the CsA-CyP complex may alter certain nuclear proteins such as NF-AT, AP-3, and NF-κB, thereby regulating gene transcription. It is thought that CsA-induced osteoporosis is mediated by the presence of T lymphocytes. In rat bone, CsA has been shown to increase mRNA expression of IL-1, a cytokine known to be involved in bone resorption. The *in vivo* effects of CsA administration on the expression of other bone cytokines have yet to be elucidated.

Tacrolimus (FK 506) has been found to be a more potent immunosuppressive than CsA and has been shown to cause trabecular bone loss in cardiac patients. *In vivo* experiments in rats have shown that tacrolimus is more toxic to bone than CsA. Tacrolimus binds to its own class of immunophilins (FKBP). Like the CsA-CyP complex, the tacrolimus-FKBP complex inhibits calcineurin, thus affecting a common pathway of inhibiting T cell activation.

Methotrexate has been found to cause osteoporosis both in the rat model and clinically. It decreased bone volume, bone formation, and osteoblast activity; it also increased osteoclast activity. Azathioprine and rapamycin have been shown to be relatively bone sparing in the rat model and results from our laboratory show that mycophenolate mofetil is not harmful to bone. Unfortunately, there is no data available on the bone effects of newer immunosuppressants such as deoxyspergualin, brequinar sodium, mizoribine, leflunomide, and azasperane. Thus, there appears to be a definite paucity of data on the cellular and molecular effects of immunosuppressants on bone. Most of the studies are related to *in vitro*, animal or human clinical studies, which have demonstrated various adverse effects. However, mechanisms of action as regards specific pathways in bone cells still need to be clarified.

REFERENCES

Abonyi, M., Prajda, N., Hata, Y., Nakamura, H., and Weber, G. (1992). Methotrexate decreases thymidine kinase activity. Biochem. Biophys. Res. Commun. 187 (1), 522-528.

Askari, A., Vignos, Jr., P.J., and Moskowitz, R.W. (1976). Steroid myopathy in connective tissue disease. Am. J. Med. 61, 485-492.

Aslam, F., Shalhoub, V., van Wijnen, A.J., Banerjee, C., Bortell, R., Shakoori, A.R., Litwack, G., Stein, J.L., Stein, G.S., and Lian, J.B. (1995). Contributions of distal and proximal promoter elements to glucocorticoid regulation of osteocalcin gene transcription. Molec. Endocrin. 9, 679-690.

Aubia, J., Masramon, J., Serrano, S., Lloveras, J., and Marinoso, L.L. (1988). Bone histology in renal transplant patients receiving cyclosporin (Letter). Lancet. May 7, 1048-1049.

Barnes, T.B., Campbell, P., Zajac, I., and Gayda, D. (1993). Toxicological and pharmacokinetic effects following coadministration of brequinar sodium and cyclosporine to Sprague-Dawley rats. Transpl. Proc. 25 (Suppl. 2), 71-74.

Bertolini, D.R., Badger, A., and High, W. (1993). Effect of immunomodulators on bone in the rapidly growing rat. JBMR 8 (Suppl. 1), S315 (Abstract, 796.)

Bierer, B.E., Jin, Y.J., Fruman, D.A., Calvo, V., and Burakoff, S.J. (1991). FK 506 and rapamycin: molecular probes of T-lymphocyte activation. Transpl. Proc. 23 (6), 2850-2855.

Blum, M., Wallenstein, S., Clark, J., and Luckey, M. (1996). Effect of methotrexate treatment on bone in postmenopausal women with primary biliary cirrhosis. JBMR. 11 (Suppl. 1), S436 (Abstract, T545).

Bockman, R.S. and Weinerman, S.A. (1990). Steroid-induced osteoporosis. Orthop. Clin. of North America 21 (1), 97-107.

Boe, J., and Skoogh, B.-E. (1992). Is long-term treatment with inhaled steroids in adults hazardous? Eur. Respir. J. 5, 1037-1039.

Bohannon, A.D. and Lyles, K.W. (1994). Drug-induced bone disease. Clin. Geri. Med. 10 (4), 611-623.

Bowman, A.R., Sass, D.A., Marshall, I., Liang, H., Ma, F., Jee, W.S.S., and Epstein, S. (1995). Raloxifene analog (LY117018 HCL) ameliorates cyclosporin A-induced osteopenia in oophorectomized rats. JBMR 10 (Suppl. 1), S350 (Abstract, M392.)

Briner, V.A., Thiel, G., Monier-Faugere, M.-C., Bognar, B., Landmann, J., Kamber, V., and Malluche, H.H. (1995). Prevention of cancellous bone loss but persistence of renal bone disease despite normal 1,25 vitamin D levels two years after kidney transplantation. Transplantation 59 (10), 1393-1400.

Bryer, H.P., Isserow, J.A., Armstrong, E.C., Mann, G.N., Rucinski, B., Buchinsky, F.J., Romero, D.F., and Epstein, S. (1995). Azathioprine alone is bone sparing and does not alter cyclosporin A-induced osteopenia in the rat. JBMR 10 (1), 132-138.

Buchinsky, F.J., Ma, Y.F., Mann, G.N., Isserow, J.A., Rucinski, B., Cvetkovic, M., Romero, D. F., Bryer, H.P., Armstrong, E.C., Jee, W.S.S., and Epstein, S. (1994). The role of T-lymphocytes in cyclosporin A (CsA) induced osteoporosis. Clinical Research 42 (2), 276A (Abstract).

Bundick, R.V., Donald, D.K., Eady, R.P., Hutchinson, R., Keogh, R.W., Schmidt, J.A., and Wells, E. (1992). FK506 as an agonist to induce inhibition of interleukin 2 production. Transplantation 53 (5), 1150-1153.

Burckhardt, P. (1984). Corticosteroids and bone: A review. Hormone Res. 20, 59-64.

Caniggia, A., Nuti, R., Lore, F., and Vattimo, A. (1981). Pathophysiology of the adverse effects of glucoactive corticosteroids on calcium metabolism in man. J. Ster. Biochem. 15, 153-161.

Cvetkovic, M., Mann, G.N., Romero, D.F., Liang, X.G., Ma, Y., Jee, W.S.S., and Epstein, S. (1994). The deleterious effects of long-term cyclosporine A, cyclosporine G, and FK 506 on bone mineral metabolism in vivo. Transplantation 57 (8), 1231-1237.

Delany, A.M. and Canalis, E. (1995). Transcriptional repression of Insulinlike Growth Factor I by glucocorticoids in rat bone cells. Endocrinology 136 (11), 4776-4781.

Delany, A.M., Gabbitas, B.Y., and Canalis, E. (1995). Cortisol downregulates osteoblast $\alpha 1$(I) procollagen mRNA by transcriptional and posttranscriptional mechanisms. J. Cell. Biochem. 57, 488-494.

del Pozo, E., Graeber, M., Elford, P., and Payne, T. (1990). Regression of bone and cartilage loss in adjuvant arthritic rats after treatment with cyclosporin A. Arthr Rheum. 33(2), 247-252.

del Pozo, E., Elford, P., Perrelet, R., Graeber, M., Casez, J.P., Modrowski, D., Payne, T., and MacKenzie, A.R. (1992). Prevention of adjuvant arthritis by cyclosporine in rats. Sem. Arth. Rheum. 21 (6, Suppl. 3), 23-29.

Dempster, D.W. (1989). Bone histomorphometry in glucocorticoid-induced osteoporosis. JBMR 4 (2), 137-141.

Dissanayake, I., Goodman, G.R., Bowman, A.R., Ma, Y., Pun, S., Jee, W.S., and Epstein, S. (1998). Mycophenolate mofetil: A promising new immunosuppressant that does not cause bone loss in the rat. Transplantation. 65, 275-278.

Doerr, P. and Pirke, K.M. (1976). Cortisol-induced suppression of plasma testosterone in normal adult males. J. Clin. Endocr. Metab. 43, 622-629.

Doherty, W.J., DeRome, M.E., McCarthy, M.-B., and Gronowicz, G.A. (1995). The effect of glucocorticoids on osteoblast function. J. Bone Joint Surgery 77-A (3), 396-404.

Emmel, E.A., Verweij, C.L., Durand, D.B., Higgins, K.M., Lacy, E., and Crabtree, G.R. (1989). Cyclosporin A specifically inhibits function of nuclear proteins involved in T-cell activation. Science 246, 1617-1620.

Ensley, R.D., Bristow, M.R., Olsen, S.L., Taylor, D.O., Hammond, E.H., O'Connell, J.B., Dunn, D., Osburn, L., Jones, K.W., Kauffman, R.S., Gay, W.A., and Renlund, D.G. (1993). The use of mycophenolate mofetil (RS-61443) in human heart transplant recipients. Transplantation 56 (1), 75-82.

Epstein, S. (1996). Posttransplantation bone disease: The role of immunosuppressive agents and the skeleton. JBMR 11 (1), 1-7.

Epstein, S., Schlosberg, M., Fallon, M., Thomas, S., Movsowitz, C., and Ismail, F. (1990). 1,25 dihydroxyvitamin D_3 modifies cyclosporine-induced bone loss. Calcif. Tissue Int. 47, 152-157.

Epstein, S., Takizawa, M., Stein, B., Katz, I.A., Joffe, I.I., Romero, D.F., Liang, X.G., Li, M., Ke, H.Z., Jee, W.S.S., Jacobs, T.W., and Berlin, J. (1994). Effect of cyclosporin A on bone mineral metabolism in experimental diabetes mellitus in the rat. JBMR 9 (4), 557-566.

Epstein, S., Shane, E., and Bilezikian, J.P. (1995). Organ transplantation and osteoporosis. Curr. Opin. Rheum. 7, 255-261.

European FK506 Multicentre Liver Study Group (1994). Randomised trial comparing tacrolimus (FK506) and cyclosporin in prevention of liver allograft rejection. Lancet 344, 423-428.

European Mycophenolate Mofetil Cooperative Study Group (1995). Placebo-controlled study of mycophenolate mofetil combined with cyclosporin and corticosteroids for prevention of acute rejection. Lancet 345, 1321-1325.

Feher, J.J. and Wasserman, R.H. (1979). Intestinal calcium-binding protein and calcium absorption in cortisol-treated chicks: Effects of vitamin D_3 and 1,25-dihydroxyvitamin D_3. Endocrinology 104 (2), 547-551.

Freeman-Narrod, M. and Narrod, S.A. (1977). Chronic toxicity of methotrexate in mice. J. Nat. Cancer Inst. 58 (3), 735-741.

Friedlaender, G.E., Tross, R.B., Doganis, A.C., Kirkwood, J.M., and Baron, R. (1984). Effects of chemotherapeutic agents on bone: I. Short-term methotrexate and doxorubicin (Adriamycin) treatment in a rat model. J. Bone Joint Surgery 66-A (4), 602-607.

Fruman, D.A., Klee, C.B., Bierer, B.E., and Burakoff, S.J. (1992). Calcineurin phosphatase activity in T-lymphocytes is inhibited by FK 506 and cyclosporin A. Proc. Nat. Acad. Sci. U.S.A. 89, 3686-3690.

Gabbitas, B. and Canalis, E. (1996). Cortisol enhances the transcription of insulinlike growth factor-binding protein-6 in cultured osteoblasts. Endocrinology 137 (5), 1687-1692.

Gennari, C. (1994). Glucocorticoid induced osteoporosis. Clinical Endocrinology 41, 273-274.

Gohel, A.R., Hand, A.R., and Gronowicz, G.A. (1995). Immunogold localization of β_1-integrin in bone: Effect of glucocorticoids and insulinlike growth factor I on integrins and osteocyte formation. J. Histochem. Cytochem. 43 (11), 1085-1096.

Gray, D.W.R. (1995). Mycophenolate mofetil for transplantation: new drug, old problems? Lancet 346, 390.

Graziani, G., Aroldi, A., Castelnovo, C., Bondatti, F., DeVecchi, A., and Ponticelli, C. (1991). Cyclosporin and calcium metabolism in renal transplanted patients. Nephron 57, 479-480.

Gronowicz, G., McCarthy, M.B., and Raisz, L.G. (1990). Glucocorticoids stimulate resorption in fetal rat parietal bones in vitro. JBMR 5 (12), 1223-1230.

Gronowicz, G.A., DeRome, M.E., and McCarthy, M.-B. (1991). Glucocorticoids inhibit fibronectin synthesis and messenger ribonucleic acid levels in cultured fetal rat parietal bones. Endocrinology 128 (2), 1107-1114.

Grotz, W., Mundinger, A., Gugel, B., Exner, V., Reichelt, A., and Schollmeyer, P. (1994). Missing impact of cyclosporine on osteoporosis in renal transplant recipients. Transpl. Proc. 26 (5), 2652-2653.

Handschumacher, R.E., Harding, M.W., Drugge, R.J., and Speicher, D.W. (1984). Cyclophilin: A specific cytosolic binding protein for cyclosporin A. Science 226, 544-547.

Hollander, A.A.M.J., van Saase, J.L.C.M., Kootte, A.M.M., van Dorp, W.T., van Bockel, H.J., van Es, L.A., and van der Woude, F.J. (1995). Beneficial effects of conversion from cyclosporin to azathioprine after kidney transplantation. Lancet 345, 610-614.

Hsueh, A.J.W. and Erickson, G.F. (1978). Glucocorticoid inhibition of FSH-induced estrogen production in cultured rat granulosa cells. Steroids 32 (5), 639-648.

Ip, M., Lam, K., Yam, L., Kung, A., and Ng, M. (1994). Decreased bone mineral density in premenopausal asthma patients receiving long-term inhaled steroids. Chest 105, 1722-1727.

Jacobs, T.W., Katz, I.A., Joffe, I.I., Stein, B., Takizawa, M., and Epstein, S. (1991). The effect of FK 506, cyclosporine A, and cyclosporine G on serum 1,25 dihydroxyvitamin D levels. Transp. Proc. 23 (6), 3188-3189.

Jaffee, B.D., Jones, E.A., Loveless, S.E., and Chen, S.F. (1993). The unique immunosuppressive activity of brequinar sodium. Transpl. Proc. 25 (3, Suppl. 2), 19-22.

Jee, W.S.S., Mori, S., Li, X.J., and Chan, S. (1990). Prostaglandin E_2 enhances cortical bone mass and activates intracortical bone remodeling in intact and ovariectomized female rats. Bone 11, 253-266.

Jenkinson, T. and Bhalla, A.K. (1993). A reappraisal of steroid-induced osteoporosis. Brit. J. Hosp. Med. 50 (8), 472-476.

Joffe, I., Katz, I., Sehgal, S., Bex, F., Kharode, Y., Tamasi, J., and Epstein, S. (1993). Lack of change of cancellous bone volume with short-term use of the new immunosuppressant Rapamycin in rats. Calcif. Tissue Int. 53, 45-52.

Kahan, B.D. (1989). Cyclosporine. New. Engl. J. Med. 321 (25), 1725-1738.

Kahan, B.D. (1992). Cyclosporin A, FK506, rapamycin: The use of a quantitative analytic tool to discriminate immunosuppressive drug interactions. J. Am. Soc. Nephr. 2 (Suppl. 3), S222-S227.

Kahan, B.D. (1995). Concentration-controlled immunosuppressive regimens using cyclosporine with sirolimus or brequinar in human renal transplantation. Transpl. Proc. 27 (1), 33-36.

Kasahara, Y., Nakai, Y., Miura, D., Yagi, K., Hirabayashi, K., and Makita, T. (1992). Mechanism of induction of micronuclei and chromosome aberrations in mouse bone marrow by multiple treatments of methotrexate. Mutation Res. 280, 117-128.

Kasahara, Y., Nakai, Y., Miura, D., Kanatani, H., Yagi, K., Hirabayashi, K., Takahashi, Y., and Izawa, Y. (1993). Decrease in deoxyribonucleotide triphosphate pools and induction of alkaline-labile sites in mouse bone marrow cells by multiple treatments with methotrexate. Mutation Res. 319, 143-149.

Katz, I.A. and Epstein, S. (1992). Perspectives: Posttransplantation bone disease. JBMR 7 (2), 123-126.

Katz, I.A., Takizawa, M., Joffe, I.I., Stein, B., Fallon, M.D., and Epstein, S. (1991). Comparison of the effects of FK506 and cyclosporine on bone mineral metabolism in the rat: A pilot study. Transplantation 52 (3), 571-574.

Katz, I.A., Jee, W.S.S., Joffe, I.I., Stein, B., Takizawa, M., Jacobs, T.W., Setterberg, R., Lin, B. Y., Tang, L.Y., Ke, H.Z., Zeng, Q.Q., Berlin, J., and Epstein, S. (1992). Prostaglandin E_2 alleviates cyclosporin A-induced bone loss in the rat. JBMR 7 (10), 1191-1200.

Katz, I., Li, M., Joffe, I., Stein, B., Jacobs, T., Liang, X.G., Ke, H.Z., Jee, W., and Epstein, S. (1994). Influence of age on cyclosporin A-induced alterations in bone mineral metabolism in the rat in vivio. JBMR 9 (1), 59-67.

Kaufmann, S., Jones, K.L., Wehrenberg, W.B., and Culler, F.L. (1988). Inhibition by prednisone of growth hormone (GH) response to GH-Releasing hormone in normal men. J. Clin. Endocrinol. Metab. 67 (6), 1258-1261.

Kawaguchi, H., Raisz, L.G., Voznesensky, O.S., Alander, C.B., Hakeda, Y., and Pilbeam, C.C. (1994). Regulation of the two prostaglandin G/H synthases by parathyroid hormone, interleukin-1, cortisol, and prostaglandin E_2 in cultures neonatal mouse calvariae. Endocrinology 135 (3), 1157-1164.

Keown, P.A. and Stiller, C.R. (1987). Cyclosporine: A double-edged sword. Hospital Practice 15 (5), 207-220.

Klein, L., Lemel, M.S., Wolfe, M.S., and Shaffer, J. (1994). Cyclosporin A does not affect the absolute rate of cortical bone resorption at the organ level in the growing rat. Calcif. Tissue Int. 55, 295-301.

Libanati, C.R. and Baylink, D.J. (1992). Prevention and treatment of glucocorticoid-induced osteoporosis. Chest 102 (5), 1426-1435.

Liu, J., Farmer, Jr., J.D., Lane, W.S., Friedman, J., and Weissman, I. (1991). Calcineurin is a common target of cyclophilin-cyclosporin A and FKBP-FK506 complexes. Cell 66, 807-815.

Loertscher, R., Thiel, G., Harder, F., and Brunner, F.P. (1983). Persistent elevation of alkaline phosphatase in cyclosporine-treated renal transplant recipients. Transplantation 36, 115-116.

Lowe, Jr., W.L. (1994). Once is not enough—promiscuity begets diversity among the insulin-like growth factor binding proteins (Editorial.) Endocrinology 135 (5), 1719-1721.

Lukert, B.P. (1992). Glucocorticoid-induced ósteoporosis. Southern Med. J. 85 (Suppl. 2), S48-S51.

Lukert, B.P. and Raisz, L.G. (1990). Glucocorticoid-induced osteoporosis: Pathogenesis and management. Ann. Int. Med. 112 (5), 352-364.

Macleod, A.M. and Thomson, A.W. (1991). FK 506: An immunosuppressant for the 1990s? Lancet 337, 25-27.

Makowka, L., Chapman, F., and Cramer, D.V. (1993). Historical development of brequinar sodium as a new immunosuppressive drug for transplantation. Transpl. Proc. 25 (3, Suppl. 2), 2-7.

Mann, G.N., Jacobs, T.W., Buchinsky, F.J., Armstrong, E.C., Li, M., Ke, H.Z., Ma, Y.F., Webster, S.S.J., and Epstein, S. (1994). Interferon-γ causes loss of bone volume in vivo and fails to ameliorate cyclosporine A-induced osteopenia. Endocrinology 135 (3), 1077-1083.

Manolagas, S.C., Anderson, D.C., and Lumb, G.A. (1979). Glucocorticoids regulate the concentration of 1,25-dihydroxycholecalciferol receptors in bone. Nature 277, 314-315.

Marcus, R. (1992). Secondary forms of osteoporosis. Disorders Bone Min. Metab. 889-904.

Marshall, I., Isserow, J.A., Buchinsky, F.J., Paynton, B.V., and Epstein, S. (1995). Expression of interleukin-1 and interleukin-6 in bone from normal and cyclosporin A-treated rats. (Abstract.) The XII International Conference on Calcium Regulating Hormones, Melbourne, Australia. Bone 16 (1, Suppl.).

May, K.P., West, S.G., McDermott, M.T., and Huffer, W.E. (1994). The effect of low-dose methotrexate on bone metabolism and histomorphometry in rats. Arth. Rheum. 37 (2), 201-206.

Mazanec, D.J. and Grisanti, J.M. (1989). Drug-induced osteoporosis. Cleveland Clin. J. Med. 56, 297-303.

McCarthy, T.L., Centrella, M., and Canalis, E. (1990). Cortisol inhibits the synthesis of insulinlike growth factor-I in skeletal cells. Endocrinology 126 (3), 1569-1575.

Moore, R., Griffin, P., Darby, C., Jurewicz, A., and Lord, R. (1995). Mycophenolate mofetil for prevention of acute rejection (Letter). Lancet 346, 253.

Morris, R.E., Hoyt, E.G., Murphy, M.P., Eugui, E.M., and Allison, A.C. (1990a). Mycophenolic acid morpholinoethylester (RS-61443) is a new immunosuppressant that prevents and halts heart allograft rejection by selective inhibition of T- and B-cell purine synthesis. Transpl. Proc. 22 (4), 1659-1662.

Morris, R.E., Wu, J., and Shorthouse, R. (1990b). Comparative immunopharmacologic effects of FK 506 and CyA in in vivo models of organ transplantation. Transpl. Proc. 22 (1, Suppl. 1), 110-112.

Movsowitz, C., Epstein, S., Fallon, M., Ismail, F., and Thomas, S. (1988). Cyclosporin-A in vivo produces severe osteopenia in the rat: Effect of dose and duration of administration. Endocrinology 123 (5), 2571-2577.

Movsowitz, C., Epstein, S., Ismail, F., Fallon, M., and Thomas, S. (1989). Cyclosporin A in the oophorectomized rat: Unexpected severe bone resorption. JBMR 4 (3), 393-398.

Movsowitz, C., Epstein, S., Fallon, M., Ismail, F., and Thomas, S. (1990a). The bisphosphonate 2-PEBP inhibits cyclosporin A-induced high-turnover osteopenia in the rat. J. Lab. Clin. Med. 115 (1), 62-68.

Movsowitz, C., Schlosberg, M., Epstein, S., Ismail, F., Fallon, M., and Thomas, S. (1990b). Combined treatment with cyclosporin A and cortisone acetate minimizes the adverse effects of either agent alone. J. Ortho. Res. 8 (5), 635-641.

Nesbit, M., Krivit, W., Heyn, R., and Sharp, H. (1976). Acute and chronic effects of methotrexate on hepatic, pulmonary, and skeletal systems. Cancer 37, 1048-1054.

Okazaki, R., Riggs, B.L., and Conover, C.A. (1994). Glucocorticoid regulation of insulinlike growth factor-binding protein expression in normal human osteoblastlike cells. Endocrinology 134 (1), 126-132.

Orcel, P., Bielakoff, J., Modrowski, D., Miravet, L., and De Vernejoul, M.C. (1989). Cyclosporin A induces in vivo inhibition of resorption and stimulation of formation in rat bone. JBMR 4 (3), 387-391.

Orwoll, E.S. (1996). Androgens as anabolic agents for bone. Trends Endocr. Metab. 7 (3), 77-84.

Peretz, A., Praet, J.-P., Bosson, D., Rozenberg, S., and Bourdoux, P. (1989). Serum osteocalcin in the assessment of corticosteroid induced osteoporosis. Effect of long- and short-term corticosteroid treatment. J. Rheum. 16 (3), 363-367.

Raisz, L.G., and Fall, P.M. (1990). Biphasic effects of prostaglandin E_2 on bone formation in cultured fetal rat calvariae: Interaction with cortisol. Endocrinology 126 (3), 1654-1659.

Raisz, L.G., Trummel, C.L., Wener, J.A., and Simmons, H. (1972). Effects of glucocorticoids on bone resorption in tissue culture. Endocrinology 90, 961-967.

Reid, I.R. (1989). Steroid osteoporosis (Editorial). Calcif Tissue Int 45, 63-67.

Reid, I.R. and Ibbertson, H.K. (1987). Evidence for decreased tubular reabsorption of calcium in glucocorticoid-treated asthmatics. Hormone Res. 27, 200-204.

Rich, G.M., Mudge, G.H., Laffel, G.L., and LeBoff, M.S. (1992). Cyclosporine A and prednisone-associated osteoporosis in heart transplant recipients .J. Heart Lung Transpl. 11 (5), 950-958.

Romero, D.F., Buchinsky, F.J., Rucinski, B., Cvetkovic, M., Bryer, H.P., Liang, X.G., Ma, Y.F., Jee, W.S.S., and Epstein, S. (1995). Rapamycin: A bone sparing immunosuppressant? JBMR 10 (5), 760-768.

RS-61443 Investigation Committee–Japan (1995). Pilot study of mycophenolate mofetil (RS-61443) in the prevention of acute rejection following renal transplantation in Japanese patients. Transpl. Proc. 27 (1), 1421-1424.

Russell, G., Graveley, R., Seid, J., Al-Humidan, A.-K., and Skjodt, H. (1992a). Mechanisms of action of cyclosporine and effects on connective tissues. Sem. Arth. Rheum. 21 (6, Suppl. 3), 16-22.

Russell, R.G.G., Graveley, R., Coxon, F., Skjodt, H., del Pozo, E., Elford, P., and Mackenzie, A. (1992b). Cyclosporin A. Mode of action and effects on bone and joint tissues. Scand. J. Rheum. 21 (Suppl. 95), 9-18.

Sambrook, P.N., Kelly, P.J., Keogh, A.M., Macdonald, P., Spratt, P., Freund, J., and Eisman, J.A. (1994). Bone loss after heart transplantation: A prospective study. J. Heart Lung Transpl. 13 (1, Pt.1), 116-121.

Scheven, B.A.A., van der Veen, M.J., Damen, C.A., Lafeber, F.P.J.G., Van Rijn, H.J.M., Bijlsma, J.W.J., and Duursma, S.A. (1995). Effects of methotrexate on human osteoblasts in vitro: Modulation by 1,25-dihydroxyvitamin D_3. JBMR 10 (6), 874-880.

Schlosberg, M., Movsowitz, C., Epstein, S., Ismail, F., Fallon, M.D., and Thomas, S. (1989). The effect of cyclosporin A administration and its withdrawal on bone mineral metabolism in the rat. Endocrinology 124 (5), 2179-2184.

Schwartz, A.M. and Leonidas, J.C. (1984). Methotrexate osteopathy. Skeletal Radiol. 11, 13-16.

Shane, E. and Epstein, S. (1994). Immunosuppressive therapy and the skeleton. Trends Endocr. Metab. 5 (4), 169-175.

Shane, E., Del C. Rivas, M., Silverberg, S.J., Sook Kim, T., Staron, R.B., and Bilezikian, J.P. (1993). Osteoporosis after cardiac transplantation. Am. J. Med. 94, 257-264.

Shaw, K.T.-Y., Ho, A.M., Raghavan, A., Kim, J., Jain, J., Park, J., Sharma, S., Rao, A., and Hogan, P.G. (1995). Immunosuppressive drugs prevent a rapid dephosphorylation of transcription factor NFAT1 in stimulated immune cells. Proc. Nat. Acad. Sci. USA 92, 11205-11209.

Siekierka, J.J. and Sigal, N.H. (1992). FK-506 and cyclosporine A: Immunosuppressive mechanism of action and beyond. Curr. Opin. Immunol. 4, 548-552.

Siekierka, J.J., Staruch, M.J., Hung, S.H.Y., and Sigal, N.H. (1989). FK-506, A potent novel immunosuppressive agent binds to a cytosolic protein which is distinct from the cyclosporin A-binding protein, cyclophilin. J. Immunol. 143 (5), 1580-1583.

Starzl, T.E., Fung, J., Jordan, M., Shapiro, R., Tzakis, A., McCauley, J., Johnston, J., Iwaki, Y., Jain, A., Alessiani, M., and Todo, S. (1990). Kidney transplantation under FK 506. JAMA 264 (1), 63-67.

Stein, B., Halloran, B.P., Reinhardt, T., Engstrom, G.W., Bales, C.W., Drezner, M.K., Currie, K.L., Takizawa, M., Adams, J.S., and Epstein, S. (1991a). Cyclosporin-A increases synthesis of 1,25-dihydroxyvitamin D_3 in the rat and mouse. Endocrinology 128 (3), 1369-1373.

Stein, B., Takizawa, M., Katz, I., Joffe, I., Berlin, J., Fallon, M., and Epstein, S. (1991b). Salmon calcitonin prevents cyclosporine-A-induced high turnover bone loss. Endocrinology 129 (1), 92-98.

Stempfle, H.U., Wehr, U., Meiser, B., Angermann, C.E., Rambeck, W.A., and Gartner, R. (1996). Effect of FK506 (tacrolimus) on trabecular bone loss shortly after cardiac transplantation (Abstract 130). JBMR 11 (Suppl. 1), S127.

Stepkowski, S.M. and Kahan, B.D. (1993). Synergistic activity of the triple combination: Cyclosporine, rapamycin, and brequinar. Transpl. Proc. 25 (3, Suppl. 2), 29-31.

Stewart, P.J. and Stern, P.H. (1988). Inhibition of parathyroid hormone and interleukin 1-stimulated bone resorption by cyclosporine A but not by cyclosporine H or F. Transpl. Proc. 20 (3, Suppl. 3), 989-992.

Stewart, P.J. and Stern, P.H. (1989). Cyclosporines: Correlation of immunosuppressive activity and inhibition of bone resorption. Calcif. Tissue. Int. 45, 222-226.

Stewart, P.J., Green, O.C., and Stern, P.H. (1986). Cyclosporine A inhibits calcemic hormone-induced bone resorption in vitro. JBMR 1(3), 285-291.

Suthanthiran, M., and Strom, T.B. (1994). Renal transplantation. New Engl. J. Med. 331(6), 365-376.

Suzuki, S., Hayashi, R., Kenmochi, T., Shimatani, K., Fukuoka, T., and Amemiya, H. (1990). Clinical application of 15-deoxyspergualin for treatment of acute graft rejection following renal transplantation. Transpl. Proc. 22 (4), 1615-1617.

Takahashi, K., Ota, K., Tanabe, K., Oba, S., Teraoka, S., Toma, H., Agishi, T., Kawaguchi, H., and Ito, K. (1990). Effect of a novel immunosuppressive agent, deoxyspergualin, on rejection in kidney transplant recipients. Transpl. Proc. 22 (4), 1606-1612.

Tepper, M.A., Petty, B., Bursuker, I., Pasternak, R.D., Cleaveland, J., Spitalny, G.L., and Schacter, B. (1991). Inhibition of antibody production by the immunosuppressive agent, 15-deoxyspergualin. Transpl. Proc. 23 (1), 328-331.

Thiebaud, D., Krieg, M.A., Gillard-Berguer, D., Jacquet, A.F., Goy, J.J., and Burckhardt, P. —(1996). Cyclosporine induces high bone turnover and may contribute to bone loss after heart transplantation. Eur. J. Clin. Invest. 26, 549-555.

Tullberg-Reinert, H. and Hefti, A.F. (1991). Different inhibitory actions of cyclosporine A and cyclosporine A-acetate on lipopolysaccharide, interleukin 1, 1,25 dihydroxyvitamin D_3

and parathyroid hormone-stimulated calcium and lysosomal enzyme release from mouse calvaria in vitro. Agents and Actions 32 (3/4), 321-332.

Unterman, T.G. and Phillips, L.S. (1985). Glucocorticoid effects on somatomedins and somatomedin inhibitors. J. Clin. Endocrinol. Metab. 61, 618-626.

The U.S. Multicenter FK 506 Liver Study Group (1994). A comparison of tacrolimus (FK 506) and cyclosporine for immunosuppression in liver transplantation. New Engl. J. Med. 331 (17), 1110-1115.

Venkataramanan, R., Jain, A., Cadoff, E., Warty, V., Iwasaki, K., Nagase, K., Krajack, A., Imventarza, O., Todo, S., Fung, J.J., and Starzl, T.E. (1990). Pharmacokinetics of FK 506: Preclinical and Clinical Studies. Transpl. Proc. 22 (1, Suppl. 1), 52-56.

Walsh, C.T., Zydowsky, L.D., and McKeon, F.D. (1992). Cyclosporin A, the cyclophilin class of peptidylprolyl isomerases, and blockade of T-cell signal transduction. J. Biol. Chem. 267, 13115-13118.

Wheeler, D.L., Vander Griend, R.A., Wronski, T.J., Miller, G.J., Keith, E.E., and Graves, J.E. (1995). The short- and long-term effects of methotrexate on the rat skeleton. Bone 16 (2), 215-221.

Wilmink, J.M., Bras, J., Surachno, S., van Heyst, J.L.A.M., and van der Horst, J.M. (1989). Bone repair in cyclosporin-treated renal transplant patients. Transpl. Proc. 21 (1), 1492-1494.

Wolinsky-Friedland, M. (1995). Drug-induced metabolic bone disease. Endocr. Metab. Clin. North America 24 (2), 395-420.

Yasunaga, C., Cramer, D.V., Chapman, F.A., Wang, H.K., Barnett, M., Wu, G.D., and Makowka, L. (1993). The prevention of accelerated cardiac allograft rejection in sensitized recipients after treatment with Brequinar Sodium. Transplantation 56 (4), 898-904.

Yuh, D.D. and Morris, R.E. (1991). 15-deoxyspergualin is a more potent and effective immunosuppressant than cyclosporine but does not effectively suppress lymphoproliferation in vivo. Transpl. Proc. 23 (1), 535-539.

Zhu, X., Imamura, M., Tanaka, J., Han, C.W., Hashino, S., Imai, K., Asano, M., Nakane, A., Minagawa, T., Kobayashi, M., Sakurada, K., and Miyazaki, T. (1994). Effects of 15-deoxyspergualin in vitro and in vivo on cytokine gene expression. Transpl. 58 (10), 1104-1109.

DEVELOPMENTAL DISORDERS OF BONE

Jay R. Shapiro, MD

Advances in Organ Biology
Volume 5C, pages 765-795.

ISBN: 0-7623-0390-5

I. INTRODUCTION

From the standpoint of structure and function, skeletal tissue is highly complex and is subject to extensive modifications during every period of life. The genetic component of skeletal development has received increasing attention during recent years as investigators have applied new technologies to gain an understanding of disorders of skeletal development recognized during early life (Ralston 1997). There are a large number of genetically-determined developmental disorders in children associated with defective skeletal formation. These have been clinically categorized by McKusick in 1992. On the other end of the age spectrum, the attainment of adequate peak bone mass during the third decade may also be determined genetically. Recent investigations have explored factors that may serve to limit bone loss with increasing age. Among these are several candidate genes associated with decreased bone mineral density in older persons (Morrison et al., 1994; Kobayashi et al., 1996).

This chapter will present a summary of findings related to the molecular biology of several of the more common developmental disorders of bone. The reader is also referred to reviews for additional information not included here because of limitations on space (Tilstra and Byers, 1994; Lacombe, 1995; Wood, 1998; Uitterlinden, 1998).

II. FACTORS REGULATING SKELETAL DEVELOPMENT

The development of skeletal tissue follows two basic pathways: intramembranous bone formation whereby mesenchymal precursor cells differentiate directly into osteoblasts, and endochondral bone formation whereby chondrocytes in cartilage templates undergo a process of cellular proliferation, hypertrophy, calcification of their extracellular matrix, and eventually cell death (apoptosis). During the evolution of this process, precursor-derived osteoblasts and osteoclasts in invading neovascular tissue replace the endochondral cartilage template with osseous matrix (Reddi, 1994a; Erlebacher et al., 1995).

Essential to normal skeletal development is the orderly sequential expression of genes regulating bone morphogenesis, as well as the timely interaction of these genes with multiple growth factors, cytokines, and hormones that serve to modulate cell differentiation. Sonic hedgehog (Shh) is a morphogenic protein that is essential for the normal development of multiple organ systems including craniofacial structures and limbs (Ming,

1998). Genes of the Shh response pathway include patched, the Sonic hedgehog receptor, gli, and gli2/4 (Bórycki, 1998). Mutations involving the Shh signaling pathway genes include holoprosencephaly, basal cell carcinoma syndromes, and the congential disorders Greig syndrome, Pallister-Hall syndrome, and postaxial polydactly (Ming, 1998). Indian hedgehog (Ihh) secreted by prehypertrophic chondrocytes modulates chondrocyte differentiation and also activates another secreted factor, parathyroid hormone–related protein (PTHrP) (Vortkamp, 1998). PTHrP, first recognized as a mediator of tumor-related hypercalcemia, has now been shown by targeted disruption studies in mice to regulate chondrocyte differentiation and endochondral bone formation (Suda, 1997). Also essential in the process of skeletal development are the homeobox genes of the *hox* family (e.g., Hoxd-4, Hoxc-8) of genes that code for transcription factors that regulate gene expression related to patterning of the head and neck as well as limb morphogenesis. (Yueh, 1998). Members of the transforming growth factor β (TGFβ) super-family of growth factors that includes several of the bone morphogenetic proteins (BMPs) that are also critical to modulating skeletal development (Reddi, 1994b; Horan et al., 1995; Martin et al., 1995). The fibroblast growth factors are also members of the TGFβ family of growth factors.

The transcription factor Cbfa1, identified initially in osteoblasts and bone, has recently been identified as "the master gene for bone growth". Treatment with BMP-7, which promotes osteoblast differentiation, led to expression of *Cbfa1* before expression of other osteoblast-specific genes while in mouse embryos, peak expression of Cbfa1 occurred before the first ossification centers appeared. When transfected into several nonosteoblastic cell strains, this transcription factor induces production of osteoblast-related proteins such as osteocalcin and bone sialoprotein. Conversely, targeted disruption of Cbfa1 results in a failure to form bone (Kimori, 1997). Heterozygous mutations involving Cbfa1 are associated with cleidocranial dysplasia (Mundlos, 1997; Otto, 1997)

III. ONCOGENES AND BONE CELL PHYSIOLOGY

Expression of the family of c-fos and c-jun proto-oncogenes by osteoblasts and osteoclasts is involved in regulating the process of bone formation and remodeling (McCabe et al., 1965; Jacenko, 1995). The proto-oncogene proteins c-fos and c-jun are part of a family of nuclear transcription factors that also includes fos B, fra-1, fra-2, jun B, and jun D. These oncoproteins are components of the AP-1 transcription complex that form heterodimers with any jun protein: dimer formation is required for binding to DNA AP-1 sites (Machwate et al., 1995). High levels of c-fos, c-jun, and jun B mRNA transcripts have been demonstrated during the period of osteoblast development (McCabe et al., 1965). Expression of c-fos, c-jun, jun B, and jun D is also significantly enhanced late in the developmental sequence when apoptosis is evident. The temporal pattern of c-fos expression during intramembranous and endochondral bone formation has been investigated in fetal and neonatal rat bone (Machwate et al., 1995; Grigoriadis, 1995). Expression of c-fos, histone H-4, and osteocalcin mRNAs were shown to follow a temporal sequence during bone formation. In fetal rat bone, c-fos mRNA levels were found to increase transiently at birth in both calvaria and femur preceeding a rise in osteocalcin mRNA. c-fos was demonstrated in osteoprogenitor cells in the perichondrium and periosteum, but not in mature osteoblasts (McCabe et al., 1965).

The TFGβ family of growth factors that includes the group of bone morphogenetic proteins (BMPs) interact with c-fos in regulating osteoblast differentiation (Ohta et al., 1992). Studies with cultured human osteoblasts have demonstrated that TGFβ rapidly increases mRNA levels of c-fos and gradually increases expression of mRNA for jun B, but decreases c-jun mRNA levels (Subramaniam et al., 1995). Noggin is a BMP antagonist expressed in condensing cartilage and immature chondrocytes. When noggin expression is inhibited in mice, cartilage hyperplasia and failed joint development ensue (Brunet, 1998).

Proto-oncogene proteins also play a critical role in osteoclastic bone resorption. The gene *src* encodes pp60c-src, a nonreceptor protein tyrosine kinase (Yoneda et al., 1993). Pp60c-src protein expression and pp60-src tyrosine kinase activity were shown to correlate with the numbers of active murine bone marrow derived osteoclasts in tissue culture (Yoneda et al., 1993). Mice carrying a disruption of this gene develop osteopetrosis due to defective osteoclast function secondary to failure of the osteoclasts to form ruffled borders (Boyce et al., 1993; Lowe et al., 1993). However, expression of a wild-type kinase-deficient Src in src-/- mice has been shown capable of reversing the src-/- phenotype (Schwartzberg, 1997). *In vitro* studies using selective inhibitors of c-src (herbimycin A, gelanamycin)decrease osteoclastic bone resorption (Hall et al., 1994). However, parathyroid hormone will increase, and calcitonin will decrease pp60-src tyrosine kinase activity and the numbers of osteoclasts in bone marrow cultures (Yoneda et al., 1993).

Definition of the role of early response genes in osteoblast and osteoclast differentiation and of the coordination of factors modulating gene expression by bone cells is critical to understanding how the cellular phenotype is altered by genetic disease as well as for defining methods by which these disorders may be treated in the future.

IV. DISORDERS RELATED TO MUTATIONS OF THE FIBROBLASTIC GROWTH FACTOR RECEPTORS

Fibroblastic growth factor receptors (FGFRs) are members of the tyrosine kinase family of cell surface receptors (Ullrich and Schlessinger, 1995). Four members have been identified that bind a group of at least 15 related mitogenic fibroblast growth factors (Johnson and Williams, 1993; Szebenyi, 1998). Three of the four FGFR genes generate multiple splice variants thus increasing the complexity of the system (Johnson and Williams, 1993). Ligand binding results in proliferation of multiple cell types, neural outgrowth from cerebral and hippocampal neurons and differentiation of preadipocytic fibroblasts into adipocytes. Different point mutations may activate fibroblast growth factor receptors via both ligand-dependent and -independent mechanisms (Neilson, 1996). Heterozygous mutations involving the fibroblast growth factor 3 have been reported in patients with achondroplasia, thanatrophoric dysplasia and hypochondroplasia. (Horton, 1997). Mutations as a cause of skeletal disorders have been reviewed by Muenke and Francomano, and Gorlin (1997). (Francomano, 1995; Muenke and Schell, 1995).

A. Achondroplasia

Achondroplasia is the most common of the chondrodysplasias in man (Tanaka, 1997). It is characterized by short limbed dwarfism, a normal sized trunk, and macrocephaly. It is inherited as a autosomal dominant trait although the majority of cases are sporadic. Achondroplasia occurs with an estimated frequency of 1/15,000–40,000 live births (Rosseau et al., 1994).

The gene for achondroplasia has been mapped by several groups using linkage techniques. Francomano localized the gene to 4p16.3 using 18 multigenerational families with achondroplasia and 10 short tandem repeat polymorphic markers from this region (Francomano et al., 1994; Le Merrer et al., 1994;Velinov et al., 1994). The fibroblast growth factor receptor 3 gene (FGFR3) was known to be situated in this area. Subsequently, two groups reported mutations affecting the FGFR3 gene in achondroplasia (Rosseau et al., 1994; Shiang et al., 1994). Unlike the variable pattern in other heritable disorders (see osteognesis imperfecta below), in 97% of cases, achondroplasia was found associated with on of two mutations: G to A transition at nucleotide 1138, or G to C mutation at nucleotide 1138 affecting the transmembrane domain of the FGFR3 gene by substituting arginine for glycine at position 380 of the protein. These apparently occur on the paternally derived chromosome in sporadic cases of achondroplasia (Wilkin, 1998). Although a G to C mutation was observed in another case (Shiang et al., 1994), Rosseau also found that 23 mutations in achondroplasia cases resulted from the same G to A. (Rosseau et al., 1994). A single case bore a novel G-T transition at codon 375 (Ikegawa et al., 1995). Ikegawa also reported a G-T transition at condon 375 which resulted in the substitution of a cysteine for a glycine. These mutations frequently occur in a cytosine phosphate (Cp) next to guanine, a CpG dinucleotide that is known to be a mutational "hotspot" (Francomano, 1995).

B. Thanatophoric Dysplasia

Thanatophoric dysplasia is a sporadic lethal disorder that occurs with a frequency of approximately 0.28–0.60 in 10,000 (Tavomina et al., 1995). (Kaufman et al., 1970). The phenotype of thanatophoric dysplasia includes micromelic shortening of limbs, macrocephaly with bossing, decreased height of vertebral bodies and shortened ribs resulting in a reduced thoracic cavity, and a bell-shaped abdomen. There is poor cellular proliferation and column formation in cartilaginous growth plates of long bones. The disorder is usually lethal due to early respiratory failure. Different mutations affecting FGFR3 have been reported in different sporadic types of thanatophric dysplasia (types I and II) (Tavomina et al., 1995).

Various mutations affecting FGFR3 have been reported in different sporadic phenotypes of thanatophric dysplasia. The classification into phenotypes types I (no cloverleaf skull deformity) and type II thanatophoric dysplasia remains uncertain. (Tavormina et al., 1995; Weber, 1998). In type I, mutations are reported to occur in the Ig2–Ig3 linker domain or in the extracellular juxtamembrane domain and involve mutation of the wild-type residue to Cys that allows abnormal dimerization and thus ligand-independent receptor activation. (d'Avis, 1998). Correlation with the phenotypes has been reported by Wilcox and colleagues. Cases with a Lys650Glu substitution had straight femora with craniostenosis and frequently the cloverleaf skull deformity. Histopathologically, cases with the Lys650Glu substitution had better preservation of the growth plate. Cases with Tyr373Cys tended to have more severe radiographic findings than those with the Arg 248Cys cases (Wilcox, 1998).

C. Craniostenosis Syndromes

Premature fusion of sutures of the skull bones can occur among a series of heterogeneous disorders, the most well-studied being the Apert's, Crouzon's, Pfeiffer's, and Jackson-Weiss syndromes. Of interest in these syndromes, as in other heritable disorders of connective tissue, is the association of different fibroblastic growth factor receptor (FGFR) mutations in patients with similar clinical phenotypes. Conversely, identical mutations have been reported in phenotypically different syndromes. An example of this is the identification in the Jackson-Weiss syndrome (craniostenosis associated with foot malformations) of missense mutations shared with Crouzon and Pfeiffer phenotypes (Tartaglia, 1997). These mutations mainly affect the *FGFR-2* gene and the *FGFR-1* and *FGFR-3* genes to a lesser extent. Fibroblast growth factor-2 (FGF-2) inhibits longitudinal bone growth by altering cartilage matrix production. (Mancilla, 1998). FGFR exon 7 mutations have been reported in the Pfeiffer, Crouzon, and Jackson-Weiss syndromes (Jabs et al., 1994; Rutland et al., 1995).

Pfeiffer Syndromes

In the Pfeiffer syndromes, (types I, II, III) bicoronal craniostenosis, midface hypoplasia and beaked nasal tip are associated with broad thumbs and toes, syndactly, and brachydactly (Moore, 1995; Robin, 1998). Mutations in both the FGFR1 and fibroblast growth factor 2 (FGF2) receptors have been reported in the Pfeiffer syndrome (Laujenie et al., 1995; Muenkde et al., 1995; Ruthland et al., 1995). Single-strand conformational polymorphism (SSCP) was employed by Rutland et al. (1995) to define point mutations in FGFR2 exon 7 in seven patients with sporadic Pfeiffer syndrome. In five patients, these were T→C transitions at nucleotide 1036, a mutation (Cys342Arg) also reported in the Crouzon syndrome (Reardon et al., 1994a). A G→A transition at nucleotide 1037, also previously reported in Crouzon syndrome, and an A→C transversion at 1033 not previously observed were identified. A serine 351-cysteine mutation in the *FGFR-2* gene has also been reported in a sporadic case (Mathijssen, 1998).

Crouzon Syndrome

This autosomal dominant syndrome is characterized by abnormal cranial shape and proptosis due to early suture closure. Although common to other craniostenosis syndromes, digital abnormalities are not present in the Crouzon syndrome. Crouzon syndrome has been mapped to chromosome 10q25-q26 (Preston et al., 1994). The FGF2 gene which is also located on chromosome 10 (10q25.3-26) was considered a putative candidate gene for this syndrome (Miki et al., 1992). Reardon and colleagues (1994b) have reported polymorphisms observed with SSCP analysis involving the coding sequence and splice junctions of exon B of FGFR2 that segregated with the disorder in three families. Direct sequencing of exon B revealed mutations not found in normals. These involved a G to A transition (cys342Tyr substitution) at nucleotide 1037. An additional sporadic case involved a cys342→ argenine substitution that predicts a serine subsitituon for glycine. Yet in other cases, mutations have been observed to predict substitution of serine for cysteine-B. A C→T mutation leading to Tyr340-His substitution in another case predicted a C to G mutation. A C→G mutation led to a Ser 354Cys substitution. A G→A transition was found at nucleotide 1044.

V. THE OSTEOPETROSIS SYNDROMES

The osteopetrosis syndromes are a heterogenous group of congential disorders characterized by the deposition of variable amounts of dense bone that typically appears radiopaque on x-ray but is frequently structurally defective (Frame et al., 1987). Osteopetrosis is the end result of defective development or defective function of osteoclasts, the cells that normally resorb bone (Marks, 1989). The phenotype varies from individuals with the autosomal dominant benign types (types I and II) of osteopetrosis who are usually asymptomatic and have a normal life span, to the rare "malignant osteopetrosis" syndrome that is inherited as an autosomal recessive trait. The clinical spectrum of osteopetrotic disorders has been reviewed by Bollerslev and Anderson (1987).

The children with recessive disease ("malignant" osteopetrosis) present in infancy with failure to thrive, anemia, pancytopenia, hepatosplenomegaly, and multiple cranial nerve palsies. This condition is usually lethal in early childhood. There is diffuse skeletal sclerosis with loss of marrow spaces. In a study of 33 patients, Gerritsen et al. (1994a) found that ocular involvement occurring at a median age of two months was found at initial examination in one-half of the patients. Retinal degeneration and neurologic complications also were common (Gerritsen et al., 1994a). Based on mapping of the murine osteosclerosis (oc) mutation, human recessive osteopetrosis has been located to 11q13 (Heaney, 1998). It has been noted that in addition to defective osteoclast formation, osteoblast function may be abnormal in these patients (Lajeunesse, 1996; Lian, 1996). Osteoblastic cells obtained before bone marrow transplantation showed defective alkaline phosphatase and osteocalcin responses to 1,25(OH) vitamin D and failed to produce macrophage colony stimulation factor in response to IL-1 and TNF. These defects normalized after bone marrow transplant repopulated the marrow with functioning osteoclasts.

Allogenic bone marrow transplantion is currently the mainstay of treatment for malignant osteopetrosis, but an acceptable donor can be found in only 40% of patients. (Coccia et al., 1980; Gerritsen et al., 1994b; Whyte, 1995). Malignant osteopetrosis has been identified in one study as an absolute indication of unrelated donor marrow transplantation (Miano, 1998). Successful marrow engraftment may lead to an increase in bone remodeling and new non-sclerotic bone formation (Solh et al., 1995). In a European study there was 47% survival two years after transplantation, with a cure rate of 62% in the survivors (Fischer et al., 1991). High dose vitamin D therapy has been administered to increase differentiation of osteoclasts and promote bone resorption, with amelioriation of osteopetrosis in 25% of patients (Key et al., 1984). However, patients may become refractory to this therapy. Long-term treatment of subjects with congenital osteopetrosis with interferon γ-1b has been found to increase bone resorption and to improve hematopoesis and leukocyte function (Key et al., 1995). The rational for this treatment was based on the observation that interferon 1β corrected a defect in the generation of superoxide by leukocytes of subjects with chronic granulomatous disease, a defect also found in the leukocytes of certain osteopetrotic subjects.

The benign forms of osteopetrosis are usually diagnosed by chance, but may be associated with anemia, facial palsy, or nerve compression syndromes such as deafness (Frame et al., 1987). The phenotypes of type I and II disease were presented by Bollerslev and Anderson (1987). Pathologic fractures may occur in benign disease and healing may be impaired. In a survey of 42 patients with benign disease, el-Tawil and Stoker (1993) found that fractures were more common in type II disease as were sclerotic metaphyseal bands. The gene for autosomal dominant osteopetrosis has been localized to chromosome 1p21 (Van Hul, 1997). Administration of thyroid hormone has recently been studied as a means of increasing bone remodeling in subjects with type I-and -II autosomal dominant osteopetrosis (Bollerslev, 1998).

Renal tubular acidosis may occur in certain patients with osteopetrosis due to a deficiency of carbonic anhydrase II. The severity of the disease is highly variable, as is the age of onset. The phenotype includes the presence of osteopetrosis, a mixed type of renal tubular acidosis, cerebral calcification and growth retardation. Mental retardation occurs in as many as 90% of affected individuals. Dental malocclusion is also common in this syndrome (Sly et al., 1985; Aramaki et al., 1993). Erythrocyte carbonic anhydrase II levels are undetectable (Whyte et al., 1980). Obligate heterozygotes have half-normal enzyme levels (Sly et al., 1985). Identification of mutations in carbonic anhydrase II may provide the opportunity for prenatal diagnosis in subsequent pregnancies. (Strisciuglio, 1998). Carbonic anhydrase II has been demonstrated in osteoclasts (Vaananen and Parvinen, 1983). Sly has proposed that failure to titrate the OH^- produced by bicarbonate limits the ability of the osteoclast to secrete H^+, and thus limits osteoclast resorption leading to osteopetrosis.

There are five isoenzymes of carbonic anhydrase. Carbonic anhydrase I, II, and III are clustered on chromosome 8q22 (Tashian, 1991). Carbonic anhydrase II deficiency is inherited as an autosomal recessive trait (Sly et al., 1985). The first reported mutation was in a Belgian patient and was a C→T transition in exon 3, resulting in a tyrosine replacing histidine at position 103 (Venta et al., 1991). Roth et al. (1992) have reported mutations affecting the structural gene for carbonic anhydrase II in the American family in which this syndrome was first described. The three affected siblings from this family were compound heterozygotes, each having inherited two different mutations. The paternal mutation was a splice acceptor mutation at the 3' end of exon 5. The maternal mutation substituted tyrosine for histidine 107 in exon 3 (Roth et al., 1992). Subsequent studies have demonstrated carbonic anhydrase II gene mutations in patients from different countries: in Japanese families, mutations involved a T→G transition at exon 2. Here paternal and maternal mutations were the same (Soda, 1994). Splice junction mutations in intron 2 have been reported for which many Arabic families were homozygous (Hu et al., 1992a). Seven Hispanic families were found to be homozygous for a single base deletion in the coding region of exon 7 (Hu et al., 1994a). Six additional patients of Caribbean ancestry with osteopetrosis and renal tubular acidosis were homozygous for the same mutation, but expression of the disorder varied widely in severity (Hu et al., 1994b). Analysis of six unrelated Arabic kindreds revealed five to be homozygous and one heterozygous for a donor splice site mutation at the 5' end of exon 3. The Arabic patients differ from their American and Belgian counterparts with the His 107→Tyr mutation in having mental retardation and infrequent fractures (Shapiro et al., 1988). Tunisian patients, who also have mental retardation, are reported to have a splice junction mutation at the 5' end of intron 2 (Fathallah et al., 1994). A frameshift mutation due to a single base deletion leading to the formation of a truncated protein was described in a Hispanic girl who had no renal tubular acidosis (Hu et al., 1992b). Decreased enzyme activity was demonstrated when the enzyme was expressed in bacteria. However, it was determined that the mutant enzymatic activity was not due to the truncated protein, but rather it was due to depressed activity of an allele product of nearly normal size. This mutation was later reported in seven other unrelated Hispanic families in whom expression of the identical mutation was found to be highly variable (Hu et al., 1992b, 1994b).

Animal models of osteopetrosis have been invaluable in permitting the definition of different abnormalities that lead to defective osteoclast function (Seifert et al., 1993). The pioneering work of Walker (1975) demonstrated that grey-lethal mice may be cured by parabiosis to normal littermates and that intravenous administration of bone marrow or spleen cell suspensions from normal littermates have the same effect (Walker, 1975). Nine genetically distinct osteopetrotic mutations have been described in three animal species (rat, mouse, and rabbit) (Hermey et al., 1995) and each is inherited as a autosomal recessive trait. (Table 1).

The heterogeneity of these animal models in regard to various osteoclast defects is illustrated by the fact that mutations have been assigned to four different chromosomes (Marks, 1989). Mice homozygous for the op mutation are characterized by defective osteoclast formation from macrophages due to lack of macrophage colony stimulating factor (M-CSF/CSF-1) (Takatsuka, 1998). Gene transfer of CSF-1 into op/op marrow stromal cells has been reported to correct in vitro osteoclastogenesis in these animals (Abboud, 1998). Studies in the toothless rat mutant (*tl/tl*) demonstrate multiple abnormalities: decreased bone

Table 1. Animal Models of Osteopetrosis

Species	*Reference*	*Defect*
Rat		
Osteopetrosis (*op/op*)	Hermey et al., 1995	
Incisors-absent (*ia/ia*)	Schneider et al., 1979	
Toothless (*tl/tl*)	Marks (73?)	M-CSF improves osteoclast number and morphology
Mouse		
Osteopetrotic (*op/op*)	Felix et al., 1990	Lack of M-CSF/CSF-1
Microophthalmia(*mi/mi*)	Steingrimsson et al., 1994	
Rabbit		
Osteosclerotic (*os/os*)	Lenhard et al., 1990	

osteoclasts and macrophages, elevated levels of osteopontin gene expression and decreased levels of osteocalcin gene expression, and elevated serum levels of $25(OH)_2D_3$ (Jackson et al., 1994). In addition, an accelerated maturational sequence of gene expression has been reported for cultured osteoblasts from the *tl/tl* mutant (Shalhoub et al., 1994). Unlike the op/op mouse, treatment with CSF-1 increased, but did not normalize, osteoclast number in the tl/tl rat (Marks, 1997). As a consequence of this heterogeneity, multiple methods such as marrow and spleen cell transplantation, or cytokines such as colony stimulating factor, have been employed in attempts to correct the osteoclast lesion in these animal mutants (Felix et al., 1990; Nesbitt et al., 1995).

Studies in transgenic mice have pointed to the significant role c-*src* plays in the process of osteoclastic bone resorption (Lowe et al., 1993; Hall et al., 1994). Targeted disruption (knockout) of the *src* proto-oncogene in mice results in the development of osteopetrosis due to the failure of the src mutant osteoclasts to form ruffled borders (Boyce et al., 1993). However, analysis of fibroblasts from three individuals with malignant osteopetrosis has demonstrated similar levels of c-*src* protein and c-*src* kinase activity between affected individuals and controls (Whyte, 1995). c-*src* is presumed to function in the translocation and/or fusion of exocytic vesicles to the border of the osteoclast ruffled membrane (Hall et al., 1994). Recent studies indicate that expression of a wild-type Src in only a limited number of tissues in transgenic mice can fully rescue the src-/- osteopetrotic phenotype. Here, Src may act to increase the function of specific tyrosine kinases (Schwartzenberg, 1997).

VI. HYPOPHOSPHATASIA

Hypophosphatasia is an inborn error of metabolism expressed as rickets in children and osteomalacia in adults (Whyte, 1994). The disorder is characterized by phenotypic and genetic heterogeneity. Biochemically, hypophosphatasia is associated with a deficiency of the tissue-nonspecific isoenzyme of alkaline phosphatase (TNSALP) in serum and tissues, and increased excretion of phosphoethanolamine in the urine (Whyte, 1994). Four clinical types of hypophosphatasia have been identified: perinatal lethal disease, infantile hypophosphatasia, childhood, and adult forms of the disease (Whyte, 1994). Odontohypophosphatasia refers to subjects with only dental manifestations. Pseudohypophosphatasia is a syndrome in which the skeletal picture resembles hypophosphatasia but serum alkaline phosphatase levels are normal. Variability in each clinical phenotype is also typical of these disorders. The severe forms of hypophosphatasia are inherited as autosomal recessive traits (Moore et al., 1990). Inheritance in the milder types of hypophosphatasia is less well-defined. Milder forms of hypophosphatasia may be transmitted as an autosomal dominant trait (Eastman and Bixler, 1983).

Serum alkaline phosphatse levels are typically subnormal in hypophophatasia. In general, the earlier the onset of disease, the more severe the disease and the lower the alkaline phosphatase. Serum calcium values are normal. By contrast, in vitamin D deficiency, vitamin D resistant rickets or osteomalacia, serum calcium is usually low and alkaline phosphatase levels are usually increased (Whyte, 1994).

The gene for hypophosphatasia is located on chromosome 1p34-36 (Smith et al., 1988; Harris, 1989). A missense mutation (Alanine162→Threonine) in exon 6 of the TNSALP gene was first noted in a lethal case by Weiss et al. in 1988. However, several other cases failed to show this mutation. Subsequently, one of eight different missense mutations were identified in 23 of 50 patients screened by Henthorn and Whyte (1992). Similar point mutations were observed in both severe and clinically mild forms of the disease. (Henthorn and Whyte, 1992). Greenberg and coworkers (1993) subsequently reported that homozygosity for a tenth mutation involving exon 10 (Gly317→Asp) was responsible for a high incidence of lethal disease in Canadian Mennonites. An interesting subject with lethal disease was found to be a compound heterozygote who inherited a missense mutation in exon 9 (G1068→A) and a silent mutation in exon 10 from the father, and a frameshift mutation secondary to a deletion of T at codon 503 in exon 12 from the mother (Orimo et al., 1994).

VII. OSTEOGENESIS IMPERFECTA

Osteogenesis imperfecta (OI) is a heritable disorder of connective tissue that has been classified into four major clinical types based on phenotypic characteristics: type 1, mild OI; type II, lethal; type III, progressive deforming OI; and type IV, moderately severe OI. (Sillence et al., 1979; Sillence, 1981; Beighton et al., 1988). The cardinal manifestations of OI are the presence of osteoporotic brittle bones that fracture easily, usually resulting in variable amounts of skeletal deformity. Scoliosis, joint laxity, a blue-grey hue to the sclera in many (but not all) patients, adult onset hearing loss and, very frequently, short stature are common phenotypic traits. The majority of OI cases are inherited as autosomal dominant traits. Approximately 6–7% are inherited as autosomal recessive traits (Beighton and Versfeld, 1985). It is recognized that many cases thought to be recessive are probably the result of germinal or somatic mosaicism in a normal appearing parent (Cohn et al., 1990; Zlotogora, 1998).

Although OI has been recognized as a clinical syndrome since 1733, it was in 1975 that a mutation affecting type I collagen defect in OI was first suspected based on the finding of extreme friability of connective tissues in a neonate with OI (Pentinnen et al., 1975). Type I collagen is the main structural protein of bone, tendon, ligament, and skin and, for this reason, investigators have focused on defining mutations affecting type I collagen α chains that could impair the normal organization and mineralization of skeletal matrix.

Type I collagen is a heterotrimer that contains two pro-α1(I) chains and one pro-α2(I) chain, assembled in a triple helical configuration (Kivirikko, 1993). Each pro-α chain contains the repeat triplet $(\text{glycine-X-Y})_{338}$ where 23% of the residues may be proline or hydroxy-proline. These polypeptide chains contain a large triple helical domain of about 1,000 amino acids, ending in a telopeptide region and N- and C-terminal propeptide extensions. Hsp47 and Hsp70 are stress-inducible heat shock glycoproteins that function in the folding and assembly of procollagen molecules in the rough endoplasmic reticulum (Naki et al., 1992; Ferreira et al., 1994). Following assembly, procollagen chains are secreted into the extracellular space where the propeptide extensions are cleaved by specific proteinases. The C-terminal proteinase has recently been identified as identical to BMP 1 (Kessler et al., 1996). Following propeptide cleavage, the triple helical molecules self-assemble into fibrils joined by lysyl and OH-lysyl cross-links. Maturation of these cross-links includes the formation of pyridinoline cross-links in the N-terminal domain.

Nineteen distinct collagen types involving 30 different pro-α chains comprise the collagen family of proteins identified to date (Mayne and Brewton, 1993). Advances in molecular biology have facilitated the identification of different mutations affecting either the COL1A1 or COL1A2 genes in subjects with OI. These genes have been localized to chromosomes 17q21.3-q22 and 7q21.3-q22.1, respectively (Retif et al., 1985; Huerre et al., 1992). The COL1A1 and COL1A2 genes are large and complex—4 kb in size—encompassing 52 exons. Mutations occur either in the portion of the gene coding for the triple helical region of the collagen molecule or, less commonly, in the regions coding the telopeptide or N- or C-terminal propeptide extensions.

In a survey of mutations affecting fibrillar collagens (types I, II, III, and XI), Kuivaniemi and colleagues reported that a majority of the mutations were single-base and changed either the codon of a critical amino acid (63%; e.g., first position glycine) or led to abnormal RNA splicing (13%) (Kuivaniemi ,1997). In OI, more than 83 mutations in 126 patients have been reported to involve different regions of the COL1A1 and COL1A2 genes. Both pro-α1(I) and pro-α2(I) mutations have been reported in each clinical phenotype (Sykes et al., 1986). Over 90% of these are heterozygous single base pair mutations usually affecting the first-position glycine, and with the majority affect the pro-α1(I) chain. Also reported are splice-site mutations associated with exon skipping and the formation of stop codons leading to the synthesis of shortened pro-α chains, as well as deletions and insertions of different sizes (Kuivaniemi et al., 1988; Tromp and Prockop, 1988). Mutations that intefere with procollagen triple helix assembly alter the stability of the procollagen chains as demonstrated by an increase in susceptibility to proteolytic digestion and thermal denaturation (Bachinger et al., 1993). Because assembly of the triple helix preceeds from the C- to N-terminal direction, mutations in the C-terminal domain are frequently, but not always, associated with more severe disease. The reader is referred to several excellent reviews for additional information about specific type I collagen mutations in OI and the type-I collagen mutation database (Rowe, 1990; Cole, 1994a; Tilstra and Byers, 1994; Prockop and Kivirikko, 1995; Shapiro et al., 1996; Kuivaniemi, 1997; Dalgleish, 1997).

A. Null Allelic Mutations

Null mutations have been most commonly identified in mild type I OI in whom structural mutations affecting the collagen helical portion are usually not observed. As reported with cultured dermal fibroblasts, the null mutation is associated with a 50% decrease in total collagen synthesis but a normal 2:1 ratio of α1(I) to α2(I) chains secreted into the extracellular matrix (Willing et al., 1994). Most cases have decreased mRNA levels for pro-α1(I) and decreased production ofα1(I)chain (Willing et al., 1994). The mechanism underlying the null allele apparently involves accumulation of the mutant mRNA in the nucleus due to a premature stop mutation rather than its transport to the cytoplasm (Redford-Badwal, 1996). In an examination of nonsense and frameshift mutations inducing premature termination of translation of the *COL1A1* gene in type-I OI, Willing and colleagues found a marked reduction in mRNA from the mutant allele suggesting that nonsense-mediated decay of COL1A1 is a nuclear event (Willing, 1996). This is illustrated by one OI patient in which, in the presence of a donor splice site mutation, the abnormally spliced product which was out of frame was retained while the product of the normal allele was transported to the cytoplasm, translated and secreted into the extracellular matirx (Stover et al., 1993). However, other mechanisms that might produce the null allele with underproduction of collagen chains include posttranscriptional, translational and posttranslational mutations where chain assembly is altered by mutations affecting the C-terminal propeptide (Shapiro et al., 1996). For example, a splice site point mutation (AG to GG) in exon 16 resulted in deletion of exon 17 and a shortened pro-α1(I) chain (Willing et al., 1994). Polymorphic markers including a 4-bp insertion in the 3' UTR of the COL1A1 gene localized downstream of a recognized MnII RFLP site have been used in combination to identify type-I patients exhibiting the null allele effect (Nuytinck, 1998).

B. Dominant-Negative Mutations in OI

The dominant-negative mutation results when the mutated pro-α chain is incorporated into the procollagen trimer and compromises the function of the other normal chains in the triple helical structure. These mutations affect either the pro-α1(I) or pro-α2(I) chains and are responsible for clinical phenotypes of varying severity ranging from lethal type II OI to the moderately severe type IV phenotype (Byers et al., 1988).

C. Helical Domain Glycine Mutations

These mutations have been reported in each OI phenotype (Bachinger et al., 1993). Point substitutions most frequently occur in the first or second nucleotides of a GGN codon for the pro-α1(I) chain. Most commonly reported are point mutations involving the substitution of a first position glycine by cysteine. Other potential substitutions could involve glycine substitutions by alanine, arginine, aspartate, cysteine, glutamic acid, serine, valine, and tryptophan. Tryptophan and glutamic acid substitutions have not been reported. The substitution of glycine by larger and more bulky amino acids may have a more disruptive effect on the assembly of the collagen pro-α chains and may be associated with more severe disease. Glycine substitutions by aspartate have been reported only in association with lethal disease (Cole, 1994b). In lethal cases, a cysteine substitution (Gly748→Cys and Gly 718→Cys) led to a kink in the triple helix (Vogel et al., 1988; Lightfoot et al., 1992) but a similar substitution in another lethal case (Gly→Cys) was shown to have only a minor effect on the stability of the triple helix. A chain-dependent stoichiometric effect of mutations on phenotype is illustrated by two patients harboring identical gly 661→ser mutations. Where this affected the pro-α1 (I) chain (e.g. α1/α2 ration =2:1), a severe phenotype resulted whereas when affecting the pro-α2(I) chain, only a mild type-I phenotype occurred (Nuytinck, 1996). Of interest are mutations reported in families exhibiting parental mosaicism—one a pro-α2(I) gly802→asp, where an infant with OI type III/IV was born to normal parents, and a second causing lethal OI due to a pro-α1(I) gly382→arg transversion where 36% of the phenotypically normal father's fibroblasts contained the mutant allele (Cohen-Solal, 1996).

D. Splicing Mutations

Several point mutations or deletions affecting donor or acceptor splice sites have resulted in exon skipping (Rowe et al., 1985; Kuivaniemi et al., 1991; Bateman et al., 1994). With most proteins, exon skipping leads to frameshifts and the introduction of premature stop condons resulting in shortened pro-α chains that may be degraded intracellularly. However, in type I collagen, each exon codes multiple complete (Gly-X-Y) triplets, thus maintaining the triple helical sequence (Cole, 1994b). Some intron mutations result in activation of cryptic splice sites that lead to insertion of intron sequences. In one interesting example, a point mutation in the donor splice site of COL1A1 intron 8 resulted in skipping of the upstream exon as well as having a secondary effect of activating a cyptic 5' splice site in the next upstream intron causing the inclusion of 96 base pairs in intron 7. This resulted in the insertion of a non-collagenous protein sequence in the triple helix in addition to the exon skip (Bateman et al., 1994). This pattern has also been observed with mutations affecting the pro-α2(I) chain (Genovese et al., 1989). Alternative splicing affecting COL1A2 exon 26 due to an A to G transition in IVS 26 donor splice site has been suggested to explain intrafamilial phenotypic variability where the father and paternal grandmother had mild OI but the proband had type-IV disease detected in utero (Zolezzi, 1997).

E. Helical Domain Deletions and Insertions

Deletions involving the pro-α1(I) and pro-α2(I) chains have been associated with lethal OI (Chu et al., 1983, 1985; Barsh et al., 1985; Willing et al., 1988). Deletion of exons 22–24 was reported in lethal disease, as was deletion of exons 34–40 in the pro-α2(I) chain (Willing et al., 1988). In a family with type-IV OI, sequencing of the pro-α(I) demonstrated a 9-bp deletion involving codons 1003–1006 resulting in the removal the three amino acids (Gly-Pro-Pro) (Lund, 1996).

F. C-Terminal Telopeptide and Propeptide Mutations

Associated phenotypes vary from severe to mild (Cohn et al., 1988). Because the pro-α(I) chain is assembled from the C- to N-terminal direction, mutations at these sites would be expected to alter chain association and disulfide bonding. As observed by Cole, variable amounts of normal and mutant α chains are produced in the presence of C-telopeptide and C-propeptide mutations (Cole, 1994b). However, as illustrated in one type III OI patient and in the murine counterpart of this patient, the *oim* mutation, a C-propeptide pro-α2(I) (exon 52) mutation may be associated with failure to synthesize the pro-α2(I) chain resulting in the formation of an α1(I) homotrimer (Nicholls et al., 1984; Chipman et al., 1993; McBride and Shapiro, 1994). Mutations affecting the C-terminal domains are associated with posttranslational overmodification of the helical domain, delayed secretion, and retention in the endoplasmic reticulum. However, Raghunath et al. (1994) have reported normal kinetics in a case with a Gly-1017→Cys mutation that was outside the triple helical region in the C-terminal telopeptide. An infant with lethal OI (type-II) was found to harbor an unusual substitution of tryptophan 94 by cysteine in the carboxy-terminal propeptide of pro- α1(I) (Cole, 1996). This is significant because the clinical phenotype differed from neonates with type-II OI resulting from mutations affecting the helical domains of type I collagen.

VIII. TYPE II COLLAGEN MUTATIONS AND HERITABLE OSTEOARTHRITIS

Type II collagen, the major collagen component of cartilage, is a homotrimer composed three identical α chains: [α1(II)$_3$]. Mutations affecting the COL2A1 gene have been reported in several syndromes including the spondyloephiphyseal dysplasias, spondyloepimetaphyseal displasia, hypochondrogenesis, the Kniest dysplasia syndrome, Stickler syndrome and the achondrogenesis syndromes (Vikkula et al., 1994; Williams and Jiminez, 1995). Early onset of osteoarthritis is a feature of several of these syndromes (Bleasel et al., 1995). COL2A1 gene mutations have been identified in several families affected with a syndrome including the precocious development of osteoarthritis involving both weight-bearing and non-weight-bearing joints associated with mild vertebral chondrodysplasia (Ritvaniemi et al., 1995). These mutations include an Agr519→Cys base substitution, a Gly976→Ser substitution, and an Arg75→Cys mutation in a kindred with early onset osteoarthritis and mild spondyloepipyseal dysplasia. A Gly493→Ser mutation has been identified in a similar family (Katzenstein et al., 1992; Ritvaniemi et al., 1995; Williams and Jaminez, 1995). These mutations focus on precocious cartilage degeneration and its role in promoting the development of osteoarthritis.

IX. TYPE X COLLAGEN AND SKELETAL DISEASE

The growth of long bones occurs by the process of endochondral bone formation at the epiphyseal growth plate. In this zone, the formation of hypertropic chondrocytes preceeds the mineralization of extracellular matrix. It is during this stage of matrix development that the hypertrophic chondrocytes secrete type X collagen. Type X collagen, a product of the COL10A1 gene, is a homotrimer of three α 1 chains, $\alpha 1(X)_3$. Transgenic mice expressing a mutated type X transgene develop several abnormalities: a decrease in newly formed bone, compression of the hypertropic growth plate cartilage, leukocyte deficiency, and a reduction in the size of the thymus, spleen, and lymphopenia (Jacenko et al., 1993). By contrast, using gene targeting procedures, type X-collagen null mice were not different from their normal littermates suggesting that type X collagen was not normal for long bone development (Rosati et al., 1994). However, this is not the case in humans.

The Schmid syndrome (metaphyseal chondrodysplasia), is associated with mutations in type X collagen (Sweetman et al., 1992). It is characterized by progressive shortening of stature, bowed legs, and a waddling gait. Major long bones show widening, cupping, and irregular metaphyseal mineralization prior to epiphyseal closure. Type X mutations appear to localize mainly in the carboxy-terminal region of the molecule (McIntosh et al., 1995). In addition to a 13 base pair deletion in one type X allele (bp 1748–1760), other reported type X collagen mutations are Gly595→ Glu, Asp617→Lys, a single base deletion (T1908), as well as single base subsitituions at positions 597, 644, and 648 (Warman et al., 1993; Bonaventure et al., 1995).

REFERENCES

Abboud, S.L., Woodruff, K.A., and Choudhury, G.G. (1998). Retroviral-mediated gene transfer of CSF-1 into op/op stromal cells to correct defective in vitro osteoclastogenesis. J. Cell. Physiol. 1786, 323-331.

Aramaki, S., Yoshida, I., Yoshino, M., Kondo, M., Sato, Y., Noda, K., Okue, A., Sai, N., and Yamashita, F. (1993) Carbonic anhydrase II deficiency in three unrelated Japanese patients. J. Inherit. Metab. Dis. 16, 982-990.

Bachinger, H., Morris, N., and Davis, J. (1993). Thermal stability and folding of the collagen triple helix and effects of mutations in osteogenesis imperfecta on the triple helix of type-I collagen am. J. Med. Genet. 45, 152-162.

Barsh, G.S., Roush, C.L., Bonadio, J., Byers, P.H., and Gelinas, R.E. (1985). Intron-mediated recombination may cause a deletion in an α I Type-I collagen chain in a lethal form of osteogenesis imperfecta. Proc. Nat. Acad. Sci. USA 82, 2870-2874.

Bateman, J.F., Chan, D., Moeller, I., Hannagan, M., and Cole, W.G. (1994). A 5' splice site mutation affecting the pre-mRNA splicing of two upstream exons in the collagen *COL1A1* gene. Biochem. J. 302, 729-735.

Beighton, P. and Versfeld, G.A. (1985). On the paradoxically high relative prevalence of osteogenesis imperfecta type-III in the Black population of South Africa. Clin. Gent. 27, 398-401.

Beighton, P., dePaepe, A., Danks, D., Finidori, G., Gedde-dahl, T., Goodman, R., Hall, T.G., Hollister, D.W., Horton, W., and McKusick, V.A. (1988). International nosology of heritable disorders of connective tissue, Berlin. (1986). Am. J. Med. Genet. 29, 581-594.

Bleasel, J.F., Bisagni-Faure, A., Holderbaum, D., Vacher-Lavenu, M.C., Haqqi, T.M., Moskowitz, RW., and Menkes, C.J. (1995). Type-II procollagen gene (*Col2A1*) mutation in exon 11 associated with spondyloepiphyseal dysplasia, tall stature, and precocious osteoarthritis. J. Rheum. 22, 255-261.

Bollerslev, J., and Anderson, P.E. (1987). Radiological, biochemical, and hereditary evidence of two types of autosomal dominant osteopetrosis. Bone 9, 7-13.

Bollerslev, J., Ueland, T., Grodum, E., Haug, E., Brixen, K., and Djoseland, O. (1998). Biochemical markers of bone metabolism in benign human osteopetrosis: A study of two types at baseline and during stimulation with triiodothyronine. Eur. J. Endocrinol. 139, 29-35.

Bonaventure, J., Chaminade, F., and Maroteaux, P. (1995). Mutations in three subdomains of the carboxy-terminal region of collagen type-X account for most of the Schmid-metaphyseal dysplasia. Hum. Genet. 96, 58-64149.

Boyce, B.F., Chen, H., Soriano, P., and Mundy, GR. (1993). Histomorphometric and immunopcytochemical studies of *src*-related osteopetrosis. Bone 14, 335-340.

Brunet, L.J., McMahon, J.A., McMahon, A.P., and Harland, R.M. (1998). Noggin, cartilage morphogenesis and joint formation in the mammalian skeleton. Science 280, 1455-1457.

Byers, P.H., Tsipouras, F.J., Bonadio, F.J., Starman, B., and Schwartz, R.C. (1988). Perinatal lethal osteogenesis imperfecta (OI type-II): A biochemically heterogeneous disorder usually due to new mutations in the genes for type-I collagen. Am. J. Hum. Genet. 42, 237-248.

Chipman, S.D., Sweet, H.O., McBride, D.J., Davisson, M.T., Marks, S.C., Shuldiner, A., Wenstrup, R.J., Rowe, D.W., and Shapiro, J.R. (1993). Defective pro-a2(I) collagen synthesis in a recessive mutation in mice: A model of human osteogenesis imperfecta. Proc. Nat. Acad. Sci. USA 90, 1701-1705.

Chu, M.L., Williams, C.J., Ramirez, F (1985). Multiexon deletion in an osteogenesis imperfecta variant with increased type-III collagen mRNA. J. Biol. Chem. 260, 691-694.

Chu, M.L., Williams, C.J., Pepe, G., Hirsch, J.L., Prockop, D.J., Ramirez, F. (1983). Internal deletion in a collagen gene in a perinatal lethal form of osteogenesis imperfecta. Nature 304, 78-80.

Coccia, P.F., Krivit, W., Cervenka, J. et al. (1980). Successful bone marrow transplant for malignant osteopetrosis. N. Eng. J. Med. 302, 701-708.

Cohen-Solal, L., Zolezzi, F., Pignatti, P.F., and Mottes, M. (1996). Intrafamilial variable expressivity of osteogenesis imperfecta due to mosaicism for a lethal G382R substitution in the COL1A1 gene. Mol. Cell. Probes. 10, 219-225.

Cohn, D.H., Apone, S., Eyre, D.H., Starman, B., Andreassen, P., Charbonneau, H., Nicholls, A.C., Pope, F.M., and Byers, P.H. (1988). Substitution of cysteine for glycine within the carboxy-terminal telopeptide of the α 1 chain of type-I collagen produces mild osteogenesis imperfecta. J. Biol. Chem. 263, 14605-14607.

Cohn, D.H., Starman, B.J., Blumberg, B., and Byers, P.H. (1990). Recurrence of lethal osteogenesis imperfecta due to parental mosaicism for a dominant mutation in a human type-I collagen (*COL1A1*) gene. Am. J. Hum. Genet. 46, 591-601.

Cole, W.G. (1994a). Collagen Genes: Mutations affecting collagen structure and expression. Prog. Nucleic Acid Res. 47, 29-79.

Cole, W.G. (1994b). Osteogenesis imperfecta as a consequence of naturally occurring and induced mutations of type-I collagen. J. Bone Min. Res. 8, 167-204.

Cole, W.G., Chow, C.W., Bateman, J.F., and Sillence, D.O. (1996). The phenotypic features of osteogenesis imperfecta resulting from a mutation of the carboxy-terminal pro α 1(I) propeptide that impairs the assembly of type-I procollagen and formation of the extracellular matrix. J. Med. Genet. 33, 965-967.

Dalgleish, R. (1997). The human type-I collagen mutation database. Nucleic Acids Res. 25, 181-187.

D' Avis, P.Y., Robertson, S.C., Meyer, A.N., Bardwell W.M., Webster, M.K., and Donoghue, D.J. (1998). Constitutive activation of fibroblast growth factor receptor-3 by mutations responsible for the lethal skeletal dysplasia thanatophoric dysplasia type I. Cell Growth Differ. 9, 71-78.

Eastman, J.R., and Bixler, D. (1983). Clinical, laboratory, and genetic investigations of hypophosphatasia: Support for autosomal dominant inheritance with homozygous lethality. J. Craniofac. Genet. Dev. Biol. 3, 213-234.

el-Tawil, T., and Stoker, D.J. (1993). Benign osteopetrosis: A review of 42 cases showing two different patterns. Skeletal Radiol. 22, 587-593.

Erlebacher, A., Filvaroff, E.H., Gitelman, S.E., and Derynck, R. (1995). Toward a molecular understanding of skeletal development. Cell 80, 371-378.

Fathallah, D.M., Bejaoui, M., Sly, W.S., Lakhoua, R., and Dellagi, K. (1994). A unique mutation underlying carbonic anhydrase II deficiency syndrome in patients of Arab descent. Hum. Genet. 94, 581-582.

Felix, R., Cecchini, M.G., and Fleisch, H. (1990). Macrophage colony stimulating factor restores in vivo bone resorption in the op/op osteopetrotic mouse. Endocrinology 127, 2592-2594.

Ferreira, L.R., Norris, K., Smith, T., Hebert, C., and Sauk, J.J. (1994). Association of Hsp47, Grp78, and Grp94 with procollagen supports the successive or coupled action of molecular chaparones. J. Cell Biochem 56, 518-526.

Fischer, A., Friedrich, W. Fasth, A., Blanche, S., Le Deist, F., Girault, D., Veber, F., Vossen, J., Lopez, M., and Griscell, C. (1991). Reduction of graft failure by a monoclonal antibody (anti-LFA-1 CD11A) after HLA nonidentical bone marrow transplantation in children with immunodeficiencies, osteopetrosis, and Fancon's anemia: A European group for immunodeficiency/European group for bone marrow transplantation report. Blood 77, 240-256.

Frame, B., Honasoge, M., and Kottamasu, S.R. (1987). Osteosclerosis In: *Hyperostosis and Related Disorders*. pp. 334-359. Elsevier, New York.

Francomano, C.A., De Luna, O., Hefferon, T.W., Bellus, G.A., Turner, C.E., Taylor, E., Meyers, D.A., Blanton, S.H., Murray, J.C., and McIntosh, I. (1994). Localization of the achondroplasia gene to the distal 2.5 Mb of human chromosome 4p. Hum. Mol. Genet. 3, 787-792.

Francomano, C.A. (1995). The genetic basis of dwarfism. N. Eng. J. Med. 332, 58-59.

Genovese, C., Brufsky, A., Shapiro, J.R., and Rowe, D. (1989). Detection of mutations in human type-I collagen mRNA in osteogenesis imperfecta by indirect RNase protection. J. Biol. Chem. 264, 9632-9637.

Gerritsen, E.J., Vossen, J.M., van Loo, I.H., Hermans, J., Helfrich, M.H., Griscelli, C., Fischer, A. (1994a). Autosomal recessive osteopetrosis: Variability of findings at diagnosis and during the natural course. Pediatrics 93, 247-253.

Gerritsen, E.J., Vossen, J.M., Fasth, A., Friedrich, W., Morgan, G., Padmos, A., Vellodi, A., Porras, G., O'Meara, A., and Porta, F. (1994b). Bone marrow transplantation for autosomal recessive osteopetrosis. A report from the Working Party on Inborn Errors of the European Bone Marrow Transplantation Study. J. Ped. 125, 896-902.

Gorlin, R.J. (1997). Fibroblast growth factors, their receptors and receptor disorders. J. Craniomaxillofac. Surg. 25, 69-79.

Greenberg, C.R., Taylor, C.L., Haworth, J.C., Seargeant, L.E., Phillips, S., Triggs-Raine, B., Chodriker, B.N. (1993). A homoallelic Gly317→Asp mutation in ALPL causes the perinatal (lethal) form of hypophosphatasia in Canadian Mennonites. Genomics 17, 215-217.

Grigoriadis, A.E., Wang, Z., and Wagner, E.F. (1995). *Fos* and bone cell development: Lessons from a nuclear oncogene. Trends Genet. 11, 436-441.

Hall, T.J., Schaeublin, M., and Missbach, M. (1994). Evidence that *c-src* is involved in the process of osteoclastic bone resorption. Biochem. Biophys. Res. Comm . 199, 1237-1244.

Harris, H. (1989). The human alkaline phosphatases: What we know and what we don't know. Clin. Chem. Acta 186, 133-150.

Heaney, C., Shalev, H., Elbredour, K.,, Carmi R., Staack,. JB., Sheffield, V.C., and Beier, D.R. (1997). Human autosomal recessive osteopetrosis maps to 11q13, a position predicted by comparative mapping of the murine osteosclerosis (oc) mutation. Hum. Mol. Genet. 7, 1407-1410.

Henthorn, P.S. and Whyte, M.P. (1992). Missense mutations of the tissue-nonspecific alkaline phosphatase gene in hypophosphatasia. Clin. Chem. 38, 2501-2505.

Henthorn, P.S., Raducha, M., Fedde, K.N., Lafferty, M.A., and Whyte, M.P. (1992). Different missense mutations at the tissue-nonspecific alkaline phosphatase gene locus in autosomal recessively inherited forms of mild and severe hypophosphatasia. Proc. Nat. Acad. Sci. USA 89, 9924-9928.

Hermey, D.C., Ireland, R.A., Zerwikh, J.E., and Popoff, S.N. (1995). Regulation of mineral homeostasis in osteopetrotic rats. Am. J. Physiol. 268, E312-E317.

Horan, G.S., Kovacs, E.N., Behringer, R.R., and Featherstone, M.S. (1995). Mutations in paralogous *Hox* genes result in overlapping homeotic transformations of the axial skeleton: Evidence for unique and redundant function. Develop. Biol. 169, 359-372.

Horton, W.A. (1997). Fibroblast growth factor receptor 3 and the human chondrodysplasias Curr. Opin. Pediatr. 9, 437-442.

Hu, P.Y., Ernst, A.R., and Sly, W.S. (1992a). UGA supression of "Hispanic Mutation" for CAII deficiency syndrome suggests novel mechanism for genetic heterogeneity. Am. J. Hum. Genet. 51, (Suppl.) A29.

Hu, P.Y., Ernst, A.T., Sly, W.S., Venta, P.J., Sakggs, L.A., and Tashina, R.E. (1994a). Carbonic anhydrase II deficiency. Single base deletion in exon 7 is the predominant mutation in Caribbean Hispanic patients. Am. J. Hum. Genet. 54, 602-608.

Hu, P.Y., Ernst, A.R., Sly, W.S., Venta, P.J., Skaggs, L.A., and Tashina, R.E. (1994b). Carbonic anhydrase deficiency: Single-base deletion in exon 7 is the predominant mutation in Caribbean Hispanic patients. Am. J. Hum. Genet. 54, 602-608.

Hu, P.Y., Roth, D.E., Skaggs, L.A., Venta, P.J., Tashian, R.E., Gibaud, P., and Sly, W.S. (1992b). A splice junction mutation in intron 2 of the carbonic anhydrase II gene of osteopetrosis patients from Arabic countries. Hum. Mutat. 1, 288-292.

Huerre, C., Junien, D., Weil, M.L., Chu, M.M., Morabito, N. et al. (1982). Proc. Nat. Acad. Sci. USA 79, 6627.

Ikegawa, S., Fukusima, Y., Isomura, M., Takeda, F., and Nakamura, Y. (1995). Mutations of the fibroblast growth factor receptor-3 gene in one familial and six sporadic cases of achondroplasia in Japanese patients. Hum. Genet. 96, 309-311.

Jabs, E.W., Li, X., Scott, A.F., Meyers, G., Chen, W., Exccles, M., Mao, J., Charnas, L.R., Jackson, C.E., and Jaye, M. (1994). Jadckson-Weiss and Crouzon syndromes are allelic with mutations in fibroblast growth factor receptor 2. Nat. Genet. 8, 275-279.

Jacenko, O. (1995). *c-fos* and bone loss: A protooncogene regulates osteoclast lineage determination. Bioassays 17, 277-281.

Jacenko, O., LuValle, P.A., and Olsen, B.R. (1993). Spondylometaphyseal dysplasia in mice carrying a dominant negative mutation in a matrix protein specific for cartilage-to bone transition. Nature 365, 56-61.

Jackson, M.E., Shalhoub, V., Lian, J.B., Stein, G.S., and Marks, Jr., S.C. (1944). Aberrant gene expression in cultured mammalian bone cells demonstrates an osteoblast defect in osteopetrosis. J. Cell Biochem. 55, 366-372.

Johnson, D.E., and Williams, L.T. (1993). Structural and functional diversity in the FGF receptor multigene family. Adv. Cancer Res. 60, 1-41.

Katzenstein, P.L., Campbell, D.F., Machado, M.A., Horton, W.A., Lee, B., and Ramirez, F. (1992). A type-II collagen defect in a new family with SED tarda and early onset osteoarthritis (OA). Arth. Rheum. 35, S41.

Kaufman, R., Rimoin, D., McAllister, W., and Kissane, J. (1970). Thanatophoric dwarfism. Am. J. Dis. Child 120, 53-57.

Kessler, E., Takahara, K., Biniaminov, L., Busel, M., and Greenspan, D. (1996). Bone Morphogenetic Protein-1: The Type I Procollagen C-Proteinase. Science 271, 360-362

Key, L., Carnes, S., Cole, S., Holtrop, M., Bar-Shavit, Z., Shapiro, F., Arceci, R., Steinberg, J., Gundberg, C., Kahn, A., Teitelbaum, S., and Anast, C. (1984). Treatment of congential osteopetrosis with high-dose calcitrol. N. Eng. J. Med. 310, 409-415.

Key, L.L., Rodriquez, R.M., Willi, S.M., Wright, N.M., Hatcher, H.C., Eyre, D.R., Cure, J.K., Griffin, P.P., and Ries, W.L. (1995). Long-term treatment of osteopetrosis with recombinant human inteferonγ. N. Eng. J. Med. 332, 1594-1595.

Kivirikko, K. (1993). Collagens and their abnormalities in a wide spectrum of diseases. Ann. Med. 25, 113-126.

Kobayashi, S., Inoue, S., Hosoi, T., Ouchi, Y., Shiraki, M., and Orimo, H. (1996). Association of bone mineral density with polymorphism of the estrogen receptor gene. J. Bone Min. Res. 11, 306-311.

Konori, T., Yagi, H., Nomura, S. et al. (1997). Targeted disruption of CBFA 1 results in a complete lack of bone formation owing to maturational arrest of osteoblasts. Cell 89, 755-764.

Kuivaniemi, H., Sippola-Thiele, M., and Prockop, D.J. (1988). A 19Bbase pair deletion in the *pro-a2(I)* gene of type-I procollagen that causes in-frame RNA splicing from exon 10 to exon 12 in a proband with a typical osteogenesis imperfecta and his asymptomatic mother. J. Biol. Chem. 263, 11407-11413.

Kuivaniemi, H., Tromp, G., and Prockop, D.J. (1991). Mutations in collagen genes: Causes of rare and some common diseases in humans. FASEB J. 5, 2052-2060.

Kuivaniemi, H., Tromp, G. and Prockop, D.J. (1997). Mutations in fibrillar collagens (types I, II, III, and XI) fibril-associated collagen (type-IX), and network-forming collagen (type-X) cause a spectrum of diseases of bone, cartilage, and blood vessels. Hum. Mutat. 9, 300-315.

Lacombe, D. (1995). Clinical dysmorphology beyond developmental genetics: Recent advances in some human developmental genes. Ann. Genet. 38, 137-144.

Lajeunesse, D., Busque, L., Menard, P., Brunettge, M.G., and Bonny, Y. (1996). Demonstration of an osteoblast defect in two cases of human malignant osteopetrosis. Correction of the phenotype after bone marrow transplant. J. Clin. Invest. 98, 1835-1842.

Laujenie, E., Ma, H.W., Bonaventure, J., Munnich, A., and Le Merrer, M. (1995). FGFR2 mutations in Pfeiffer syndrome. Nat. Genet. 9, 108.

Le Merrer, M., Rousseau, F., Legeai-Mallet, L. Et al. (1994). A gene for achondroplasia-hypochondroplasia maps to chromosome 4p. Nat. Genet. 6, 318-321.

Lenhard, S., Popoff, S.N., and Marks, Jr., S.C. (1990). Defective osteoclast differentiation and function in the osteopetrotic (os) rabbit. Am. J. Anat. 188, 438-444.

Lian, J.B. (1996). Mysterious cross talk between bone cells. J. Clin. Invest. 98, 1697-1698.

Lightfoot, S.J., Holmes, D.F., Brass, A., Grant, M.F., Byers, P.H., and Kadler, K.E. (1992). Type-I procollagens containing substitutions of aspartate, arginine, and cysteine for glycine in the proa1 (I) chain are cleaved slowly by N-proteinase but only the cysteine substitution introduces a kink in the molecule. J. Biol. Chem. 267, 25521-25528.

Lowe, C., Yoneda, T., Boyce, BF., Chen, H., Mundy, G.R., and Soriano, P. (1993). Osteopetrosis in src-deficient mice is due to an autonomous defect of osteoclasts. Proc. Nat. Acad. Sci. USA 90, 4485-4489.

Lund, A.M., Skovby, F., and Schwartz, M.. (1996). Deletion of a Gly-Pro-Pro repeat in the pro-α 2(I) chain of procollagen I in a family with dominant osteogenesis imperfecta type-IV. Hum. Genet. 97, 287-290.

Machwate, M., Jullienne, A., Moukhtar, M., and Marie, P.J. (1995). Temporal variation of c-fos protooncogene expression during osteoblast differentiation and osteogenesis in developing rat bone. J. Cell Biochem. 57, 62-70.

Mancilla, E.E., DeLuca, F., Uyeda, J.A., Czerwiec, F.S., and Baron, J. (1998). Effects of fibroblast growth factor-2 on longitudinal bone growth. Endocrinol. 139, 2900-2904.

Marks, Jr., S.C. (1989). Osteoclast biology: Lessons from mammalian mutations. Am. J. Med. Genet. 34, 43-54

Marks, S.C. (1977). Osteopetrosis in the toothless (tl) rat: Presence of osteoclasts but failure to respond to parathyroid extract or to be cured by infusion of spleen or bone marrow cells from normal littermates. Am. J. Anat. 149, 289-297.

Marks, Jr., S.C. (1989). Osteoclast biology: Lessons from mammalian mutations. Am. J. Med. Genet. 34, 43-54.

Marks, S.C. Jr., Iizuka, T., MacKay, C.A., Mason-Savas, A, and Cielinski, M.J. (1997). The effects of colony-stimulating factor-i on the number and ultrastructure of osteoclasts in toothless (tl) rats and osteoporotic (op) mice .Tissue Cell 29, 589-595.

Martin, J.F., Bradley, A., and Olson, E.N. (1995). The pairedlike homeo box gene *MHox* is required for early events of skeletogenesis in multiple lineages. Genes Develop. 9, 1237-1249.

Mathijssen, I.M., Vaanrager, J.M., Hoogeboom, A.J, Hesseling-Janssen, A.L., and van den Ouweland, A.M. (1998). Pfeiffer's syndrome resulting from an S351C mutation in the fibroblast growth factor receptor-2 gene. J. Craniofac. Surg. 9, 207-209.

Mayne R., and Brewton, R.G. (1993). New members of the collagen superfamily. Curr. Opin. Cell Biol. 5, 883-890.

McBride, D.J., and Shapiro, J.R: (1994). Confirmation of a G nucleotide deletion in the Cola-2 gene of mice with the osteogenesis imperfecta mutation. Genomics 20, 135-137.

McCabe, L.R., Kockx, M., Lian, J., Stein, J., and Stein, G. (1995). Selective expression of fos- and jun-related genes during osteoblast proliferation and differentiation. Exptl. Cell Res. 218, 255-262.

McIntosh, I., Abbot, M.H., and Francomano, C.A. (1995). Concentration of mutations causing Schmid metaphyseal chondrodysplasia in the C-terminal noncollagenous domain of type X collagen. Human Mutat. 5, 121-125.

McKusick, V., Francomano, C.A., and Antonarakis, S.E. (1992). *Mendelian Inheritance in Man: Catalogs of Autosomal Dominant, Autosomal Dominant, Autosomal Recessive, and X-Linked phenotype*, 10th edn. Johns Hopkins University Press, Baltimore.

Miano, M. Porta, F., Locatelli, F., Miniero, R., La Nasa, G., Di Bartolomeo, P., Giardini C., Messina, C., Balduzzi, A., Testi, A.M., Garbarino, L., Lanino, E., Crescenzi, F., Zecca, M., and Dini, C. (1998). Unrelated donor marrow transplantation for inborn errors. Bone Marrow Transplant. 21, S37-S41.

Miki, T. et al. (1992). Determination of ligand binding specificity by alternative splicing: Two distinct growth factor receptors encoded by a single gene. Proc. Natl. Acad. Sci. USA 89, 246-250.

Moore, C.A., Ward, J.C., Rivas, M.L., Magill, H.L., abd Whyte, M.P. (1990). Infantile ypophosphatasia, autosomal recessive transmission to two related sibships. Am. J. Med. Genet 36, 15-22.

Moore, M.H., Cantrell, S.B., Trott, J.A., and David, D.J. (1995) Pfeiffer Syndrome: A clinical review. Cleft Palate Craniofac. J. 32, 62-70.

Morrison, N.A., Cheng, J., Tokita, A., Kelly, P.J., Crofts, L., Nguyen, T.V., Sambrook, P.N., and Eisman, J.A. (1994). Prediction of bone density from vitamin D receptor alleles. Nature 367, 284-287.

Muenkde, M., Schell, U., Hehr, A., Robin, N.H., Losken, H.W., Schnizel, A. et al. (1995). A Common mutation in the fibroblaśt growth factor receptor 1 gene in Pfeiffer Syndrome. Nat. Genet. 8, 269-274.

Muenke, M., and Schell, U. (1995). Fibroblast-growth-factor receptor mutations in human skeletal disorders. Trends Genet. 11, 308-313.

Mundlos, S., Otto, F., Mundlos, C. et al. (1997). Mutations involving the transcription factor CBFA 1 cause cleidocranial dysplasia. Cell 89, 773-779.

Naki, A., Satoh, M., Hirayoshi, K., and Nagata, K. (1992). Involvement of the stress protein HSP47 in procollagen processing in the endoplasmic reticulum. J. Cell Biochem. 117, 903-914.

Nesbitt, T., Marks, Jr., S.C., Jackson, M.E., Mackay, C., and Drezner, M.K. (1995). Normalization of mineral homeostasis after reversal of osteopetrosis. J Bone Min Res 10, 1116-1121.

Nicholls, A.C., Osse, G., Schloon, G., Deak, S., Myers, J.C., Prockop, D.J., Weigel, W.R.F., Fryer, P., and Pope, F.M. (1984). The clinical features of homozygous α-2(I)-collagen deficient osteogenesis imperfecta. J. Med. Genet. 21, 257-262.

Nielson, K.M. and Friesel, R. (1996). Ligand-independent activation of fibrobast growth factor receptor by point mutation in the extracellular, transmembrane and kinase domains. J. Biol. Chem. 271, 25049-25057.

Nutytinck, L., Dagleish, R., Spotila, L., Renard, J.P., Van Regemorter, N, and De Paepe, A. (1996). Substitution of glycine-661 by serine in the α1(I) and α 2(I) chains of type-I collagen results in different clinical and biochemical phenotypes. Hum. Genet. 324-329.

Nuytinck, L., Coppin, C., and De Paepe, A. (1998). A four base-pair insertion polymorphism in the 3' untranslated region of the COL1A1 gene is highly information for null-allele testing in patients with osteogenesis imeprfecta type-I. Matrix Biol. 16, 349-352.

Ohta, S., Hiraki, Y., Shigeno, C., Suzuki, F., Kasai, R., Ikeda, T., Kohno, H., Lee, K., Kikuchi, H., Konishi, J. et al. (1992). Bone morphogenetic proteins (BMP-2 and BMP-3) induce the late- phase expression of the protooncogene c-fos in murine osteoblastic MC3T3-E1 cells. FEBS Lett. 314, 356-360.

Orimo, H., Hayashi, Z., Watanabe, A., Hirayama, T., Hirayama, T., and Shimada, T. (1994). Novel missense and frameshift mutations in the tissue-nonspecific alkaline phosphatase gene in a Japanese patient with hypophosphatasia. Hum. Mol. Genet. 9, 1683-1684.

Otto, F., Thornell, A.P., Crompton, T. et al. (1997). Cfba 1 a candidate gene for cleidocranial dysplasia syndrome is essential for osteoblast differentiation and bone development. Cell 89. 765-771.

Pentinnen, R., Lichtenstein, J.R., Martin, G.R., and McKusick, V.A. (1975). Abnoraml collagen metabolism in cultured cells in osteogenesis imperfecta. Proc. Natl. Acad. Sci. USA 72, 586-589.

Preston, R.A. et al. (1994). A gene for Crouzon craniofacial dystosis maps to the long arm of chromosome 10. Nat. Genet. 7, 149-153.

Prockop, D.J., and Kivirikko, K.I. (1995). Collagens: Molecular biology, diseases, and potentials for therapy. Ann. Rev. Biochem. 64, 403-434.

Raghunath, M., Bruckner, P., and Steinmann, B. (1994). Delayed triple helical formation of mutant collagen from patients with osteogenesis imperfecta. J. Mol. Biol. 236, 940-949.

Reardon, W., Winter, R.M., Rutland, P., Pulleyn, L.J., Jones, B.M., and Calcom, S. (1994a). Mutation in the fibroblast growth factor receptor 2. Nat. Genet. 8, 275-279.

Reardon, W., Winter, R.M., Rutland, P., Pulleyn, L.J., Jones, B.M., and Malcom, S. (1994b). Mutations in the fibroblast growth factor receptor II gene cause Crouzon syndrome. Nat. Genet. 8, 98-103.

Reddi, A.H. (1994a). Bone and cartilage differentiation. Curr. Opin. Genet. Dev. 4, 737-744.

Reddi, A.H. (1994b). Cartilage morphogenesis: Role of bone and cartilage morphogenetic proteins, homeobox genes, and extracellular matrix. Matrix Biol. 14, 599-606.

Redford-Badwal, D.A., Stover M.L., Valli, M., McKinstry, M.B., and Rowe, D.W. (1996). Nuclear retention of COL1A1 messenger RNA identifies null alleles causing mild osteogenesis imperfecta. J. Clin. Invest. 97, 1035-1040.

Retief, E., Parker, M.I., and Retief, A.E. (1985). Regional Chromosome mapping of human collagen genes α-2 (I and α-1) COPLIA-1 and COLIA 2 1A2. Hum. Genet. 69, 304-398.

Ritvaniemi, P., Korkko, J., Bonaventure, J., Vikkula, M. et al. (1995). Identification of *COL2A1* mutations in patients with chondrodysplasias and familial osteoarthritis. Arthr. Rheum. 38, 999-1004.

Robin, N.H., Scott, J.A., Arnold, J.E., Goldstein, J.A., Shilling, B.B, Marion, R.W, and Cohen, M.M., Jr. (1998). Favorable prognosis for children with Pfeiffer syndrome types-2 and -3: Implications for classification. Am. J. Med. Genet. 75, 240-244.

Rosati, R., Horan, G., Pinero, G.J., Garafolo, S., Keene, D.R., Horton, W., Vuorio, E., deCrombrugghe, B., and Behringer, R.R. (1994). Normal long bone growth and development in type-X collagen-null mice. Nat. Genet. 8, 129-135.

Rosseau, F., Bonaventure, J., Legeai-Mallet, L., Pelet, A., Rozet, J.M., Maroteaux, P., Le Merrer, M., and Minnich, A. (1994). Mutations in the gene encoding fibroblast growth factor receptor-3 in achondroplasia. Nature 3721, 252-254.

Roth, D.E., Venta, P.J., Tashian, R.E., and Sly, W.S. (1992). Molecular basis of human carbonic anhydrase deficiency Proc. Natl. Acad. Sci. USA 89, 1804-1808.

Rowe, D.W., Poirer, M., and Shapiro, J.R. (1985). Type- I collagen biosynthesis in osteogenesis imperfecta. J. Clin. Invest. 76, 604-611.

Rowe, D.W. (1990). Osteogenesis imperfecta. J. Bone Miner. Res. 7, 209-241.

Rutland, P., Pulleyn, I.J., Reardon, W., Baraitzer, M., Hayward, R., Jones, B. et al. (1995). Identical mutations in the *FGFR2* gene cause both the Pfeiffer and Crouzon syndrome phenotypes. Nat. Genet. 9, 173-176.

Schell, U., Hehr, A., Feldman, G.J., Robin, N.H., Zackai, E.H., de Die Smulders, C. et al. (1995). Mutations in *FGFR1* and *FGFR2* cause familial and sporadic Pfeiffer syndrome. Hum. Molec. Genet. 4, 321-328.

Schneider, G.B., Cuenoud, M.L., and Marks, Jr., S.C. (1979). The diagnosis and cure for neonatal osteopetrosis: Experimental evidence from congenitally osteopetrotic (ia) rats. Metab. Bone Dis. Rel. Res. 1, 335-339.

Schwartzberg, P.L., Xing, L., Hoffman, O., Lowell, C.A., Garrett, L., Boyce, B.F., and Varmus, H.E. (1997). Rescue of osteoclast function by transgenic expression of kinase-deficient Src in src-/- mutant mice. Genes Dev. 11, 2835-2844.

Seifert, M., Popoff, S.N., Jackson, M.E., Mackay, C.A., Cielinski, M., and Marks, Jr., S.C. (1993), Experimental studies of osteopetrosis in laboratory animals. Clin. Ortho. Rel. Res. 294, 23-33.

Shalhoub, V., Bettancourt, B., Jackson, M.E., MacKay, C., Glincher, M., Marks, S.C., Stein, G., and Lian, J. (1994). Abnormalities of phosphoprotein gene expression in three osteopetrotic rat mutations: Elevated mRNA transcripts, protein synthesis, and accumulation in bone of mutant animals. J. Cell Physiol. 158, 110-120.

Shapiro, F., Key, L., and Anast, C. (1988). Variable osteoclast appearance in human infantile osteopetrosis. Calcif. Tissue Int. 43, 67-76.

Shapiro, J.R., Primorac, D., and Rowe, D.W. (1996). Osteogenesis Imperfecta: Current Concepts. In: *Principles of Bone Biology*. (Bilezikian, J, Raisz, L, and Rodan, G., Eds.), pp. 889-902. Academic Press, Orlando, FL.

Shiang, R., Thompson, L.M., Zhu, Y.Z., Church, D.M., Fielder, T.J., Bocian, M., Winokur, S.T., and Wasmuth, J.J. (1994). Mutations in the transmembrane domain of FGFR3 cause the inmost common genetic form of dwarfism, Achondroplasia. Cell 78, 335-342.

Sillence, D.O., Senn, A., and Danks, D.M. (1979). Genetic heterogeneity in osteogenesis imperfecta. J. Med. Genet. 16, 101-116.

Sillence, D.O. (1981). Osteogenesis imperfecta: An expanding panorama of variants. Clin. Orthop. 159, 11-25.

Sly, W.S., Whyte, M.P., Sundaram, V., Tashian, R.E., Hewett-Emmett, D., Guibard, P., Vainsel, M., Baluarte, H.J., Gruskin, A., Al-Mosawi, M., Sakati, N., and Ohlsson, A. (1985). Carbonic anhydrase II deficiency in 12 families with the autosomal recessive syndrome of osteopetrosis with renal tubular acidosis and cerebral calcification. N. Eng. J. Med. 313, 139-145.

Smith, M., Weiss, M.J., Griffin, C.A., Murray, J.C., Buetow, K.H., Emanuel, B.S., Henthorn, P.S., and Harris, H. (1988). Regional assignment of the gene for human liver/bone/kidney alkaline phosphatase to chromosome 1p36-1p34. Genomics 2, 139-143.

Soda, H. (1994). Carbonic anhydrase II deficiency syndrome—clinico-pathological, biochemical and molecular studies. Kurume Med. J. 41, 233-240.

Solh, H., DaCuhna, A.M., Giri, N., Padmos, A., Spence, D., Clink, H., Ernst, F., and Sadati, N. (1995). Bone marrow transplantation for infantile malignant osteopetrosis. J. Pediatr. Hematol. Oncol. 17, 350-355.

Steingrimsson, E., Moore, K.J., Lamoreux, M.L., Ferre-D'Amare, A.R., Burley, S.K., Sanders Zimring, D.C., Skow, L.C., Hodgkinson, C.A., Arnheiter, H., Copeland, N.G., and Jenkins, N.A. (1994). Molecular basis of mouse Micropthalmia (mi) mutation helps explain their developmental and phenotypic consequences. Nat. Gen. 8, 256-263.

Stover, M.L., Primorac, D., Liu, S.C., McKinstry, M.B., and Rowe, D.W. (1993). Defective splicing of mRNA from one *COL1A1* allele of type-I collagen in nondeforming (type-I) osteogenesis imperfecta. J. Clin. Invest. 92, 1994-2002.

Strisciuglio, P., Hu, P.Y, Lim, E.J., Ciccolella, J., and Sly, W.S. (1998). Clinical and molecular heterogeneity in carbonic anhydrase II deficiency and prenatal diagnosis in an Italian family. J. Pediatr. 132, 717-720.

Subramaniam, M., Oursler, M.J., Rasmussen, K., Riggs, B.L., Spelberg, T.C. (1995). TGFβ regulation of nuclear protooncogenes and TGFβ gene expression in normal human osteoblastlike cells. J. Cell Biochem. 57, 52-61.

Suda, N. (1997). Parathyroid hormone-related protein (PTHrP) as a regulating factor of endochondral bone formation. Oral Dis. 3, 229-231.

Sweetman, W.A., Rash, B., Sykes, B., Beighton, P., Hecht, J.T., Zabel, B., and Thomas, J.T. (1992). SSCP and segregation analysis of the human type-X collagen gene (*COL10A1*) in heritable forms of chondrodysplasia. Am. J. Hum. Genet. 51, 841-849.

Sykes, B., Oglive, D., Wordsworth, P., Anderson, J., and Jones, N. (1986). Osteogenesis imperfecta is linked to both type-I collagen structural genes Lancet 2, 69-72.

Szebenyi, G. and Fallon, J.F. (1998). Fibroblast growth factors as multifunctional signaling factors. Int. Rev. Cytol. 185, 45-106.

Takatsuka, H., Umezu, H., Hasegawa, G., Usuda, H., Ebe, Y., Naito, M., and Shultz, L.D. (1998). Bone remodeling and macrophage differentiation in osteopetrosis (op) mutant mice defective in the production of macrophage colony-stimulating factor. J. Submicrosc. Cytol. Pathol. 30, 239-247.

Tanaka, H. (1997). Achondroplasia: Recent advances in diagnosis and treatment. Acta Pediatr. Jpn. 39, 514-520.

Tashian, R.E. (1991). Genetics of the mammalian carbonic anhydrases. Adv. Genet 30, 321-356.

Tavormina, P.L., Shiang, R., Thompson, L.M., Zhen-Zhu, Y., Wilkin, D.J., Lachman, R.S., Wilcox, W.R., Rimoin, D.L., Cohn, D.H., and Wasmuth, J.J. (1995). Thanatophoric dysplasia (types I and II) caused by distinct mutations in fibroblast growth factor receptor 3. Nature Gen. 9, 321-328.

Tilstra, D.J., and Byers, P.H. (1994). Molecular basis of hereditary disorders of connective tissue. Ann. Rev. Med. 45, 149-163.

Tromp, G.P., and Prockop, D.J. (1988). Single base mutation in the propalpha a(1) collagen gene that causes efficient splicing of the RNA from exon 27 to exon 29 and synthesis of shortened but in-frame proalpha 2(I) chain. Proc. Natl. Acad. Sci. USA 85, 5254-5258.

Uitterlinden, A.G., Burger, H., Hyang, Q., Yue, F., McGuigan, F.E., Grant, S.F., Hofman, A., van Leeuwen, J.P., Pols, H.A., and Ralston, S.H. (1998). Relation of alleles of the collagen type-I α 1 gene to bone density and the risk of osteoporotic fractures in postmenopausal women. New Eng. J. Med. 338, 1016-1021.

Ullrich, A., and Schlessinger, J. (1995). Signal transduction by receptors with tyrosine kinase activity. Cell 61, 203-212.

Van Hul, W., Bollerslev, J., Gram J., Van Hul, E., Wuyts, W., Benichou, O., Vanhoenacker, F, and Willems, P.J. (1997). Localization of a gene for autosomal dominant osteopetrosis (albers Schonberg disease) to chromosome 1p21. Am. J. Hum. Genet. 61, 363-369.

Vaananen, H.K., and Parvinen, E.K. (1983). High activity isoenzyme of carbonic anhydrase in rat calvaria osteoclasts: Immunohistochemical study. Histochemistry 78, 481-486.

Velinov, M., Slaughenhaupt, S.A., Stoilov, I., Scott, Jr., CL, Gusella, J.F., and Tsipouras, P. (1994). The gene for achondroplasia maps to the telomeric region of chromosome 4p. Nature Genet. 6, 314-317.

Venta, P.J., Welty, R.J., Johnson, TM., Sly, WS., and Tashian, R.E. (1991). Carbonic anhydrase II deficiency syndrome in a Belgian family is caused by a point mutation at an invariant histidine residue (107 His-Tyr): Complete structure of the normal human CA II gene. Am. J. Hum. Genet. 49, 1082-1090.

Vikkula, M., Metsaranta, M., and Ala-kokko, L. (1994). Type-II collagen mutations in rare and common cartilage diseases. Ann. Med. 26, 107-114.

Vogel, B., Doelz, R., Kadler, K., Hoijima, Y., Engel, J., and Prockop, D. (1988). A substitution of cysteine for glycine 748 of the a1 chain produces a kink at this site in the procollagen I molecule and an altered N-proteinase cleavage site over 225 nm away. J. Biol. Chem. 263, 18249-18255.

Vortkamp, A., Pathi, S., Peretti G.M., Caruso, E.M., Zaleske, D.J., and Tabin, C.J. (1998). Recapitulation of signals regulating embryonic bone formation during postnatal growth and in fracture repair. Mech. Dev. 71, 65-76.

Walker, D.G. (1975). Bone resorption restored in osteopetrotic mice by transplants of normal bone marrow and spleen cells. Science 190, 784-785.

Warman, M.L., Abbott, M., Apte, S.S., Heffernon, T., McIntosh, I., Cohn, D.H., Hecht, J.T., Olsen, B.R., and Francomano, C.A. (1993). A type-X collagen mutation causes Schmid metaphyseal chondrodysplasia. Nature Genet 5, 79-82.

Weber, M., Johannisson, R., Carstens, C., Pauschert, R., and Niethard, F.U. (1998) Thanatophoric dysplasia type II: New entity? J. Pediatr. Orthop. 7, 10-22.

Weiss, M.J., Ray, K., Henthorn, P.S., Lamb, B., Kadesch, T., and Harris, H. (1988). Structure of the human liver/bone/kidney alkaline phosphates gene. J. Biol. Chem. 263, 12002-12010.

Whyte, M.P. (1995a). Chipping away at marble bone disease. N. Eng. J. Med. 332, 1639-1640.

Whyte, M.P., Murphy, W.A., Fallon, M.D., Sly, WS., MacAlister, W.H., and Avioli, L.V. (1980). Osteopetrosis, renal tubular acidosis and basal ganglia calcification in three sisters. Am. J. Med. 69, 64-74.

Whyte, M.P. (1994). Hypophosphatasia and the role of alkaline phosphatase in skeletal mineralization. Endo. Rev. 15, 439-461.

Whyte, M.P. (1995b). Hypophosphatasia. (1995). In: *The Metabolic and Molecular Basis of Inherited Disease*, 7th ed. (Scriver, C.R., Beaudet, A.L., Sly, W.S., and Valle, D., Eds.), pp. 4095-4111. McGraw Hill, New York.

Wilcox, W.R., Tavormina, P.L., Krakow, D., Kitoh, H., Lachman, R.S., Wasmuth, J.J., Thompson, L.M., and Rimoin, D.L. (1998). Molecular, radiologic and histopathologic correlations in thanatophoric dysplasia. Am. J. Mol., 274-281.

Williams, C.J., and Jiminez, S.A. (1995). Heritable diseases of cartilage caused by mutations in collagen genes. J. Rheum 22, (Suppl.43), 28-33.

Wilkin, D.J., Szabo, J.K., Cameron, R., Henderson, S., Bellus, G.A., Mack, M.L., Kaitila, I., Loughlin, J., Munnich, A., Sykes, B., Bonaventure, J., and Francomano, C.A. (1998). Mutations in fibroblast growth-factor receptor 3 in sporadic cases of achondroplasia occur exclusively on the paternally derived chromosome. Am. J. Hum Genet. 63, 711-716.

Willing, M.C., Cohn, D.H., Starman, B., Holbrook, K.A., Greenberg, C.R., and Byers, P.H. (1988). Heterozygosity for a large deletion in the α2 (I) collagen gene has a dramatic effect on type-I collagen secretion and produces perinatal lethal osteogenesis imperfecta. J. Biol. Chem. 263, 8398-8404.

Willing, M.C., Deschenes, S.P., Scott, D.A., Byers, P.H., Slayton, R.L., Pitts, S.H., Arikata, H., and Roberts, E.J. (1994). Osteogenesis imperfecta type-I: Molecular heterogeneity for *COL1A1* null alleles of type-I collagen. Am. J. Hum .Genet. 55, 638-647.

Willing, M.C., Deschenes, S.P., Slayton, R.L., and Roberts, E.F. (1996). Premature chain termination is a unifying mechanism for *COL1A1* null alleles in osteogenesis imperfecta type-I cell strains. Am. J. Hum. Genet. 59, 799-809.

Wood, R.J. and Fleet, J.C. (1998). The genetics of osteoporosis: Vitamin D receptor polymorphisms. Annu. Rev. Nutr. 18, 233-258.

Yeuh, Y.G., Gardner, D.P., and Kappen, C. (1998). Evidence for regulation of cartilage differentiation by the homeobox gene Hoxc-8. Proc. Natl. Acad. Sci. USA 95, 9956-9961.

Yoneda, T., Niewolna, M., Lowe, C., Izbicka, E., and Mundy, G.R. (1993). Hormonal regulation of *pp60-src* expression during osteoclast formation in vitro. Mol. Endocrinol. 7, 1313-1318.

Zlotogora J. (1998). Germ line mosaicism.. Human Genet 102, 381-386.

Zolezzi, F., Valli, M., Clementi, M., Mammi, I., Cetta, G., Pignatti, P.F., and Mottes M. (1997).Mutation producing alternative splicing of exon 26 in the COL1A2 gene causes type-IV osteogenesis imperfecta with intrafamilial clinical variability. Am. J. Med. Genet. 71, 366-379.

SKELETAL FLUOROSIS: MOLECULAR ASPECTS

Ambrish Mithal

I. INTRODUCTION

Fluoride is a highly reactive electronegative element that is widely distributed in nature, although it is rarely found in its free form. Minute amounts of

Advances in Organ Biology
Volume 5C, pages 797-808.

ISBN: 0-7623-0390-5

fluoride are necessary for many enzymatic actions in the human body. Fluoride, in low doses, helps in dental caries prevention. There has recently been a resurgence of interest in its role in the treatment of osteoporosis (Pak et al., 1995). Large doses of fluoride can, however, cause adverse effects on several body tissues, especially the dental and skeletal systems.

The first reports of "skeletal fluorosis" appeared more than half a century ago (Shortt et al., 1937; Roholm, 1937), yet the condition continues to be endemic in many parts of the world, leading to considerable morbidity and loss of human resources. While this chapter deals with skeletal fluorosis in general, the description largely pertains to endemic fluorosis resulting from chronic consumption of fluoride contaminated drinking water ("hydric fluorosis") since it is the best studied and accounts for the majority of cases worldwide.

II. CELLULAR ACTION OF FLUOROSIS

A. Effect on Osteoblasts

Although several decades have elapsed since it was shown that fluoride has profound effects on the skeleton, the molecular mechanism of action of this ion on bone remains unclear. Farley et al. (1983) used embryonic chick calvaria to show for the first time that sodium fluoride can increase the rate of proliferation of bone cells as assessed by [^{3}H]-thymidine incorporation. In addition, sodium fluoride increased the alkaline phosphatase content of bone cells, and enhanced the growth and mineralization of embryonic bone. Importantly, the effects of sodium fluoride on [^{3}H] thymidine incorporation could be modulated by parathyroid hormone (PTH) and human skeletal growth factor (Farley et al., 1983). Fluoride has similar effects on human osteoblasts and it has been suggested that this action may be dependent on the calcium influx (Khoker and Dandona, 1990). It has also been shown that preosteogenic mesenchyme of chicken can be stimulated by sodium fluoride to differentiate into osteoblasts and deposit bone matrix, indicating that fluoride may actually initiate osteogenesis (Hall, 1987).

The mechanism by which fluoride induces osteoblast proliferation has been the subject of several studies. Lau et al. (1989) showed that low, clinically effective ("osteogenic") concentrations of fluoride inhibited phosphotyrosyl acid phosphatase and enhanced tyrosyl phosphorylation of cellular proteins in cultured chicken calvarial cells. This action was skeletal tissue specific. In this study mitogenic concentrations of fluoride potentiated the mitogenic actions of growth factors with tyrosyl kinase receptors, for example, insulin, epidermal growth factor and insulin-like growth factor (IGF-1) to a greater degree than they did for others like basic fibroblast growth factor. In a separate study the same group showed that systemic bone cell mitogens like PTH and calcitonin increased the maximum response to fluoride, and that the presence of phosphate was important for this action (Farley et al., 1988). These studies led the authors to hypothesize that fluoride directly induced inhibition of an osteoblastic phosphatase, which increased overall cellular tyrosyl phosphorylation, resulting in bone cell proliferation (Lau et al., 1989).

Fluoride has also been shown to elicit a rapid increase in human osteoblast intracellular calcium level and activate the phosphatidylinositol pathway (Zerwekh et al., 1990). Reed et al. (1993) found that in the presence of fluoride, cells from the human clonal osteoblast cell line HOS TE85 were more sensitive to stimulation by transforming growth factor β (TGFβ), a potent stimulator of bone growth.

However, unlike estrogen, fluoride did not increase mRNA for TGFβ. The importance of TGFβ in mediating fluoride induced osteoblast proliferation was reinforced by the finding that an antibody to TGFβ abolished the effects of fluoride. The authors concluded that fluoride activates osteoblast proliferation by modulating the cellular sensitivity to TGFβ (Reed et al., 1993). It has been reported that fluoride, with traces of aluminum, enhances a signal transduction cascade triggered by integrin and/or G protein receptors involving activation of focal adhesion kinase and stimulation of mitogen activated protein kinase activity in osteoblast-like cells. This could possibly occur via protein synthesis promoting Ras signal transduction pathways (Caverzasio et al., 1995; 1996; 1997).

The concept of fluoride interaction with growth factors in producing its bone enhancing effects is further strengthened by recent information that bone mass increases induced by fluoride in rats are significantly enhanced by simultaneous administration of IGF-1 (Ammann et al., 1995; 1998). Fluoride has also been shown, in a different study, to enhance tyrosine phosphorylation of several cellular proteins including extracellular signal-regulated kinases by inhibition of a specific fluoride-sensitive tyrosine phosphatase (Thomas et al., 1996). Several lines of evidence also suggest that the action of fluoride on bone cells involves tryosine kinase activation and interaction with growth factors. Despite all this evidence, however, there is still no proof that these ("tyrosine phosphorylation") actions of fluoride are actually responsible for the enhanced osteoblast proliferation.

Several lines of evidence suggest that fluoride may have a biphasic action on bone forming cells. It may stimulate osteoblasts at low concentrations, and may cause suppression at high concentrations leading to flat, inactive osteoblasts (Boivin et al., 1989). In addition, high doses of fluoride may lead to cross-link deficient collagen, possibly due to changes in its amino acid composition, and increased accumulation of glycosaminoglycans, proteoglycans, and dermatan sulfate (Susheela et al., 1985). Impaired mineralization is also a consequence of fluoride toxicity (doses >25 mg/day) in humans (Boivin et al., 1989). Thus cortical and trabecular bone mass increases at "therapeutic" doses, but there is reduced trabecular bone mass at higher (toxic) doses. Electron microscopy reveals disordered and non-lamellar bone formation (Fratzel et al., 1994). Fluoride also seems to have an effect on osteocytes, as evidenced by the presence of mottled periosteocytic lacunae surrounded by an eccentric, hypomineralized zone, as well as enlarged lacunae due to perilacunar resorption (Baud and Boivin, 1978).

B. Effect on Osteoclasts

The direct effect of fluoride on osteoclasts is still poorly understood, although there is histomorphometric and biochemical evidence of increased bone resorption in fluorosis (Faccini and Teotia, 1974; Boivin et al., 1989). In vitro studies with osteoclasts, however, have shown a decrease in resorptive activity (Okuda et al., 1990).

Functional quality of bone is also affected by fluoride in a biphasic manner, with improvement at low doses but deleterious effects at high doses. In the presence of fluoride, fluorapatite is preferentially formed over hydroxyapatite. Fluorapatite, which has a large crystal size, is less soluble than hydroxyapatite (Eanes and Reddi, 1979). Interestingly, fluoride also inhibits dissolution of preformed hydroxyapatite (Christofferson and Christofferson, 1984).

III. FLUORIDE TOXICITY

Fluorosis is a term used to describe the syndrome resulting from chronic exposure to excess fluoride. The disorder principally affects the skeletal and dental systems, although reports have highlighted metabolic and soft tissue manifestations such as glucose intolerance (Trivedi et al., 1993) and "gastritis /dyspepsia" like syndromes (Susheela et al., 1993). Traditional descriptions of skeletal fluorosis emphasize osteosclerosis, particularly of the axial skeleton (Singh and Jolly, 1961). Recently, however, it has been recognized that osteosclerosis is not necessarily the predominant feature in skeletal fluorosis (Christie, 1980; Krishnamachari, 1986; Mithal et al., 1993).

A. Clinical Correlates

Fluorosis due to consumption of drinking water containing toxic amounts of fluoride, or hydric fluorosis, continues to be endemic in vast areas of the developing world particularly in Asia and Africa. In India alone, 25 million people are thought to be at risk for development of fluoride toxicity. Fluorosis results from presence of excessive amounts of fluoride in the soil, which leeches into the well water and is subsequently consumed (Krishnamachari, 1986; Mithal et al., 1992). Hydric fluorosis resulting from consumption of fluoride-rich bottled mineral waters has been reported from France (Boivin et al., 1989). Inhalational (air borne) fluorosis could be telluric fluorosis due to inhalation of fluoride-rich dusts (Haikel et al., 1986) or industrial fluorosis due to inhalation of fluorinated gases that was first described in cryolite workers (Roholm, 1937). Further studies have emphasized the occurrence of fluorosis occurring due to coal burning in China (Jianxue and Cao, 1994). Iatrogenic fluorosis may occur with the prolonged consumption of fluoride containing drugs, e.g., niflumic acid, or sodium fluoride for the treatment of osteoporosis. Other causes include wine fluorosis (described from Spain, where fluoride was added to stop the fermentation process in certain local wines) (Soriano and Manchon, 1966), and dentrifice induced fluorosis (Ste-Marie et al., 1995).

Early skeletal involvement in fluorosis may produce only mild or nonspecific musculoskeletal symptoms, despite radiological evidence of significant disease. Overt skeletal fluorosis is characterized by generalized bone and joint pains, stiffness and restriction of movement of large joints and the spine. Fixed deformities, especially of the spine, hips, and knees are common in the late stages and may ultimately lead to incapacitation either due to the deformities themselves or as a result of the complicating compressive myeloradiculopathy. Some of these patients may present with bowing of the femur, tibia, radius, or ulna while others present with genu valgum or varum deformities (Krishnamachari and Krishnaswamy, 1973; Mithal et al., 1992). The latter typically occur in children with poor calcium intake, and have been described primarily from regions and populations where calcium undernutrition is common, such as parts of India, Tanzania, China, and South Africa (Krishnamachari, 1986; Mithal et al., 1992). Typical forms of fluorosis result in spinal deformity in adulthood.

However, variant forms with long bone involvement may lead to crippling deformities in children and adolescents. Over the years it has been shown that very young children can also suffer from severe skeletal fluorosis (Krishnamachari, 1986). In fact the rapid bone turnover state of the growing long bones may make them particularly vulnerable to the deleterious effects of toxic amounts of fluoride, especially in the absence of adequate calcium. Most patients of fluorosis, especially of the endemic hydric variety, also show evidence of dental involvement which is characterized (in increasing severity) by chalky white opaque areas over the enamel surface, marked attrition of surfaces subjected to wear, brown staining, discrete or confluent pitting, and hypoplasia of enamel.

Traditional descriptions of fluorosis have emphasized the presence of osteosclerosis (especially spinal) as the hallmark of skeletal involvement (Roholm, 1937; Shortt et al., 1937; Singh and Jolly, 1961). Calcification of the interosseous membrane of the forearm has been regarded as the *sine qua non* of skeletal fluorosis (Singh and Jolly, 1961). In addition, there is a marked tendency for calcification of ligaments and muscular attachments, irregular osteophytoses and exostoses. However, later descriptions have highlighted the wide spectrum of radiological findings including the simultaneous presence of osteosclerosis, osteoporosis, and osteomalacia (Christie, 1980; Mithal et al., 1993). A report from India showed that in addition to osteosclerosis there may be evidence of a coarse trabecular pattern, axial osteosclerosis with distal osteopenia, and diffuse osteopenia. In addition, occasional patients show the changes of hyperparathyroidism, and rickets or osteomalacia. Interestingly, in this report subjects with osteopenic changes had a significantly lower dietary calcium and protein intake than those with typical osteosclerosis (Mithal et al., 1993). It thus seems that osteosclerosis is the predominant lesion in fluorosis patients with adequate dietary calcium intake, while the osteopenic/osteomalacic picture predominates in people with suboptimal or poor calcium intake. However other as yet undefined factors, including toxins or nutrients, could also be modifying the skeletal picture.

B. Changes in Bone Metabolism Parameters

Most studies have found essentially normal levels of calcium (total and ionized) and phosphate in fluorosis, although there is evidence that these patients may retain more calcium than normal subjects (Srikantia and Siddiqui, 1965). In a study of South African children, however, hypocalcemia was reported (Pettifor et al., 1989). Several cases of hyperparathyroidism ("secondary hyperparathyroidism") have been reported (Teotia and Teotia, 1973; Sivakumar and Krishnamachari, 1976; Srivastava et al., 1989) but others reported normal parathyroid function in most patients (Boillat et al., 1978). Recent data, using the intact PTH assay, suggest that hyperparathyroidism may be less common than previously thought (Mithal et al., 1995). The factors determining the occurrence of hyperparathyroidism in a particular patient or group of patients remain unclear, although it has been suggested that calcium deficient genu valgum patients may show greater degree of secondary hyperparathyroidism (Krishnamachari, 1986). One study has suggested the occurrence of renal resistance to PTH in children with skeletal fluorosis (Pettifor et al., 1989). There is, however, evidence of rapid bone turnover, regardless of parathyroid function. Skeletal scintigraphy shows evidence of increased tracer uptake and turnover in virtually all patients of overt skeletal fluorosis (Gupta et al., 1993). Bone formation markers like serum alkaline phosphatase and osteocalcin are elevated, as are markers of resorption like urinary hydroxyproline and tartarate resistant acid phosphatase (Krishnamachari, 1986; Srivastava et al., 1989; Mithal et al., 1991, 1995).

Levels of 25(OH) vitamin D have generally been found to be normal (Krishnamachari, 1986; Srivastava et al., 1989; Mithal et al., 1995). There are reports of radiological osteomalacia in patients with fluorosis, but in these patients the 25(OH) vitamin D level was not estimated (Christie, 1980). Interestingly, a patient of osteoporosis treated with fluoride, vitamin D, and calcium developed osteomalacia despite high levels of 25(OH) vitamin D (Compston et al., 1980). There is a report of elevated serum 1,25 $(OH)_2$ vitamin D levels in children with endemic fluorosis (Pettifor et al., 1989). There does not seem to be a major alteration in serum calcitonin levels in skeletal fluorosis although reports of both high and low values are available (Krishnamachari, 1986; Mithal et al., 1995).

C. Histological Considerations

Histomorphometric analysis of bone from typical "osteosclerotic" fluorosis patients shows significant increases in cortical thickness and porosity but no reduction of compact bone mass. There is a significant increase in trabecular bone volume. Parameters reflecting bone formation (trabecular osteoid volume, trabecular osteoid surfaces, and thickness of osteoid seams) are markedly elevated. The increase in osteoid surfaces is significantly greater than that seen on trabecular resorption. There is a greater number of osteoblasts, although there is a greater proportion of flat osteoblasts. There is evidence of mineralization defects in the form of mottled periosteocytic lacunae. Thus, there is evidence of an unbalanced coupling in favor of bone formation (Boivin et al., 1989). One study has shown an enhanced bone resorptive activity in severe skeletal fluorosis (Faccini and Teotia, 1974).

D. Diagnostic Evaluation

The diagnosis of skeletal fluorosis is made by the presence of typical clinical and radiological features as described above along with the demonstration of an increased urinary fluoride level. Typical cases of skeletal fluorosis show urinary fluoride levels ranging from 1.2 to 10 mg per 24 hours. Urinary excretion of fluoride continues to be elevated for prolonged periods (years) following termination of fluoride exposure, whereas serum fluoride level is only an indication of the current fluoride consumption and is less useful in diagnosis (Krishnamachari, 1986). Bone (dry ash) fluoride levels are not generally required to make the diagnosis except in rare cases where the other tests are inconclusive. The residence of the patient in a known endemic area is an obvious clue.

E. Prevention and Treatment

No satisfactory treatment is available for the full-blown syndrome of skeletal fluorosis. However, removing the source of fluoride, for example by changing the source of water, results in some reduction in aches and pains, although the restricted mobility and deformities remain. There is some evidence in literature that there may be actual radiological improvement in skeletal changes several years after withdrawl of the fluoride insult, but this remains to be proven on a larger scale (Grandjean and Thomsen, 1983).

For most forms of skeletal fluorosis, prevention is the only effective measure to escape the disease. For hydric fluorosis, change or modification of the water supply is the only effective measure. One way is to dig deep bore wells: deeper wells have lower, although not necessarily appropriate fluoride content (Teotia et al., 1987). Several measures, including activated alumina technique and "Nalgonda" technique (alkalinization and chlorination followed by addition of aluminum salt) have been used to remove fluoride from water and are currently under trial (Bulusus, 1984). However, there is continuing controversy about the ideal level of fluoride in water.

While the World Health Organization recommends an upper limit of 1.5 mg per liter, it has been suggested that this may not be appropriate for hot climates where there is much greater water loss and consumption (Brouwer et al., 1988). In addition, since the diet, especially calcium intake, plays a major role in determining disease severity, poorly nourished populations may develop disease manifestations at much lower fluoride load. Thus in countries like India most experts feel that the acceptable level should be less than 1 part per million.

ACKNOWLEDGMENTS

The author gratefully acknowledges the contributions of his former students Sushil K. Gupta, M.D., D.M. and Nitin Trivedi, M.D., D.M., his colleague Madan M. Godbole, Ph.D., as well as all other members of the Fluoride Collaborative Study Group, who were instrumental in initiating and conducting field studies on endemic fluorosis under most trying circumstances. Grant support received from Japan International Cooperation Agency (India), and Sanjay Gandhi Post Graduate Institute of Medical Sciences is also gratefully acknowledged. Most of all, however, thanks are due to residents of endemic villages from Unnao District, Uttar Pradesh who willingly allowed themselves to be subjected to tedious investigations for the sake of science.

REFERENCES

Ammann, P., Rizzoli, R., Caverzasio, J., and Bonjour, J-P. (1995). Fluoride potentiates the osteogenic effects of IGF-1 in aged ovariectomized rat. Bone 17, 557 (Abstract).

Ammann, P., Rizzoli, R., Caverzasio, J., and Bonjour, J.P. (1998). Fluoride potentiates the osteogenic effects of IGF-I in aged ovariectomized rats. Bone. 22, 39-43.

Baud, C.A., and Boivin, G. (1978). Modifications of the perilacunar walls resulting from the effect of fluoride on osteocytic activity. Metab. Bone Dis. Relat. Res. 2S, 231-237.

Bellows, C.G., Heersche, J.N.M., and Aubin, J.E. (1991). The effects of fluoride on osteoblast progenitors in vitro. J. Bone Miner. Res. 5 (Suppl.1), 101-105.

Boivin, G., Chavassieux, P., Chapuy, M.C., Baud, C.A., and Meunier, P.J. (1989). Skeletal fluorosis: Histomorphometric analysis of bone changes and bone fluoride content in 29 patients. Bone 10, 89-99.

Boillat, M.A., Garcia, J., Dettweiler, W., Burckhardt, P., and Courvoisier, B. (1978). Clinical aspects of industrial fluorosis. In: *Fluoride and Bone*. (Courvoisier, B., Donath, A., Baud, C.A., Eds.), pp. 155-162, Huber, Bern.

Brouwer, I.D., Dirks, O.B., De Bruin, A., and Hautvast, J.G.A.J. (1988). Unsuitability of World Health Organisation Guidelines for fluoride concentrations in drinking water in Senegal. Lancet 1, 223-225.

Bulusus, K.R. (1984). Fluoride removal from potable water by various methods : Operating and capital cost estimation based on laboratory and plant data. In: *Fluoride Toxicity*. (Susheela, A.K., Ed.), pp. 91-101, International Society for Fluoride Research, New Delhi.

Caverzasio, J., Imai, T., Ammann, P., Burgener, D., and Bonjour. J-P. (1995). Mitogenic signaling process induced by fluoride in osteoblastlike cells involves focal adhesion kinase, GRB2, and mitogen activated protein kinase. J. Bone Miner. Res. 10 (Suppl.1) S143, #20 (Absract.)

Caverzasio, J., Imai, T., Ammann, P., Burgener, D., and Bonjour, J-P. (1996). Aluminium potentiates the effects of fluoride on tyrosine phosphorylation and osteoblast replication in vitro and bone mass in vivo. J. Bone Miner. Res. 11, 46-55.

Caverzasio, J., Palmer, G., Suziki, A., and Bonjour, J.P. (1997). Mechanism of the mitogenic effects of fluoride on osteoblastlike cells: Evidences for a G protein–dependent tyrosine phosphorylation process. J. Bone Miner. Res. 12, 1973-1983.

Caverzasio, J., Palmer, G., and Bonjour, J.P (1998). Fluoride: Mode of action. Bone. 22, 585-589.

Christie, D.P. (1980). The spectrum of radiographic bone changes in children with fluorosis. Radiology 136, 85-90.

Christofferson, J., and Christofferson, M.R. (1984). The effect of bendroflumethiazide and hydrochlororthiazide on the rate of dissolution of calcium hydroxyapatite. Calcif. Tissue Int. 36, 431-434.

Compston, J.E., Chadha, S., and Merett, A.L. (1980). Osteomalacia developing during treatment of osteoporosis with sodium fluoride and vitamin D. Brit. Med. J. 281, 910-911.

Dasrathi, S., Das, T.K., Gupta, I.P., Susheela, A.K., and Tandon, R.K. (1996). Gastroduodenal manifestations in patients with skeletal fluorosis. J. Gastroenterol. 31, 333-337.

Eanes, E.D., and Reddi, A.H. (1979). The effect of fluoride on bone mineral apatite. Metab. Bone Dis. Rel. Res. 2, 3-10.

Faccini, J.M., and Teotia, S.P.S. (1974). Histopathological assessment of endemic skeletal fluorosis. Calc. Tiss. Res. 16, 45-57.

Farley, J.R., Wergedal, J.E., and Baylink, D.J. (1983). Fluoride directly stimulates proliferation and alkaline phosphatase activity of bone forming cells. Science 222, 330-332.

Farley, J.R., Tarbaux, N., Hall, S., and Baylink, D.J. (1988). Evidence that fluoride stimulated ^{3}[H]-thymidine incorporation in embryonic chick calvarial cultures is dependent on the presence of a bone cell mitogen, sensitive to changes in the phosphate concentration, and modulated by systemic skeletal factors. Metabolism 37, 988-995.

Fratzel, P., Roschger, P., Eschberger, J., Abendroth, B., and Klaushofer, K. (1994). Abnormal bone mineralization after fluoride treatment in osteoporosis: A small angle x-ray scattering study. J. Bone Miner. Res. 9, 1541-1549.

Grandjean, P., and Thomsen, G. (1983). Reversibility of skeletal fluorosis. Brit. J. Industr. Med. 40, 456-461.

Gupta, S.K., Gambhir, S., Mithal, A., and Das, B.K. (1993). Skeletal scintigraphic findings in endemic skeletal fluorosis. Nucl. Med. Commun. 14, 384-390.

Haikel, Y., Voegel, J.C., and Frank, R.M. (1986). Fluoride content of water, dust, soils, and cereals in the endemic dental fluorosis area of Khouribga (Morocco). Arch. Oral Biol. 31, 279-286.

Hall, B.K. (1987). Sodium fluoride as an initiator of osteogenesis from embyronic mesenchyme in vitro. Bone 8, 111-116.

Jianxue, L., and Cao, S. (1994). Recent studies on endemic fluorosis in China. Fluoride 27, 3, 125-128.

Khoker, M.A., and Dandona, P. (1990). Fluoride stimulates ^{3}[H] thymidine incorporation and alkaline phosphatase production by human osteoblasts. Metabolism 39, 1180-1121.

Krishnamachari, K.A.V.R., and Krishnaswamy, K. (1973). Genu valgum and osteoporosis in an area of endemic fluorosis. Lancet 2, 877-879.

Krishnamachari, K.A.V.R. (1986). Skeletal fluorosis in humans : A review of the recent progress in the understanding of the disease. Prog. Food Nutr. Sci. 10, 279-314.

Lau, W.K.H., Farley, J.R., Freeman, T.K., and Baylink, D.J. (1989). A proposed mechanism of the mitogenic action of fluoride on bone cells: Inhibition of the activity of osteoblastic acid phosphatase. Metabolism 38, 858-863.

Mithal, A., Gupta, S.K., Kumar, S., Gupta, R.K, Godbole, M.M., Moonga, B.S., and Zaidi, M. (1991). Endemic skeletal fluorosis in India: Spectrum of the disease and a preliminary clinical work up of 100 patients. J. Bone Miner. Res. 6 (Suppl.), (Abstract.) 192.

Mithal, A., Trivedi, N., and Gupta, S.K. (1992). Endocrinology of fluorosis: An overview and a preliminary study of one hundred patients. In: *Endocrinology, Metabolism and Diabetes*, Vol. I. (Kochupillai, N., Ed.) pp. 308-316, MacMillan India Ltd., Delhi.

Mithal, A., Trivedi, N., Gupta, S.K., Kumar, S., and Gupta, R.K. (1993). Radiological spectrum of endemic fluorosis: Relationship with calcium intake. Skeletal. Radiol. 22, 257-261.

Mithal, A., Trivedi, N., Gupta, S.K., Dabadghao, P., Godbole, M.M., and Zaidi, M. (1995). Skeletal fluoride toxicity in children: Clinical, radiologic, and metabolic profile J. Bone Miner. Res. 10 (Suppl.1), P246.

Okuda, A., Kanehisa, J., and Heersche, J.N.M. (1990). The effects of sodium fluoride on resorptive activity of isolated osteoclasts. J. Bone Miner. Res. 5 (Suppl.1), 115, 1990.

Pak, C.Y.C., Sakhaee, K., Adams-Huet, B., Piziak, V., Peterson, R.B., and Poindexter, J.R. (1995). Treatment of post menopausal osteoporosis with slow-release sodium fluoride. Final update of a randomized controlled trial. Ann. Int. Med. 123, 401-408.

Pettifor, J.M., Schnitzler, C.M., Ross, F.P., and Moodley, G.P. (1989). Endemic skeletal fluorosis in children: Hypocalcemia and the presence of renal resistance to parathyroid hormone. Bone Miner. 7, 275-288.

Reddy, G.S., and Srikantia, S.G. (1970). Effect of dietary calcium, vitamin C, and protein in the development of experimental skeletal fluorosis. I. Growth, serum chemistry, and changes in composition, and radiological appearance of bones. Metabolism 20, 642-649.

Reed, B.Y., Zerwekh, J.E., Antich, P.P., and Pak, C.Y.C. (1993). Fluoride-stimulated ^{3}H thymidine uptake in a human osteoblastic osteosarcoma cell line is dependent on transforming growth factor β. J. Bone Miner. Res. 8, 19-25.

Roholm, K. (1937). *Fluoride Intoxication: A Clinical Hygienic Study*. Nordisk Forlog, Arnold Busck, Copenhagen, Denmark.

Shortt, H.E., Pandit, C.C., and Raghavachari, T.N.S. (1937). Endemic fluorosis in the Nellore district of South India. Ind. Med. Gazette 72, 396-403.

Singh, A., and Jolly, S.S. (1961). Endemic fluorosis. Epidemiological, clinical, and biochemical study of chronic fluorine intoxication in Punjab (India). Medicine 42, 229-246.

Sivakumar, B., and Krishnamachari, K.A.V.R. (1976). Circulating levels of immunoreactive parathyroid hormone in endemic genu valgum. Horm. Metab. Res. 8, 317-319.

Soriano, M., and Manchon, F. (1966). Radiological aspects of a new type of bone fluorosis, periosteitis deformans. Radiology 87, 1089-1094.

Srikantia, S.G., and Siddiqui, A.H. (1965). Metabolic studies in skeletal fluorosis. Clin. Sci. 28, 477-485.

Srivastava, R.N., Gill, D.S., Moudgil, A., Menon, R.K., Thomas, M., and Dandona, P. (1989). Normal ionized calcium, parathyroid hypersecretion, and elevated osteocalcin in a family with fluorosis. Metabolism 38, 120-124.

Ste-Marie, L.G., Froment, D.H., Bourelle, D., Rivest, C., and Boivin, G. (1995). Skeletal fluorosis induced by dentrifice. J. Bone Miner. Res. 10 (Suppl.1), S499, (Abstract, T580.)

Susheela, A.K., Koacher, J., Jain, S.K., Sharma, K., and Jha, M. (1985). Fluoride toxicity: A molecular approach. In: *Fluoride Toxicity*. (Susheela, A.K., Ed.), pp. 78-80. Kalpana Printing House, New Delhi.

Teotia, S.P.S., and Teotia, M. (1973). Secondary hyperparathyroidism in patients with endemic skeletal fluorosis. Brit. Med. J. 1, 637-640.

Teotia, S.P.S., Teotia, M., Singh, D.P., and Nath, M. (1987). Deep bore drinking water as a practical approach for eradication of endemic fluorosis in India. Indian J. Med. Res. 85, 699-705.

Thomas, A.B., Hashimoto, H., Baylink, D.J., and Lau, K.H.W. (1996). Fluoride at mitogenic concentrations increases the steady-state phosphotyrosyl phosphorylation level by cellular proteins in human bone cells. J. Clin. Endocrinol. Metab. 81, 2570-2578.

Trivedi, N., Mithal, A., Gupta, S.K., and Godbole, M.M. (1993). Reversible impairment of glucose intolerance in endemic fluorosis. Diabetologia 36, 826-830.

Zerwekh, J.E., Morris, A.C., Padalino, P.K., Gottschalk, F., and Pak, C.Y.C.(1990). Fluoride rapidly and transiently increases intracellular calcium in human osteoblasts. J. Bone Miner. Res. 5 (Suppl.1), 131-136.

MOLECULAR PHARMACOLOGY OF ANTIRESORPTIVE THERAPY FOR OSTEOPOROSIS

Olugbenga A. Adebanjo, Edna Schwab, Li Sun, Michael Pazianas, Baljit Moonga, and Mone Zaidi

Advances in Organ Biology
Volume 5C, pages 809-834.

ISBN: 0-7623-0390-5

I. INTRODUCTION

Skeletal remodeling requires a balance between the resorption and formation of bone. Highly specialized multinucleated osteoclasts resorb bone, while osteoblasts form or lay down new bone. These two processes are closely coupled, but are regulated independently by a variety of hormones, cytokines, and other local influences. Furthermore, both the activity as well as the formation of osteoclasts and osteoblasts can be regulated independent of the other.

It is well known that excessive osteoclast activity causes inappropriately high levels of bone destruction in several bone and joint diseases including osteoporosis, Paget's bone disease, tumor-induced osteolysis, hyperparathyroidism, and rheumatoid arthritis. In these conditions, the aim is to prevent further bone loss. This can be achieved by either inhibiting osteoclast activity or number or by restoring bone through an increase in formation. While appropriate therapeutic agents can increase the thickness of individual trabeculae, and hence trabecular bone volume, no agent has yet been developed that could bridge trabeculae that have undergone microfracture. Here we review the cellular and molecular pharmacology of established agents that are known to inhibit osteoclastic activity, as well as those that are currently under evaluation.

II. ESTABLISHED AGENTS FOR RESORPTIVE BONE DISEASE

Table 1 lists some of those bone-forming and antiresorptive agents that are currently in use clinically or are undergoing clinical investigation. Bone-forming agents are discussed in the chapter by Reid in volume 5C. We will limit our discussion, however, to antiresorptive compounds.

Table 1. Inhibitors and Stimulators of Bone Formation

Inhibitors of Bone Resorption	*Stimulators of Bone Formation*
Estrogens and estrogen agonists-antagonists	Parathyroid hormone
Calcitonin	Fluoride
Bisphosphonates	Vitamin D metabolites (under test)
Ipriflavone	Prostanoids (under test)
	Electromagnetic fields (under test)
	Growth hormone (selected conditions)

A. Estrogen

Estrogen is one of the main therapeutic modalities for treating and preventing post-menopausal bone loss. It reduces both cortical and trabecular bone loss in the axial as well as the peripheral skeleton (Gotfredsen et al., 1986; Riis and Christiansen, 1988). In women with an intact uterus, a progestogen must be added to protect against unopposed endometrial actions of estrogen. The latter are thought not to interfere with the skeletal effects of estrogens (Christiansen et al., 1980, 1981), but result in withdrawal bleeding.

The four broad classes of estrogens are shown in Table 2. Synthetic estrogens without a steroid ring structure, such as tamoxifen and clomiphene, are no longer in use except as inhibitors of estrogen action. Synthetic estrogens, such as ethinyloestradiol, with a steroid ring structure are used in oral contraceptives. In the U.S., conjugated equine estrogens are most commonly used for post-menopausal therapy, while in Europe, the mainstay of therapy centers around the use of human estrogens, such as 17β-estradiol or its esters. Note that there is also a marked variation in the potency of estrogens when different bioassays are used to provide potency estimates (Hammond and Maxson, 1982; Maschchak et al., 1982). Currently, all estrogens are thought not to differ in their side effect profile.

It remains unclear how estrogens interact with bone cells. Earlier hypotheses have suggested indirect mechanisms. These include the induction of a reduced sensitivity of bone to the resorptive effects of parathyroid hormone (Teitelbaum and Kahn, 1980), enhanced secretion of calcitonin (Stevenson et al., 1983), and an increase in calcium absorption from the intestine (Selby, 1990). Note that estrogen receptors have not been identified on osteoclasts. Nevertheless, bone-forming osteoblasts (Eriksen et al., 1988; Lin et al., 1991) do possess estrogen receptors at a density of around 1,600 receptors per cell. The receptor number is therefore one quarter of the total receptor number in reproductive tissues. Estrogen acts on its receptors on osteoblasts to produce a spectrum of effects, only some of which are specific to bone. Others are nonspecific in that they occur in all estrogen-sensitive tissues. Some effects of estrogen include progesterone receptor induction (Eriksen et al., 1988),

Table 2. Classification of Estrogen Preparations
Synthetic estrogen analogues (–steroid skeleton)
Synthetic estrogen analogues (+ steroid skeleton)
Non-human estrogens
Human estrogens

enhanced procollagen production (Gray et al., 1987; Ernst et al., 1989; Lin et al., 1991) and increased alkaline phosphatase expression. In addition, it has become obvious that estrogen may modulate the synthesis of certain growth factors and cytokines. For example, it inhibits the production of interleukin-1 (IL-1), tumor necrosis factor a (TNF-α) and interleukin-6 (IL-6) (Tabibzadeh et al., 1989; Girasole et al., 1992). Their release from peripheral blood mononuclear cells is also suppressed. Notably, higher levels of IL-1, TNF-α, and granulocyte-macrophage colony-stimulating factor (GM-CSF) are released from monocytes derived from untreated post-menopausal women with documented bone loss than from cells obtained from ones treated with estrogen (Pacifici et al., 1989). Also, 17β-estradiol at circulating concentrations suppresses monocytic TNF production (Ralston et al., 1990). Indeed, that estrogen-sensitive inhibition of TNF release may be absent post-menopausally may contribute to the increased bone loss (Kimble et al., 1996; 1997).

Estrogen, when applied to osteoblasts, induces mitogenesis and enhances insulin-like growth factor-1 (IGF-1) messenger ribonucleic acid (mRNA) expression (Ernst et al., 1989). The autocrine production of IGF-1 thus plays a role in the responsiveness of osteoblasts to estradiol. Since both resorptive and antiresorptive cytokines are produced from osteoblasts, the regulation of their production by estrogens may offer a means by which the hormone may affect the balance between bone formation and resorption. More recently, it has been shown that estrogen application results in a dramatic reduction in the levels of steady-state mRNA for osteocalcin, prepro α-2(I) chain of type I collagen, osteonectin, and osteopontin (Turner et al., 1990). The latter are known bone-matrix proteins upon which the effects of estrogen could be exerted.

B. Calcitonin

Calcitonin is an established agent in the pharmacotherapy of osteoporosis and had been licensed for use in osteoporosis treatment. High turnover osteoporosis, such as the post-menopausal form, responds better to calcitonin than does the low turnover form (Overgaard et al., 1992). There is evidence that calcitonin reduces fracture frequency (Gruber et al., 1988). Synthetic preparations of salmon, human, eel, and asusuberic$^{1-7}$-eel calcitonins are available for human use (Zaidi et al., 1990b). Apart from osteoporosis, the use of calcitonin has also been established for Paget's disease of bone and hypercalcemia of malignancy. The major drawback is that the peptide has bone conserving actions and lowers plasma calcium in hypercalcemia, but that there is an escape from the drug's action shortly following its initiation (Pecile, 1992). An advantage of calcitonin, particularly of fish calcitonin, is its analgesic action, which allows for its use in bone diseases associated with severe pain, including the vertebral crush syndrome (Seiber et al., 1986). The analgesic action of the compound is thought to be mediated via endorphins.

The primary amino acid sequences of various calcitonins are shown in Figure 1. There is an amino-terminal disulphide bridge between positions 1

Amino acid sequences of the CALCITONINS

	Man1	Man2*	Rat	Sa.2	Sa.3	Sa.1	Eel	Chick.	Porc.	Bov.	Ov.
	H										
1	Cys	Tyr	—	—	—	—	—	—	—	—	—
2	Gly	Ser		Ser	Ser	Ser	Ser	Ala	Ser	Ser	Ser
3	Asn							Ser			
4	Leu	—	—	—	—	—	—	—	—	—	—
5	Ser	—	—	—	—	—	—	—	—	—	—
6	Thr	—	—	—	—	—	—	—	—	—	—
7	Cys	—	—	—	—	—	—	—	—	—	—
8	Met	Leu			Val	Val	Val	Val	Val	Val	Val
9	Leu	Gln	—	—	—	—	—	—	—	—	—
10	Gly								Ser	Ser	Ser
11	Thr			Lys	Lys	Lys	Lys	Lys	Ala	Ala	Ala
12	Tyr			Leu	Leu	Leu	Leu	Leu			
13	Thr	Leu		Ser	Ser	Ser	Ser	Ser	Trp	Trp	Trp
14	Gln								Arg	Lys	Lys
15	Asp	Tyr				Glu	Glu	Glu	Asn		
16	Phe	Leu	Leu	Leu	Leu	Leu	Leu	Leu	Leu	Leu	Leu
17	Asn	Lys		His	His	His	His	His			
18	Lys	Asn							Asn	Asn	Asn
19	Phe			Leu	Leu	Leu	Leu	Leu		Tyr	Tyr
20	His			Gln	Gln	Gln	Gln	Gln			
21	Thr	Met							Arg	Arg	Arg
22	Phe					Tyr	Tyr	Tyr			Tyr
23	Pro							Ser	Ser	Ser	
24	Gln	Gly		Arg	Arg	Arg	Arg	Arg	Gly	Gly	Gly
25	Thr	Ile							Met	Met	Met
26	Ala	Asn	Ser	Asn	Asn	Asn	Asp	Asp	Gly	Gly	Gly
27	Ile	Phe		Thr	Thr	Thr	Val	Val	Phe	Phe	Phe
28	Gly	—	—	—	—	—	—	—	—	—	—
29	Val	Pro		Ala	Ala	Ser	Ala	Ala	Pro	Pro	Pro
30	Gly	Gln						Glu	Glu	Glu	Glu
31	Ala	Ile		Val	Val	Thr	Thr	Thr	Thr	Thr	Thr
32	Pro	—	—	—	—	—	—	—	—	—	—
	NH_2										

* Man2 - predicted

Figure 1. Amino acid sequences of calcitonins from eight species including human calcitonin (Man 1 and Man 2*). Abbreviations: Sa, Salmon; Chick, chicken; Porc, porcine; Bov, bovine; Ov, ovine. The invariant residues are shown by horizontal lines; two of these have been altered in the predicted (*) second human calcitonin-like sequence (Man 2*). Reproduced with kind permission from the Quart. J. Exp. Physiol. 72, 371-408, 1987.

and 7 and a carboxyl-terminal prolineamide residue. Note that the methionine residue at position 8 in human calcitonin also appears essential for biological activity (Maier et al., 1974). Its oxidation at position 25 in the porcine, bovine, and ovine calcitonin, however, does not affect their biological activity. Amino acid residues at the two ends of the molecule are highly conserved suggesting that they are essential for biological activity. The middle portion of the molecule (positions 10 to 27) is variable and is thought to control potency and action duration. The crystal structure of calcitonins computer predictions suggest all calcitonins have an α-helix, a turn, and a β-sheet.

The ability of calcitonins to assume numerous conformations, or its conformational flexibility, is thought to affect its biological activity (Epand et al., 1986). More flexible molecules usually attain a greater number of conformations at the receptor-membrane complex. This is why ultimobranchial or fish (salmon and eel) calcitonins may be more active than their mammalian counterparts (Galante et al., 1971). Potency is enhanced when amino acids having bulky functional groups are substituted with those with less bulky side chains. Note that three glycine residues are highly conserved both in salmon and eel calcitonin, and that glycine constitutes the least bulky functional group. Both peptides therefore possess a low potential for forming a rigid helix allowing for the molecule to access the receptor more favorably or to avoid a less active, but more stable, conformation (Epand et al., 1986).

Conformational flexibility alone nevertheless does not predict the relative potency of some calcitonins. It must be considered together with the peptide's hydrophobic moment as well as overall hydrophobicity. For example, *des*-Leu^{16}-salmon calcitonin and human calcitonin have less helix-forming potential and are consequently more flexible than salmon calcitonin, but are much less potent (Epand et al., 1985). In addition, long range interactions between the amino and carboxyl regions of the molecules at least in the receptor-bound form can result in conformational coupling between different regions of the molecule. Calcitonin binding to its receptor is also coupled to cooperative conformational interactions similar to those proposed for enzyme-substrate binding (Tanuichi, 1984).

A disulphide bridge may or may not be for calcitonin action. The salmon calcitonin analogue, *des*-ser^{2}-salmon calcitonin, for example, has fewer amino-terminal amino acids, yet it is fully active (Schwarrtz et al., 1981). In line with this notion, it has been noted that an intact hexapeptide ring structure is not essential for the binding of antagonistic arginine vasopressin (AVP) analogues to vasopressor or antidiuretic AVP receptors (Manning et al., 1987). Furthermore, when the disulphide bridge is lost in a carba-type analogue of eel calcitonin, asusubéric$^{1-7}$-eel calcitonin, full biological activity is retained (Morikawa et al., 1976; Zaidi et al., 1987a). Finally, when the disulphide bridge in salmon calcitonin between positions 1 and 7 is replaced by an S-acetamidomethylcysteinyl linkage, neither is the hypocalcemic activity nor is the renal cAMP-stimulating activity of the peptide abolished (Orlowski et al., 1987). However, a similar modification in the human calcitonin molecule removes the peptide's hypocalcemic action, while the renal effects remain intact. Such modifications may form the basis for the design of linear analogues of cyclic peptides as pharmacological probes and for therapeutic use.

The main action of calcitonin is on bone, although it may, at pharmacological concentrations, increase renal calcium and phosphate excretion (Agus et al., 1981) and 1,25-dihydroxycholecalciferol production. When infused into experimental animals with a high bone turnover, or into human subjects with Paget's bone disease, the peptide causes a marked acute fall of plasma calcium. This action results from the inhibition of osteoclast activity and hence a diminished Ca^{2+} flux from bone to blood (MacIntyre et al., 1967).

That calcitonin inhibits basal and stimulated resorption of intact bone in organ culture has long been established (Aliopoulious et al., 1965; Freidman and Raisz, 1965; Milhaud et al., 1965; Robinson et al., 1967; Reynolds et al., 1968; Holtrop et al., 1974). In bone sections, calcitonin application results in a rapid loss of ruffled borders (Kallio et al., 1972; Holtrop et al., 1974). When applied for a longer term, there is a reduction in the number of osteoclasts in bone, an action that appears relevant to its use in Paget's bone disease (Macintyre et al., 1987). Thus, calcitonin acts directly on the osteoclast (Chambers and Magnus, 1982). When applied *in vitro* to isolated osteoclasts, femtomolar concentrations of salmon calcitonin result in an acute cessation of cytoplasmic motility followed by a gradual pseudopodial retraction. Furthermore, over a 24-hour period, the excavation, by osteoclasts, of slices of devitalized cortical bone is inhibited again by very low concentrations of salmon calcitonin (EC_{50} = 0.1 pg/ml) (Chambers et al., 1985a,b; Zaidi et al., 1987a,b, 1988). Porcine and human calcitonin are equally effective only at 30-fold higher concentrations, a ratio that correlates well with their *in vivo* hypocalcemic potency. Calcitonin also inhibits acid hydrolase secretion (Anderson et al., 1982; Akisaka and Gay, 1986; Chambers et al., 1987; Moonga et al., 1990) and Na^+-K^+-ATPase activity (Akisaka and Gay, 1986) and alters the localization of carbonic anhydrase (Anderson et al., 1982) in osteoclasts.

Receptors for calcitonin have been cloned and sequenced. The first cloned receptor, a porcine kidney epithelial cell receptor (Lin et al., 1991), comprised 482-amino acids and displayed a high affinity for salmon calcitonin (K_d = 6 nM). This receptor appears to be coupled functionally to increases in intracellular cyclic adenosine 3', 5'-monophosphate (cAMP) and to intracellular Ca^{2+} release. It is homologous to the PTH/parathyroid hormone-related peptide (PTHrP) and secretin receptor (Jüppner et al., 1991). Note also that the renal receptor has been shown to couple to separate G proteins during different phases of the growth cycle (Chakarborty et al., 1991).

We developed a novel method to quantitate osteoclast motility. This first shed some light on the post-receptor effect of calcitonin on the osteoclast by showing that two components of calcitonin action separately affect osteoclastic bone resorption (Zaidi et al., 1990b). The more rapid quiescence (Q) component ($t_{1/2}$ ~ 15 minutes) is characterized by a cessation of motility. The late retraction (R) component ($t_{1/2}$ ~ 27 minutes) is characterized by pseudopodial retraction which may itself reduce resorption by decreasing the area of osteoclastic contact with the bone surface. Thus, the anti-secretory action of calcitonin (Anderson et al., 1982) may result from reduced granule movement as part of the Q component. We subsequently found that the Q and R effects of calcitonin were mediated by separate transduction pathways, and distinct receptor subtypes (Zaidi et al., 1990b). Thus, two G proteins seem to be coupled to distinct surface receptors that become activated in response to calcitonin action. A G_s-like G protein was found to be sensitive to cholera toxin stimulation and to mediate the cAMP-dependent Q component. The pertussis toxin-sensitive G protein was shown to mediate R via a rise of intracellular [Ca^{2+}]. A less-specific receptor subtype was activatable by calcitonin and its related peptides, amylin and calcitonin gene-related peptide (CGRP). In contrast, the subtype presumably mediating the R effect was highly specific for calcitonin (Alam et al., 1991). Thus distinct surface receptor subtypes appear to mediate the two components, Q and R, of calcitonin action at least pharmacologically. Both receptor isoforms in the rat, C1a and C1b, were subsequently cloned and sequenced.

C. Bisphosphonates

Bisphosphonates are used therapeutically in conditions characterized by increased bone remodeling, namely Pagets's disease of bone, hypercalcemia of malignancy, and osteoporosis (Fleisch, 1991; 1998; see the chapter in volume 5C by Fleisch). All bisphosphonates increase bone mineral density and reduce the frequency of new vertebral fractures in post-menopausal women. Their main effect is to inhibit bone resorption, although some compounds reduce mineralization with similar potency. Bisphosphonates have relatively few adverse side-effects, although esophageal irritation has been documented with the newer bisphosphonate, alendronate. The half-life of bisphosphonates is different both in plasma and bone; it is in the order of minutes to hours in plasma but prolonged in bone depending on skeletal turnover. The skeleton takes up to 50% of a single dose administered and the rest is excreted in the urine.

Essentially, in bisphosphonates, the oxygen atom of the P-O-P (pyrophosphate), is replaced with a carbon atom resulting in a P-C-P bond. Thus the physicochemical effects of most bisphosphonates are similar to those of P-O-P and polyphosphates. Their avidity to bone, similar to that of fluoride and tetracyclines, seems related to their marked affinity for solid phase calcium phosphate. They bind to the surface by chemical adsorption to calcium and then act as crystal poisons for both bone formation and resorption. The two P-C bonds are resistant to enzymatic hydrolysis and allow variations either by changing the two lateral side-chains or by esterification of the phosphate group. Small changes in bisphosphonate structure lead to extensive alterations in their physicochemical, biological, therapeutic, and toxicological properties. This is the basis of the pharmaceutical missions development of new compounds.

The critical issue in pharmaceutical research has been to develop a high potency bisphosphonate that does not inhibit bone mineralization. The antiresorptive potency of various bisphosphonates, assessed in the rat, are shown in Table 3. Note that etidronate inhibits resorption and mineralization in the same concentration range. Clodronate is about 10 times more potent as an anti resorptive agent, but does not significantly inhibit mineralization (Fleisch et al., 1969; Russell et al., 1971). Furthermore, several chemical modifications, such as the addition of a hydroxyl group to the

Table 3. Antiresorptive Potency of Bisphosphonates

Etidronate	1
Clodronate	10
Tiludronate	10
Pamidronate	100
Alendronate	> 100, < 1,000
Dimethyl APD	> 100, < 1,000
EB-1053	> 100, < 1,000
YM-175	> 100, < 1,000
BM-21.0955	> 1,000, < 10,000
Risedronate	> 1,000, < 10,000
CGP42'446	> 10,000

carbon atom at position 1 (Shinoda et al., 1983) or the addition of an amino group to a side-chain, or an increase in its length significantly enhance antiresorptive potency. The 3-amino derivative, pamidronate, is therefore 10 times more active than clodronate (Bijvoet et al., 1978; Reitsma et al., 1980). In turn, the 4-carbon bisphosphonate, alendronate, is about 10 times more active than pamidronate (Schenk et al., 1986). The potency of pamidronate is enhanced by the demethylation of the amino group (Boonekamp et al., 1987). Such activity is further increased by adding a methyl and a pentyl group to the nitrogen atom as in BM 21.0955. The latter is 10,000-fold more potent than etidronate (Muhlbauer et al., 1989, 1991) and is also among the most powerful bisphosphonate synthesized to date. Cyclic bisphosphonates are also highly active, especially those containing a nitrogen atom in the ring, such as risedronate (Bevan et al., 1988; Sietsema et al., 1989). Thus, all highly active compounds contain a nitrogen atom. This rank order of potency shown in rats is similar in humans. Furthermore, other compounds used only in animals are CGP 42'446 and YM-529 which have an imidazole ring in their structure. In addition, there are those with a ring within their structure such as in YM 175 and EB-1053.

Early suggestions on the modes of action of bisphosphonates on osteoclastic bone resorption are shown in Table 4. Although bisphosphonates inhibit resorption (Fleisch, 1983), the exact mechanism is not fully understood. One suggestion is that bisphosphonates inhibit bone resorption by the physicochemical inhibition of hydroxyapatite dissolution. However, this appears unlikely as their antiresorptive potency does not correlate with their ability to inhibit hydroxyapatite dissolution. Nevertheless, there is no doubt that bisphosphonates are avidly adsorbed onto calcified bone matrix at high local concentrations (Powell and De Mark, 1985; Monkkonen, 1988). The latter effects would explain both the specificity of bisphosphonate action on skeletal tissues and the delay in their onset of action (Muhlbauer and Fleisch, 1990). In addition, bisphosphonates alter cellular metabolism in various ways (Fast et al., 1978; Felix et al., 1984) (Table 4). These cellular actions however do not correlate with their *in vivo* antiresorptive effects.

Clodronate is thought to inhibit osteoclastic function by a direct cytotoxic action (Flanagan and Chambers, 1989); this has been ruled out for etidronate

Table 4. Cellular Pharmacology of Bisphosphonates

Inhibition of lysosomal enzymes	Inhibition of osteoclastic proton pump
Inhibition of pyrophosphatase	Reduction of protein synthesis
Inhibition of prostaglandin synthesis	Cytotoxicity

and pamidronate. The latter may, through their avid binding to bone mineral, disturb interactions between osteoclasts and regulatory cytokines, thus interupting bone remodeling (Reitsma et al., 1983; Valkema et al., 1989). The drugs may then inhibit the activity and multiplication of cytokine-producing macrophages (Cecchini et al., 1987; Cecchini and Fleisch, 1990). Another hypothesis is that bisphosphonates could inhibit the catalytic activity of the secreted acid phosphatase (Moonga et al., 1990). In isolated rat osteoclast cultures, we found that this action correlated with the antiresorptive activity of these compounds. It is therefore possible that bisphosphonates competitively inhibit acid phosphatase action on bone, thus preventing the removal of pyrophosphate, a natural inhibitor of resorption.

New insights into the action of bisphosphonates has been provided by data from Fleisch's group (Sahni et al., 1993; Vitte et al., 1994). The osteoblast is now thought to mediate the antiresorptive effects of bisphosphonates, even those actions seen in isolated osteoclast cultures. Bisphosphonates, at concentrations as low as 10^{-10} and 10^{-11}, have been shown to inhibit resorption, possibly by reducing the secretion of the yet uncharacterized osteoclast resorption-stimulating activity (ORSA). Nevertheless, very recent results from the same group have indicated that another novel osteoclast resorption-inhibiting activity may be released in response to bisphosphonate action on osteoblasts (Vitte et al., 1994). Contrary to what one might expect, this activity is not due to one of the known, osteoclast-inhibitory, prostaglandins. Furthermore, it is also unclear as to how bisphosphonates exert their actions on osteoblasts, as being highly charged molecules, they would not be expected to enter the cell. The possibility of there being a cell surface receptor is being investigated.

III. PHYSIOLOGICAL CONSIDERATIONS

The thinning of trabeculae by increased osteoclastic activity results in a complete destruction of bone architecture. As indicated previously, we do not have the means of restoring the microstructure of bone once it is destroyed. The best we can achieve with the existing therapies is increasing the width of disconnected trabeculae. This does not alter the basic strength of the overall structure of bone. Thus, prevention of destruction is the best approach available at the moment. Two strategies should guide our future approach to therapeutic intervention: inhibition of bone resorption and stimulation of bone formation. The cellular basis of future developments of osteoclast-inhibitory therapies will be discussed below.

An osteoclast forms a segregated area of close adhesion between the cell membrane and the bone surface, then isolates this newly created area and develops within it a unique microenvironment. The osteoclast plasma membrane adjacent to the bone is next thrown into complex folds, the ruffled border, a specialization that secretes resorptive enzymes and protons. The resorptive enzymes hydrolases and cysteine-proteinases digest protein matrix at a low pH (3 to 4 units) and bone mineral is dissolved by the action of the secreted acid. Together with these are free radicals, mainly $O_2^{\cdot-}$, which may aid collagen digestion (Garrett et al., 1990; Key et al., 1990; Oursler et al., 1991).

Completion of the resorptive episode is closely followed by osteoclast detachment (or de-adhesion) and retraction (Zaidi et al., 1990a). We believe that this event is triggered by a change in local Ca^{2+} concentration that results from hydroxyapatite dissolution. The response is mediated by a membrane Ca^{2+} receptor lying on the osteoclast surface (Datta et al., 1989; Malgaroli et al., 1989; Zaidi et al., 1989, 1991, 1995; Miyauchi et al., 1990; Moonga et al., 1990; Zaidi, 1990b; Zaidi et al., 1991; 1995; Shankar et al., 1992a). The latter, which is believed to be a cell surface type II ryanodine receptor, senses changes in extracellular Ca^{2+} that are subsequently transduced into intracellular signals, which in turn cause osteoclast retraction. Notably, the cells retain the ability to move away from the resorption site and to resorb at a different location. We also believe that cell migration may be facilitated by the local production of the stimulatory reactive oxygen species, hydrogen peroxide (H_2O_2). Which is produced within the hemivacuole by two successive protonations of locally secreted $O_2^{\cdot-}$ (Bax et al., 1992).

Osteoclast activity is also regulated by calcitonin while locally produced factors, including several bone- and endothelium-derived factors, influence osteoclast function either directly or indirectly. Prostaglandins are among the most well studied. We have shown that the autocoid, nitric oxide (NO) (MacIntyre et al., 1991) and the peptide, endothelin (Alam et al., 1992), have profound inhibitory effects on osteoclastic bone resorption. The osteoclast is also influenced by a complex matrix of cytokines, including various interleukins and colony-stimulating factors (Roodman, 1991). Cytokines mainly regulate osteoclast formation and recruitment, but may influence osteoclast function indirectly via the osteoblast. Recent data from our group demonstrate that locally produced IL-6 attenuates extracellular Ca^{2+} sensing, thereby releasing osteoclasts from the inhibitory effect of high Ca^{2+}. The latter may explain the mechanism by which high IL-6 production bone resorption (e.g., in multiple myeloma) is sustained. (Adebanjo et al., 1998). Several hormones, including

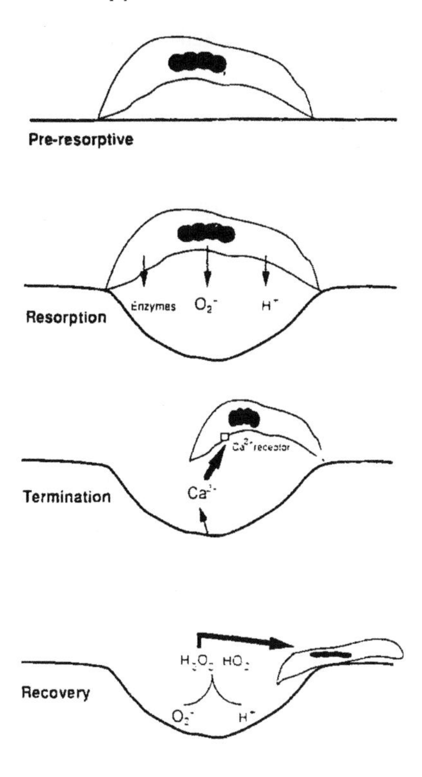

Figure 2. A hypothetical model for the bone resorptive process and its control (with kind permission from Current Opinion in Therapeutic patents).

PTH and 1, 25-dihydroxyvitamin D, also enhance osteoclastic bone resorption indirectly (McSheehy and Chambers, 1986).

IV. EXPERIMENTAL APPROACHES

We can intervene at three different stages in the bone resorption episode. One can affect osteoclast formation and recruitment, prevent osteoclast activation, or interfere with the resorptive episode per se. In order to achieve the latter objective, it is essential that we either perturb certain modes of osteoclastic activity (mode-specific inhibition) or known cellular targets (target-specific inhibition). Osteoclastic activities essential for the resorptive process are listed in Table 5. Moreover, pharmacological examples relating these activities to the stimulation/inhibition of biochemical targets in the osteoclast are presented in Figure 3.

Table 5. The Mode of Osteoclastic Activity and their Subclassification*

Abbreviation	*Mode*	*Measurement*
Motile processes		
A	Cell-matrix-adhesion	Adhesion assays
S	Cell spreading	Morphometry
M1	Pseudopodial retraction and protrusion	Digitization and analysis
M2	Cell migration	Morphometry
M3	Margin raffling	Subjective (time-lapse)
Secretory activity		
E	Enzyme secretion	Enzyme activity assays
H	Proton release	Acidification assays
F	Free radical production	Staining and densitometry

Note: * From Bioscience Reports 10, 547-556

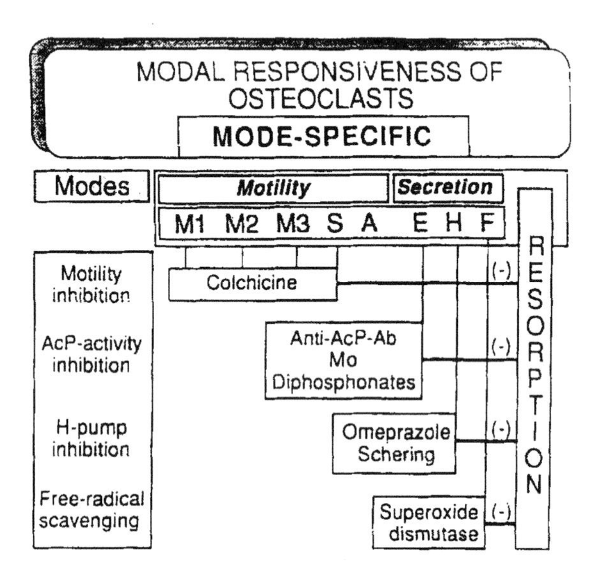

Figure 3. Mode-specific inhibition of osteoclast function. Interaction of various pharmacological agents with specific activity modes (M1, M2, M3, S, A, E, H, F) as shown in Table 5. Abbreviations: AcP, acid phosphatase; Mo, molybdate.

The attachment and subsequent sealing of osteoclast to bone depends upon the interaction of integrin $\alpha_V\beta_3$ to bone matrix proteins containing the Arg-Gly-Asp (RGD) recognition sequence (McSheehy and Chambers, 1987). This was previously termed, osteoclast functional antigen (OFA). Its expression on bone cells was found to be restricted to osteoclasts and preosteoclasts

(Davies et al., 1989; Horton and Davies, 1989; Simpson and Horton, 1989). Subsequently, it was shown that OFA is a member of the vitronectin receptor family. Its ligand(s) in bone that contain the RGD tripeptide recognition sequence include osteopontin, thrombospondin, and bone sialoprotein. It is logical to assume that any agent that can inhibit ligand-receptor interactions, by preventing the formation of the resorptive microcompartment, may have therapeutic usage in the treatment of osteoclastic bone disease. Several small molecules that would impair $\alpha_V\beta_3$-RGD interaction in bone are now in development.

Another area of exploration involves inhibition of proton secretion by the osteoclast. In an experimental system, such intervention almost certainly abolishes osteoclastic resorption. It is clear that hemivacuole acidification primarily occurs via an electrogenic vacuolar type (or V-type) proton pump (H^+-ATPase), not via a P-type plasma membrane pump (H^+-K^+-ATPase) (Webber et al., 1990). Although similarities between the osteoclast pump and the gastric mucosal and kidney H^+-K^+-APTase have been observed both immunochemically and pharmacologically (Parvinen et al., 1987; Baron, 1989;Webber et al., 1990), drugs such as omeprazole that are gastric-pump specific do block proton secretion from the osteoclast (Tuukkanen and Vaananen, 1986). Baron and colleagues have shown that the putative H^+-ATPase in the osteoclast is both vanadate- and nitrate-sensitive, and is probably novel. Drugs such as bafilomycin are potent inhibitors of H^+ secretion and bone resorption and are now being used as lead molecule for drug development.

Other potential sites for pharmacological intervention involve certain biochemical targets in the osteoclast. These include membrane receptors for peptides, such as calcitonin and endothelin, surface membrane Ca^{2+} receptor or stimulatory reactive oxygen species.

We (Zaidi et al., 1989) and others (Malgaroli et al., 1989) reported that the exposure of osteoclasts to millimolar concentrations of Ca^{2+} (Zaidi, 1990b) led to an immediate rise of cytosolic [Ca^{2+}] (Silver et al., 1988; Malgaroli et al., 1989; Zaidi et al., 1989) resulting in rapid cell retraction (Moonga et al., 1991). There was marked inhibition of bone resorption (Nicholson et al., 1986; Malgaroli et al., 1989; Zaidi et al., 1989) accompanied by reduction in the secretion of resorptive enzymes (Malgaroli et al., 1989; Zaidi et al., 1989; Moonga et al., 1990). These morphological and functional responses seen in osteoclasts when exposed to elevated extracellular [Ca^{2+}] may represent the basis for a short-range control mechanism. The effect of elevated extracellular [Ca^{2+}] in producing osteoclastic inhibition was mimicked by a range of divalent and trivalent cations of the alkaline earth, transition and lanthanide series (Bax et al., 1992b; Shankar et al., 1992a,b). These include Mg^{2+}, Ba^{2+} (Zaidi et al., 1990c), Ni^{2+} (Bax et al., 1992b), Cd^{2+} (Shankar et al., 1992a), and La^{3+} (Shankar et al., 1992b). A sensitivity to di- and trivalent cations is consistent with the existence of a surface membrane cation receptor, popularly termed the Ca^{2+} receptor. Cytosolic [Ca^{2+}] elevation has been implicated as the intracellular signal mediating the functional inhibition seen in response to extracellular [Ca^{2+}] elevation (Malgaroli et al., 1989; Zaidi et al., 1989). Thus, the functional effects of extracellular [Ca^{2+}] elevation and those of the Ca^{2+} ionophore, ionomycin, are additive. Furthermore, agents such as ionic perchlorate (Moonga et al., 1991), verapamil (Zaidi et al., 1990c), Ni^{2+} (Bax et al., 1992), Cd^{2+} (Shankar et al., 1992a), and La^{3+} (Shankar et al., 1992b) that elevate cytosolic [Ca^{2+}] inhibit osteoclastic bone resorption. We have also reported that activation of the Ca^{2+} receptor triggers a rapid mobilization of Ca^{2+} from intracellular stores (Shankar et al., 1992a,b).

Certain cell types including the osteoclast have Ca^{2+} sensors on their plasma membrane (Zaidi et al., 1993). The function of these cells is thus regulated through changes in their ambient Ca^{2+} concentration (Zaidi et al., 1993; Brown et al., 1995). For example, parathyroid cells, renal cells, and thyroid C cells possess a G protein-coupled seven-pass membrane receptor for Ca^{2+} (hence, the widely used term, Ca^{2+}-sensing receptor) (Brown et al., 1993; Mithal et al., 1994; Ricardi et al., 1995). The human Ca^{2+}-sensing receptor gene has been cloned and sequenced; point mutations in it result in the abnormal Ca^{2+} sensing that is seen in certain familial hypercalcemic syndromes (Brown et al., 1995) However, it should be noted that Ca^{2+} sensors do not belong to a family of related molecules. Instead, a diverse set of unrelated structures appear to subserve a somewhat similar biological function. In the trophoblast, for example, Ca^{2+} sensing occurs through a low density lipoprotein (LDL) receptor–like molecule (Lundgren et al., 1994). In the osteoclast, the Ca^{2+} sensor is a plasma membrane-resident ryanodine receptor (Shankar et al., 1992b; Zaidi et al., 1995) analogous to the Ca^{2+} release channel found in microsomal membranes (Meissner, 1994). When characterized in more detail, the putative Ca^{2+} receptor may be a useful target for pharmacological intervention.

We have shown that NO inhibits bone resorption. Since endothelial cells are present in abundance in the bone marrow and in close proximity to the osteoclast, the autociod may therefore play a role in downregulating osteoclast activity. The osteoclast produces NO itself, giving rise to the possibility of a feedback control loop for osteoclast activity. Moreover NO will combine with O_2^-. with the formation of the peroxynitrite radical (Yang et al., 1992). This will have the effect of diminishing peroxide formation by the osteoblast given that peroxides, in particular H_2O_2 stimulate osteoclast bone resorption (Bax et al., 1992a). Thus, an attractive hypothesis of bone turnover is that it is a peroxide driven process of osteoclast bone resorption, under the control of local NO production by endothelial cells and by the osteoclast itself. The excess production of reactive oxygen species observed in inflammatory joint disease may shift this balance towards excess osteoclast activity with the resulting effects of periarticular osteoporosis and bone erosion. It is interesting to speculate that conditions such as osteoarthritis and osteoporosis whose prevalence increases with age may be a consequence of an age-related decline in cellular NO production resulting in an imbalance of bone cell metabolism.

Therefore, a potential area for new drug development could be the use of agents that are able to release NO at sites of excessive bone resorption. However, it may be difficult to succeed as marked hypotension is known to result from the systemic administration of osteoclast-inhibitory doses of drugs, such as 3 morpholinosyndnonimine (or SIN-1) (Collin-Osdoby et al., 1991). An indirect way of achieving increased local NO activity may be to employ drugs such as phospholipase A_2 inhibitors which inhibit the oxidative modification of low density lipoprotein (Parthasarathy et al., 1985) a process thought to occur in the rheumatoid joint (Selley et al., 1992). This is relevant since it has been demonstrated that oxidized lipoproteins inactivate endothelial derived relaxing factor/NO activity (Parthasarathy et al., 1985). In addition, a promising approach that clearly deserves study is the development of novel NO-generating agents that could be targeted to sites of potential bone destruction. Finally, drugs could be developed based on their ability to decrease the production of H_2O_2 from osteoblasts and endothelial cells.

ACKNOWLEDGMENTS

M.Z. acknowledges grant support from the National Institute on Aging, National Institutes of Health (RO1 AG141917), the Department of Veterans Affairs (Merit Review Award), and Amgen Corporation, Thousand Oaks, CA.

REFERENCES

Adebanjo, O.A., Moonga, B.S., Yamate, T., Sun, L., Minkin, C., Abe, E., and Zaidi, M. (1998). Mode of action of interleukin-6 on mature osteoclasts. Novel interaction with extracellular Ca^{2+} sensing in the regulation of osteclastic bone resorption. J. Cell. Biol. 142, 1347-1356.

Agus, Z.S., Wasserstein, S., and Goldfarb, S. (1981). PTH, calcitonin cyclic nucleotides, and the kidney. Ann. Rev. Physiol. 43, 583-595.

Akisaka, T. and Gay, C.V. Ultracytochemical evidence for a proton-pump adenosine triphosphatase in chick osteoclasts. Cell Tissue Res. (1986). 245, 507-512.

Alam, A.S.M.T., Gallagher, A., Shankar, V.S., Ghatei, M.A., Datta, H.K., Huang, CL-H., Moonga, B.S., Chambers, T.J., Bloom, S.R., and Zaidi, M. (1992). Endothelin inhibits osteoclastic bone resorption by a direct effect on cell motility: Implications for the vascular control of bone resorption. Endocrinology 130, 3617-3624.

Alam A.S.M.T., Moonga, B.S., Bevis, P.J.R., Huang, C.L-H., and Zaidi, M. (1991). Selective antagonism of calcitonin-induced osteoclastic quiescence (Q effect) by human calcitonin gene-related peptide (Val^8Phe^{37}). Biochem. Biophys. Commun. 179, 134-139.

Aliopoulious, M.A., Gildhaber, P., and Munson, P.L. (1965). Thyrocalcitonin inhibition of bone resorption induced by parathyroid hormone in tissue culture. Science 151, 330-331.

Anderson, R.E., Schraer, H., and Gay, C.V. (1989). Ultrastructural immunocytochemical localisation of carbonic anhydrase in normal and calcitonin-treated chick osteoclasts. Anat. Rec. 204, 9.

Baron, R. (1989). Molecular mechanisms of bone resorption by the osteoclast. Anat. Rec. 224, 317-324.

Bax, .E., Alam, A.S.M.T., Banerji, B., Bax, C.M.R., Bevis, P.J.R., Stevens, C.R., Moonga, B.S., Blake, D.R., and Zaidi, M. (1992a). Stimulation of osteoclastic bone resorption by hydrogen peroxide (H_2O_2). Biochem. Biophys. Res. Commun. 183, 1153-1158.

Bax, C.M.R., Shankar, V.S., Huang, C.L-H., and Zaidi, M. (1992b). Is the osteoclast calcium "receptor" a receptor-operated calcium channel? Biochem. Biophys. Res. Commun. 183, 619-625.

Bevan, J.A., Franks, A.F., Mcosker, J.E., Boyce, R.W., and Buckingham, K.W. (1988). Bisphosponate action in the oophorectomized rat: The effects of 2-(-pyridinyl)-1-hydroxyethlidene-bis(phosphonate) on skeletal metabolism; comparison with etidronate. J. Bone Miner. Res. 3 (Suppl 1), S193.

Bijvoet, O.L.M., Hosking, D.J., Lemkes, H.H.P.J., Reitsma, P.H., and Frijlink, W.B. (1978). Development in the treatment of Paget's disease. In: *Endocrinology of Calcium Metabolism*. (Copp et al., Eds.), pp. 421, 48-54, Excerpta Medica, Amsterdam.

Boonekamp, P.M., Lowik, C.W.G.M., Van der Wee-pals, L.J.A., Van WijkBVan Lennep, M.L.L., and Bijvoet, O.L.M. (1987). Enhancement of the inhibitory action of APD on the transformation of osteoclast precursors into resorbing cells after dimethylation of the amino group. Bone Miner. 2, 29-42.

Brown, E.M., Gamba, G., Ricardi, I.D., Lombardi, M., Butters, R., Kifor, O., Sun, A., Hediger, M.A., Lytton, J., and Hebert, S.C. (1993). Cloning and characterization of an extracellular calcium-sensing receptor from bovine parathyroid. Nature (London) 366, 575-579.

Brown, E.M., Pollak, M., Seidman, C.E., Seidman, J.G., Chou, Y-H.W., Ricardi, D., and Hebert, S.C. (1995). Calcium-ion-sensing cell-surface receptors. N. Eng. J. Med. 333, 234-240.

Carano, A., Teitelbaum, S.L., Konsek, J.D., Schlesinger, P.H., and Blair, H.C. (1990). Bisphosphonates directly inhibit the bone resorption activity of isolated avian osteoclasts in vitro. J. Clin. Invest. 85, 456-461.

Cecchini, M.G. and Fleisch, H. (1990). Bisphosphonates in vitro specifically inhibit, among the hemopoietic series, the development the mouse mononuclear phagocyte linage. J. Bone Miner. Res. 5, 1019-1027.

Cecchini, M.G., Felix, R., Fleisch, H., and Cooper, P.H. (1987). Effects of bisphosphonates on proliferation and viability of mouse bone marrowBderived macrophages. J. Bone Miner. Res. 2, 135-142.

Chakarborty, M., Chatterji, D., Kellokumpus, S.H., and Baron, R. (1991). Cell cycle-dependent coupling of calcitonin receptor to different G proteins. Science 251, 1078-1081.

Chambers, T.J. and Magnus, C.J. (1982). Calcitonin alters behavior of isolated osteoclasts. J. Pathol. 136, 27-39.

Chambers, T.J., Fuller, K., Mcsheehy, P.M.J., and Pringle, J.A.S. (1985a). The effect of calcium-regulating hormones on bone resorption by isolated osteoclastoma cells. J. Pathology 145, 297-305.

Chambers, T.J., Mcsheehy, P.M.J., Thomson, B.M., and Fuller, K. (1985). The effect of calcium-regulating hormones and prostaglandins on bone resorption by osteoclasts disaggregated from rabbit long bones. Endocrinology 116, 234-239.

Chambers, T.J., Fuller, K., and Darby, J.A. (1987). Hormonal regulation of acid phosphatase release by osteoclasts disaggregated from neonatal rat bone. J. Cell Physiol. 132, 92-96.

Christiansen, C., Christiansen, M.S., and Transbol, I. (1981). Bone mass in postmenopausal women after withdrawal of oestrogen / gestagen replacement therapy. Lancet 1, 459-461.

Christiansen, C., Christiansen, M.S., McNair, P., Hagen, C., Stocklund, K-E., and Ansbol, I. (1980). Postmenopausal bone loss: Controlled two-year study in 315 normal females. Eur. J. Clin. Invest. 10, 273-279.

Collin-Osdoby, P., Oursler, M.J., Webber, D., and Osdoby, P. (1991). Osteoclast-specific monoclonal andibodies coupled to magnetic beads provide a rapid and efficient method of purifying avian osteoclasts. J. Bone Min. Res. 6, 11353-11365.

Datta, H.K., Macintyre, I., and Zaidi, M. (1989). The effect of ionised calcium on the morphology and function of isolated rat osteoclasts. Biosci. Rep. 9, 247-251.

Davies, J., Warwick, J., Tutty, N., Phillip, R., Helfrich, M., and Horton, M. (1989). The osteoclast functional antigen, implicated in the regulation of bone resorption, is biochemically related to the vitronectin receptor. J. Cell Biol. 109, 1817-1826.

Epand, R.M., Epand, R.F., Orlowski, R.C., Seyler, J.K., Colescott, R.L. (1986). Conformational flexibility and biological activity of salmon calcitonin. Biochemistry 25, 1964-1968.

Epand, R.M., Epand, R.F., Orlowski, R.C., Lanigan, E., and Stahl, G.L. (1985). A comparison of the interaction of glucagon, human parathyroid hormone -(1-34)-peptide, and calcitonin with dimyristoylphosphatidyl-glycerol and with dimyristoylphosphatidylcholine. Biophys. Chem. 23, 39-48.

Eriksen, E.F., Colvard, D.S., Berg, N.J., Graham, M.L., Mann, K.G., Spelsberg, T.C., Riggs, B.L. (1988). Evidence of estrogen receptors in normal human osteoblastslike cells. Science 241, 84-86.

Ernst, M., Heath, J.K., and Rodan, G.A. (1989). Estradiol effects on proliferation, messenger ribonucleic acid for collagen, and insulinlike growth factor-I and parathyroid hormone-stinmulated adenylate cyclase activity in osteoblastic cells from calvariae and long bones. Endocrinology 125, 825-833.

Fast, D.K., Felix, R., Dowse, C., Neuman, W.F., and Fleisch, H. (1978). The effects of disphosphonates on the growth and glycolysis of connective-tissue cells in culture. Biochem. J. 172, 97-107.

Felix, R., Bettex, J.D., and Fleisch, H. (1981). Effect of disphosphonates on the synthesis of prostaglandins in cultured calvaria cells. Calcif. Tissue. Int. 33, 549-552.

Felix, R., Guenther, H.L., and Fleisch, H. (1984). The subcellular distribution of ^{14}C dichloromethylene-bisphosphonate and ^{14}C 1-hydroxy-ethylidene-1,1-bisphosphonate in cultured calvaria cells. Calcif. Tissue Int. 36, 108-113.

Felix, R., Russell, R.G.G., and Fleisch, H. (1976). The effect of several disphosphonates on acid phosphohydrolases and other lysosomal enzymes. Biochim. Biophys. Acta 429, 429-438.

Flanagan, A.M. and Chambers, T.J. (1989). Dichloromethylenebisphosphonate (Cl_2MBP) inhibits bone resorption through injury to osteoclasts that resorb Cl_2MBP-coated bone. Bone. Miner. 6, 33-43.

Fleisch, H., Russell, R.G.G., and Fransis, M.D. (1969). Diphosphonate inhibits the hydroxyapatite dissolution in vitro and bone resorption in tissue culture and in vivo. Science 165 , 1262-1264.

Fleisch, H. (1983). Bisphosphonates: Mechanisms of action and clinical application. In: Bone and Mineral Research Annual 1. (Peck, W.A., Ed.), pp. 319-357. Excerpta Medica, Amsterdam.

Fleisch, H. (1991). Bisphosphonates: Pharmacology and use in the treatment of tumor-induced hypercalcaemic and metastatic bone disease. Drugs 42, 919-944.

Fleisch, H. (1998). Bisphosphonates: Mechanisms of action. Endocrine Review 19, 80-100.

Freidman, J. and Raisz, L.G. (1965). Thyrocalcitonin: Inhibitor of bone resorption in tissue culture. Science 150, 1465-1467.

Galante, L., Horton, R., Joplin, G.F., Woodhouse, and N.J.F., MacIntyre. (1971). Comparison of human, porcine, and salmon synthetic calcitonins in man and in the rat Clin. Sci. 40, 9-10.

Garrett, I.R., Boyce, B.F., Oreffo, R.O., Bonewald, L., Poser, J., and Mundy, G.R. (1990). Oxygen-derived free radicals stimulate osteoclastic bone resorption in rodent bone in vitro and in vivo. J. Clin. Invest. 85, 632-639.

Girasole, G., Jilka, R.L., Passeri, G., Boswell, S., Boder, G., Williams, D.C, and Manolagas, S.C. 17β-Estradiol inhibits interleukin-6 production by bone marrow stromal-derived stromal cells and osteoblasts in vitro: Apotential mechanism for the antiosteoporotic effect of estrogen. J. Clin. Invest. 89, 883-891.

Gotfredsen, A., Nilas, L., Riis, B.J., and Christiansen, C. (1986). Bone changes occurring spontaneously and caused by oestrogen postmenopausal women: A local or generalized phenomenon? Br. Med. J. 292, 1098-1100.

Gray, T.K., Flyn, T.C., Gray, K.M., and Nabell, L.M. 17β-estradiol acts directly on the clonal osteoblast line UMR 106. Proc. Nat. Acad. Sci. USA 84, 6267-6271.

Gruber, H.E., Ivey, J.L., Baylink, D.J., Matthews, M., Nelp, W.B., Sosom, K., and Chestnut, C.H. (1988). Long-term calcitonin therapy in postmenopausal osteoporosis. Metabolism 33, 295-303.

Hammond, C.B. and Maxson, W.S. (1982). Current status of estrogen therapy. Fertil. Sterol 37, 5-25.

Hammond, C.B. and Maxson, W.S. (1986). Estrogen replacement therapy. Clin. Obst. Gyn. 29, 407-430.

Holtrop, N.E., Raisz, L.J., and Simmons, H.A. (1974). The effects of parathyroid hormone, colchicine and calcitonin on the ultrastructure and the activity of osteoclasts in organ culture. J. Cell Biol. 60, 346-355.

Horton, M.A. and Davies, J. (1989). Perspectives: Adhesion receptors in bone. J. Bone Miner. Res. 4, 803-808.

Jüppner, H., Abou-Samra, A-B., Freeman, M., Kong, F.X., Schipani, E., Richards, J., Kolakowski, Jr., L.F., Hock, J., Potts, Jr., J.T., Kronenberg, H.M., and Segre, G.V. (1991). A G proteinBlinked receptor for parathyroid hormone-related peptide. Science 254, 1024-1026.

Kallio, D.M., Garant, P.R., and Minkin, C. (1972). Ultrastructural effects of calcitonin on osteoclasts in tissue culture. J. Ultrastruct. Res. 39, 205-216.

Key, Jr., L.L., Reis, W.L., Taylor, R.G., Hays, B.D., and Pitzer, B.R. (1990). Oxygen-derived free radicals in osteoclasts: The specificity and location of the nitroblue tetrazolium reaction. Bone 11, 115-119.

Kimble, R., Bain, S., and Pacifici, R. (1997). The functioned block of TNF but not of IL-6 prevent bone loss in ovariectomized mice. J. Bone Miner. Res. 12, 935-491.

Kimble, R.B.S., Srivastava, S., Bellone, C.J., and Pacifici R. (1996). Estrogen inhibits IL1 and TNF gene expression in transfected murine macrophogic cells. J. Bone Miner. Res. 11, Suppl.1. P235.

Lin, H.Y., Harris, T.L., Flannery, N.S., Aruffo, A., Kajiehgorn, A., Kolakowski, Jr., .F., Lodish, H.F., and Goldring, S.R.(1991). Expression cloning of an adenylate cyclase-coupled calcitonin receptor. Science 254, 1022-1024.

Lundgren, S., Hjalm, G., Hellman, P. et. al. (1994). A protein involved in calcium sensing of the human parathyroid and placental cytotrophoblast cells belongs to the LDL-receptor protein superfamily. Exp. Cell Res. 212, 344-350.

Macintyre, I., Alevizaki, M., Bevis, P.J.R., and Zaidi, M. (1987). Calcitonin and peptides from the calcitonin genes. Clin. Orthop. 21, 45.

MacIntyre, I., Parsons, J.A., and Robinson, C.J. (1967). The effect of thyrocalcitonin on blood bone calcium equilibrium in the perfused tibiae of the cat. J. Physiol. 191, 393-405.

MacIntyre, I., Zaidi, M., Alam, A.S.M.T., Datta, H.K., Moonga, B.S., Lidbury, P.S., Hecker, M., Vane, J.R. (1991). Osteoclastic inhibition: An action of nitric oxide not mediated by cyclic GMP. Proc. Natl. Acad. Sci. USA 88, 2936-2940.

Maier, R., Riniker, B., and Rittel, W. (1974). Analogues of human calcitonin. I. Influence of modifications in amino acid posiotions 29 and 31 and hypocalcaemic activity in the rat. FEBS Lett. 48, 68-71.

Malgaroli, A., Meldolesi, J., Zambonin-Zallone, A., and Teti, A. (1989). Control of cytosolic free calcium in rat and chicken osteoclasts. J. Biol. Chem. 264, 14342-14347.

Manning, M., Przybylski, A., Olma, A., Klis, W.A., Kruszynski, M., Wo, N.C., Pelton, G.H., Sawyer, W.H. (1987). No requirements of cyclic conformation of antagonists in binding to vasopressin receptors. Nature 329, 839-840.

Maschchak, C.A., Lobo, R.A., Dozono-Takono, R., Eggena, P., Nakamura, R.M., Brenner, P.F., Mishell, D.R. (1982). Comparison of pharmacodynamic properties of various estrogen formulations. Am. J. Obstet. Gynecol. 144, 511-518.

Mcsheehy, P.M.J. and Chambers, T.J. (1986). Osteoblastlike cells in the presence of parathyroid hormone release soluble factor that stimulates osteoclastic bone resorption. Endocrinology 119, 1654-1659.

McSheehy, P.M.J. and Chambers, T.J. (1987). 1,25-dihydroxyvitamin D_3 stimulates rat osteoblastic bone resorption. J. Clin. Invest. 80, 425-429.

Meissner, G. (1994). The ryanodine receptor: Structure and function. Ann. Rev. Physiol. 56, 485-508.

Milhaud, G., Perault, A.M., and Moukhtar, M.S. (1965). Etude du mecanisme de l'action hypocalcemiante de la thyrocalcitonine. C.R. Academy of Science, Paris D261, 813.

Mithal, A., Kifor, O., Thun, R., Krapcho, K., Fuller, F., Hebert, S.C., Brown, E.M., and Tamir, H. (1994). Highly purified sheep C-cells express an extracellular Ca receptor similar to that present in parathyroid. J. Bone Miner. Res. 9 (Suppl.1), B209.

Miyauchi, A., Hruska, K.A., Greenfield, E.M., Duncan, R., Alvarez, J., Barattolo, R., Colucci, S., Zambonin-Zallone, A., Teitelbaum, S.L., and Teti, A. (1990). Osteoclast cytosolic calcium, regulated by voltage-gated calcium channels extracellular calcium, controls podosome assembly and bone resorption. J. Cell Biol. 111, 2543-2552.

Monkkonen, J.A. (1988). One year follow up study of the distribution of 14C-clodronate in mice and rats. Pharmacol. Toxicol. 62, 51-53.

Moonga, B.S., Datta, H.K., Bevis, P.J.R., Huang, C.L-H., MacIntyre, I., Zaidi, M. (1991). Correlates of osteoclast function in the presence of perchlorate ions in the rat. Exp. Physiol. 76, 923-933.

Moonga, B.S., Moss, D.W., Patchell, A., and Zaidi, M. (1990). Intracellular regulation of enzyme release from rat osteoclasts and evidence for a functional role in bone resorption. J. Physiol. 429, 29-45.

Morikawa, T., Muniketa, E., Sakakibara, S., Noda, T., and Otani, M. (1976). Synthesis of eel-calcitonin and (Asu ^{1,7})-eel-calcitonin: Cntribution of the disulfide bond to the hormonal activity. Experientia 32, 1104-1106.

Muhlbauer, R.C. and Fleisch, H. (1990). A method for continual monitoring of bone resorption in rats : Evidence for a diurnal rhythm. Am. J. Physiol. 259, R679-R689.

Muhlbauer, R.C., Bauss, F., Schenk, R., Janner, M., Bosies, E., Strein, K., Fleisch, H. (1991). BM 21.0955, a potent new bisphosphonate to inhibit bone resorption. J. Bone Miner Res. 6, 1003-1011.

Muhlbauer, R.C., Stutzer, A., Schenk, R., Janner, M., Fleisch, H. (1989). 1-hydroxy-3-(methylpentylamino) propylidene-bisphosphonate (BM 21.0955), a potent new inhibitor of bone resorption. J. Bone Miner. Res. 4 (Suppl.1), S168.

Nicholson, C.G., Moseley, J.M., Sexton, P.M., Mendelson, F.A.O., Martin, T.J. (1986). Abundant calcitonin receptors in isolated rat osteoclasts. Biochemical and autoradiographic characterisation. J. Clin. Invest. 78, 355-360.

Orlowski, R.C., Epand, R.M., and Stafford, A.R. (1987). Biologically potent analogues of salmon calcitonin which do no contain an N-terminal disulfide-bridged ring structure. Eur. J. Biochem. 162, 399-402.

Oursler, M.J., Collin-Osdoby, P., Li, L., Schmitt, E., and Osdoby, P. (1991). Evidence for an immunological and functional relationship between superoxide dismutase and a high molecular weight osteoclast plasma membrane glycoprotein. J. Cell. Biochem. 46, 331-344.

Overgaard, K., Hansen, M.A, Jensen, S.B., Christiansen, C. (1992). Effect of salcalcitonin given intranasally on bone mass and fracture rates in established osteoporosis: a dose response study. Br. Med. J. 305, 556-561.

Pollak, M.R., Brown, E.M., Chou, Y-H.W., Hebert, S.C., Marx, S.J., Steinmann, B., Levy, T., Seidman, C.E., and Seidman, J.G. (1993). Mutations in the human Ca^{2+}-sensing receptor gene cause familial hypocalciuric hypercalcemia and neonatal severe hyperparathyroidism. Cell 75, 1297-1303..

Pacifici, R., Rifas, L., Mccracken, R., Vered, I., McMurtry, C., Avioli, L.V., and Peck, W.A. Ovarian steroid treatment blocks a postmenopausal increase in blood monocyte interleukin -1 release. Proc. Nat. Acad. Sci. USA 86, 2398-2402.

Parthasarathy, S., Steinbrecher, U.P., Barnett, J., Witzum, J.L., Steinberg, D. (1985). Essential role of phospholipase A_2 activity in endothelial cell-induced modification of low density lipoprotein. Proc. Nat. Acad. Sci. USA 82, 3000-3004.

Parvinen, E.K., Slot, J.W., and Vaarianen, H.K. (1987). Osteoclast acidifying enzymes, carbonic anhydrase II and H^+-K^+-ATPase immunogold localisation on ultracryosections. Calcif. Tissue Int. (Suppl.2) 41, 20.

Pecile, A. (1992). Calcitonin and relief of pain. Bone Miner. 16, 187-189.

Powell, J.H. and De Mark, B.P. (1985). Clinical pharmacokinetics of diphosphonates. In: *Bone Resorption, Metastasis, and Diphosphonates.* (Garrafini, S., Ed.), pp. 41-49. Raven Press, New York.

Ralston, S.H., Russell, R.G.G., and Gowen, M. (1990). Oestrogen inhibits release of tumor necrosis factor from peripheral blood mononuclear cells in postmenopausal women. J. Bone Miner. Res. 983-988.

Reitsma, P.H., Bijvoet, O.L.M., Verlinden, O.O.M.S.H, Van der Wee-pals, L.J.A. (1980). Kinetic studies of bone and mineral metabolism during treatment with (3-amino-1-hydroxy-propylidene)-1,1-bisphosphonate (APD) in rats. Calcif. Tissue Int. 32 , 145-147.

Reitsma, P.H., Bijvoet, O.L., Potokar, M., Van der Wee-pals, L.J., Van WijkBVan Lennep, M.M. (1983). Apposition and resorption of bone during oral treatment with (3-amino-1-hydroxypropylidene)-1,1-bisphosphonate (APD). Calcif. Tissue Int. 35, 357-361.

Reynolds, J.J, Dingle, J.T., Gudmundsson, T.V., MacIntyre, I. (1968). Bone resorption in vitro and its inhibition by calcitonin. In: *Calcitonin, Proc. Int. Symposium on Thyrocalcitonin and C-Cells.* (Taylor, S.F., Ed.), pp. 223. Heinemann, London.

Ricardi, D., Pak, J., Lee, W-S., Gamba, G., Brown, E.M., and Hebert, S.C. (1995). Cloning and functional expression of a rat kidney extracellular calcium/polyvalent cation-sensing receptor. Proc. Nat. Acad. Sci. USA 9, 131-135.

Rickard, D., Russell, G., Gowen, M. (1992). Oestradiol inhibits the release of tumor necrosis factor but not interleukin 6 from adult human osteoblasts in vitro. Osteoporosis Int. 2, 94-102.

Riis, B.J. and Christiansen, C. (1988). Measurement of spinal or peripheral bone mass to estimate early postmenopausal bone loss? Am. J. Med. 84, 646-653.

Robinson, C.J., Martin, T.J., Mathews, E.W., and Macintyre, I. (1967). Mode of action of thyrocalcitonin. J. Endocrinol. 39, 71-79.

Roodman, G.D. (1991). Osteoclast differentiation. Crit. Rev. Oral Biol. Med. 2, 389-409.

Russell, R.G.G., Bisaz, S., Donath, A., Fleisch, H. (1971). Inorganic pyrophosphate in plasma in normal persons and in patients with hypophosphatasia, osteogenesis imperfecta, and other disorders of bone. J. Clin. Invest. 50, 961-969.

Sahni, M., Guenther, H.L., Fleisch, H., Collin, P., and Martin, T.J. (1993). Bisphosphonates act on rat bone resorption through the mediation of osteoblasts. J. Clin. Invest. 91, 2004-2011.

Schenk, R., Eggli, P., Felix, R., Fleisch, H., and Rosini, S. (1986). Quantitative morphometric evaluation of the inhibitory activity of new aminobisphosphonates on bone resorption in the rat. Calcif. Tissue. Int. 38, 342-349.

Schwarrtz, K.E., Orlowski, R.C., and Marcus, R. (1981). des-Ser2 salmon calcitonin: A biologically potent synthetic analog. Endocrinology 108, 831-835.

Seiber, P., Brugger, M., Kamber, P., Riniker, B., and Rittel, W. Menschliches calcitonin. III. Die synthese von calcitonin M. Helv. Chim. Acta. 51, 2057-2061.

Selby, P.L. (1990). Oestrogens and bone. In: *Osteoporosis, Pathogenesis and Management.* (Francis, R.M., Ed.), pp 81-102. Kluwer Academic Publishers, Boston.

Selley, M.L., Bourne, D.J., Bartlett, M.R., Tymms, K.E., Brook, A.S, Duffield, A.M., and Ardlie, N.G. (1992). Occurence of (E)-4-hydroxy-2-nonenal in plasma and synovial fluid of patients with rheumatoid arthritis and osteoarthritis. Ann. Rheum. Dis. 51, 481-484.

Shankar, V.S., Alam, A.S.M.T., Bax, C.M.R., Bax, B.E., Pazianas, M., Huang, C.L.-H., and Zaidi, M. (1992a). Activation and inactivation of the osteoclast Ca^{2+} receptor by the trivalent cation, La^{3+}. Biochem. Biophys. Res. Commun. 187, 907-912.

Shankar, V.S., Bax, C.M.R., Alam, A.S.M.T., Bax, B.E., Huang, C.L.-H., and Zaidi, M. (1992b). The osteoclast Ca^{2+} receptor is highly sensitive to activation by transition metal cations. Biochem. Biophys. Res. Commun. 187, 913-918.

Shinoda, H., Adamek, G., Felix, R., Fleisch, H., Schenk, R., and Hagan, P. (1983). Structure-activity relationships of various bisphosphonates. Calcif. Tissue Int. 35, 87-99.

Sietsema, W.K., Ebetino, F.H., Salvagno, A.M., Bevan, J.A. (1989). Antiresorptive dose-response relationship across three generations of bisphosphonates . Drugs Exp. Clin. Res. 15, 389-396.

Silver, I.A., Murrills, R.J., and Etherington, D.J. (1988). Microelectrode studies on acid microenvironment beneath adherent macrophages and osteoclasts. Exp. Cell Res. 175, 266-276.

Simpson, A. and Horton, M.A. (1989). Expression of the vitronectin receptor during embryonic development: An immunohistological study of the ontogeny of the osteoclast in the rabbit. Br. J. Exp. Pathol. 70, 257-265.

Smirnova, I.N., Kudryavtseva, N.A., Komissarenko, S.V., Tarusova, N.B., Baykov, A.A. (1988). Disphosphonates are potent inhibitors of mammalian inorganic pyrophosphatase. Arch. Biochem. Biophys. 267, 280-284.

Stevenson, J.C., Abeyasekera, G., Hillyard, C.J., Phang, K.G., Macintyre, I., Campbell, S., Lane, G., Townsend, P.T., Young, O., and Whitehead, M.I. (1983). Regulation of calcium-regulating hormones by exogenous sex steroids in early post menopause. Eur. J. Clin. Invest. 13, 481-487.

Stutzer, A., Fleisch, H., and Trechsel, U. (1988). Short- and long-term effects of a single dose of bisphosphonates on retinoid-induced bone resorption in thyroparathyroidectomised rats. Calcif. Tissue. Int. 43, 294-299.

Tabibzadeh, S.S, Santhanan, U., May, L., and Sehgal, P. (1989). Cytocine-induced production of IFN-β_2/IL-6 by freshly explanted human endometrial stroma cells: Modulation by oestradiol-17β. J. Immunol. 142, 3134-3139.

Tanuichi, H. (1984). In: *The Impact of Protein Chemistry on Biomedical Sciences.* (Schechter, A.N., Dean, A., and Goldberg, R.F., Eds.), Academic Press, Orlando, FL.

Teitelbaum, S.L. and Kahn, T.J. (1980). Mononuclear phagocytes, osteoclasts, and bone resorption. Miner. Electrolyte Metab. 3, 2-7.

Thiebaud, D., Jeager, P., and Burcckhardt, P. (1986). A single-day treatment of tumor-induced hypercalcaemia by intravenous amino hydroxypropylidene bisphosphonate. J. Bone. Miner. Res. 1, 555-562.

Turner, R.T., Colvard, DS., Spelsberg, T.C. (1990). Estrogen inhibition of periosteal bone formation in rat long bones, down-regulation of gene expression for bone matrix proteins. Endocrinology 127, 1346-1351.

Tuukkanen, J. and Vaananen, H.K. (1986). Omeprazole, a specific inhibitor of H^+-K^+-ATPase inhibits bone resorption in vitro. Calcif. Tissue Intern. 38, 123-125.

Valkema, R., Visman, S.F.-J.F.E., Papapoulos, S.E. et al. (1989). Effects of long-term low-dose bisphosphonate (APD) on calcium balance and bone mineral content in patients with osteoporosis Bone Min. 5, 183-192.

Vitte, C., Fleisch, H., and Guenther, H.L. (1994). Osteoblasts mediate the bisphosphonate inhibition on bone resorption through synthesis of an osteoclast-inhibiting activity. J. Bone Min. Res. 9 (Suppl. 1), S142.

Webber, D., Osdoby, P., Hauchka, P., and Krukowski, M. (1990). Correlation of an osteoclast antigen and ruffled border on giant cells formed in response to resorbable substrates. J. Bone Miner. Res. 5, 401-410.

Yang, G., Condy, T.E., Boaro, M., Wilkaen, H.E., Jones, P., Nazhat, N.B., Saadalla-Nazhat, R.A., and Blake, D.R. (1992). Free radical yield from the homolysis of peroxynitrous acid. Free Rad. Biol. Med. 12, 327-330.

Zaidi, M., Kerby, J., Huang, C.L-H., Alam, A.S.M.T., Rathod, H., Chambers, T.J., and Moonga, B. (1991). Divalent cations mimic the inhibitory "effect of extracellular ionized calcium on bone resorption by isolated rat osteoclasts: Further evidence for a "calcium receptor@. J. Cell Physiol. 149, 422-427.

Zaidi, M., Datta, H.K., Bevis, P.J.R., Wimalawansa, S.J., and MacIntyre, I. (1990a) Amylin amide: A new bone-conserving peptide from the pancreas. Exp. Physiol. 75, 529-536.

Zaidi, M., Datta, H.K., Moonga, B.S., and MacIntyre, I. (1990b). Evidence that the action of calcitonin on rat osteoclasts is mediated by two G proteins acting via separate postreceptor pathways. J. Endocrinol. 126, 473-481.

Zaidi, M., Datta, H.K., Patchell, A., Moonga, B.S., MacIntyre, I. (1989). "Calcium-activated" intracellular calcium elevation: A novel mechanism of osteoclast regulation. Biochem. Biophys. Res. Commun. 163, 1461-1465.

Zaidi, M., Macintyre, I., and Datta, H.K. (1990c). Intracellular calcium in the control of osteoclast function. II. Paradoxical elevation of cytosolic free calcium by verapamil. Biochem. Biophys. Res. Commun. 167, 807-812.

Zaidi, M., Chambers, T.J., Bevis, P.J.R., Beacham, J.L., Gaines, D.A.S.R.E., MacIntyre, I. (1988). Effects of peptides from the calcitonin genes on bone and bone cells. Q. J. Exp. Physiol. 73, 471-485.

Zaidi, M., Breimer, L.H., and MacIntyre, I. (1987a). The biology of the peptides from the calcitonin genes. Q. J. Exp. Physiol. 72, 371-408.

Zaidi, M., Chambers, T.J., Gaines, D.A.S.R.E., Morris, H.R., and MacIntyre, I. (1987b). A direct effect of human calcitonin gene-related peptide on isolated osteoclasts. J. Endocrinol. 115, 511-518.

Zaidi, M., Fuller, K., Bevis, P.J.R., Gaines, D.A.S.R.E., Chambers, T.J., MacIntyre, I. (1987c). Calcitonin gene-related peptide inhibits osteoclastic bone resorption: A comparative study. Calcif. Tissue. Int. 40, 149-154.

Zaidi, M., Moonga, B.S., Bevis, P.J.R., Bascal, Z.A., and Breimer, L.H. (1990d). Calcitonin gene peptides: biology and clinical relevance. Crit. Rev. Clin. Lab. Sci. 28, 109-174.

Zaidi, M., Shankar, V.S., Huang, C.L-H., Rifkin, B.R., and Pazianas, M. (1994). Molecular mechanisms of calcitonin action. Endocrine 2, 459-467.

Zaidi, M., Shankar, V.S., Tunwell, R., Abebanjo, O.A., Mackvile, J., Pazianas, M., O'Connell, D., Simon, B.J., Rifkin, B.R., Venkitaranian, A.R., Huang, C.L.-H., and Lui, F.A. (1995). A ryanodine receptor like molecule expressed in the osteoclast plasma membrane functions in extracellular Ca^{2+} sensing. J. Clin. Invest. 96, 1582-1590.

Zaidi, M. (1990a). Modularity of osteoclast behaviour and mode-specific inhibition of osteoclast function. Biosci. Rep. 10, 547-556.

Zaidi, M.(1990b). Calcium "receptors" on eukaryotic cells with special reference to the osteoclast. Biosci. Rep. 10, 493-507.

Zaidi, M., Alam, A.S.M.T., Huang, C.L.-H., Pazianas, M., Bax, C.M.R., Bax, B.E., Moonga, B.S., Bevis, P.J.R., and Shankar, V.S. (1993). Extracellular Ca^{2+} sensing by the osteoclast. Cell Calcium 14, 271-277.

BISPHOSPHONATES: MECHANISMS OF ACTION

Herbert Fleisch

I. INTRODUCTION

The bisphosphonates, erroneously called diphosphonates in the past, are a class of drugs which have been developed in the past three decades for use in

Advances in Organ Biology
Volume 5C, pages 835-850.

ISBN: 0-7623-0390-5

various diseases of bone, tooth, and calcium metabolism. This chapter will deal essentially with their mechanisms of action. For information on the other aspects of these compounds, the reader is referred to a monograph (Fleisch, 1997), as well as to various reviews (Geddes et al., 1994; Fleisch, 1996; Rodan and Fleisch, 1997a)

II. CHEMISTRY

Geminal bisphosphonates, generally called bisphosphonates, are pyrophosphate analogues in which the oxygen bridge has been replaced by a carbon

[illegible]	
NH_2–$(CH_2)_3$; O^-, O^-; $O=P-C-P=O$; O^- OH O^- (4-Amino-1-hydroxybutylidene)-bis-phosphonate alendronate* *Gentili; Merck Sharp & Dohme*	O^- Cl O^-; $O=P-C-P=O$; O^- Cl O^- (Dichloromethylene)-bis-phosphonate clodronate* *Astra; Boehringer Mannheim; Gentili; Leiras; Rhône-Poulenc Rorer*
pyrrolidinyl–$(CH_2)_2$; O^-, O^-; $O=P-C-P=O$; O^- OH O^- [1-Hydroxy-3-(1-pyrrolidinyl)-propylidene]bis-phosphonate EB-1053 *Leo*	O^- CH_3 O^-; $O=P-C-P=O$; O^- OH O^- (1-Hydroxyethylidene)-bis-phosphonate etidronate* *Gentili; Procter & Gamble*
CH_3–N($(CH_2)_4$–CH_3)–$(CH_2)_2$; O^-, O^-; $O=P-C-P=O$; O^- OH O^- [1-Hydroxy-3-(methylpentylamino)propylidene]bis-phosphonate ibandronate* *Boehringer Mannheim*	cycloheptyl–NH; O^-, O^-; $O=P-C-P=O$; O^- H O^- [(Cycloheptylamino)-methylene]bis-phosphonate incadronate *Yamanouchi*
NH_2–$(CH_2)_5$; O^-, O^-; $O=P-C-P=O$; O^- OH O^- (6-Amino-1-hydroxyhexylidene)bis-phosphonate neridronate *Gentili*	H_3C–N–CH_3; $(CH_2)_2$; O^-, O^-; $O=P-C-P=O$; O^- OH O^- [3-(Dimethylamino)-1-hydroxypropylidene]bis-phosphonate olpadronate *Gador*

continued

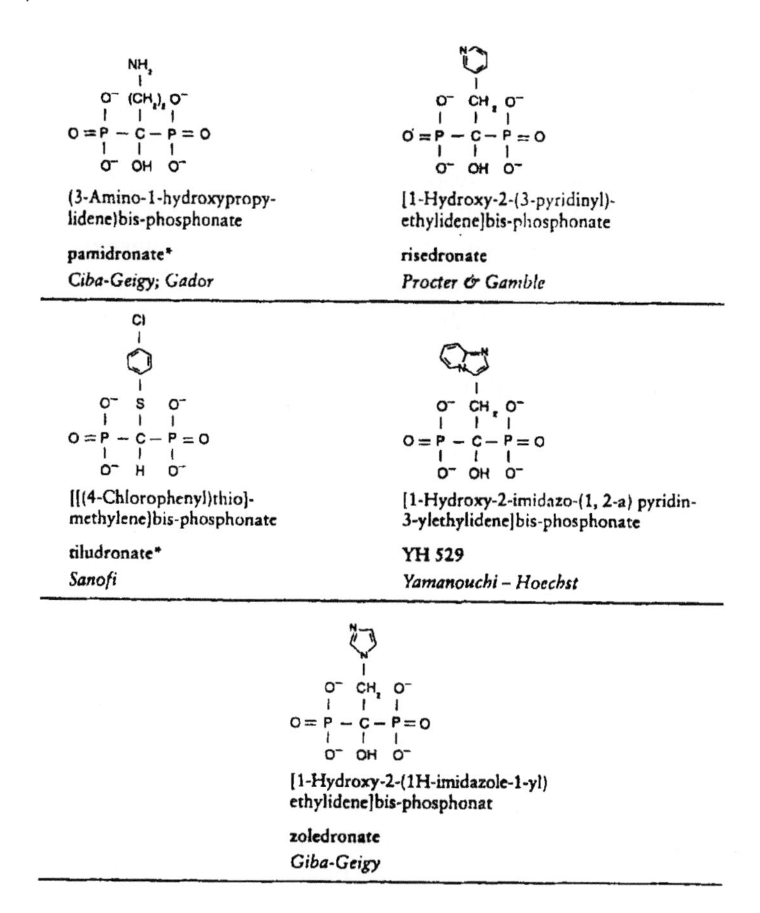

Figure 1. Chemical structure of bisphosphonates investigated in humans. (*Commercially available.)

with various side chains (P-C-P). These compounds have been known to the chemists since the 19th century, the first synthesis dating back to 1865 (Menschutkin, 1865). They were first used in various industrial procedures, among others as anticorrosive and antiscaling agents (Blomen, 1995). After the discovery that they can effectively control calcium phosphate formation and dissolution *in vitro* as well as mineralization and bone resorption *in vivo* (Fleisch et al., 1969; Francis et al., 1969), they were developed and used in the treatment of bone diseases, mostly Paget's disease, tumor bone disease, and lately osteoporosis. The following bisphosphonates have been used in man up to now, seven of them being commercially available in some countries (Figure 1).

III. GENERAL PROPERTIES

Each bisphosphonate has its own physicochemical and biological characteristics, which implies that it is not possible to extrapolate from the results of one compound to others with respect to its actions. The P-C-P bond of the bisphosphonates is stable to heat and most chemical reagents and completely resistant to hydrolysis by enzymes found in the body. The latter explains why these compounds are not metabolized *in vivo* and excreted unaltered.

The bisphosphonates have a strong affinity for metal ions such as calcium and iron. This affinity to calcium explains why they bind strongly to hydroxyapatite and deposit virtually only in calcified tissues. In bone, they deposit where bone mineral is exposed to the surrounding fluids, especially where it is formed and resorbed. Once deposited, the bisphosphonates are liberated again, some by physicochemical mechanisms, but in most cases only when the bone in which they were deposited is resorbed, which explains their long half-life (Kasting and Francis, 1992). The bisphosphonates have a low lipophilicity which hampers transcellular transport, and a high negative charge, hampering paracellular transport. This explains their low intestinal absorbability.

IV. PHYSICOCHEMICAL EFFECTS

The physicochemical effects of most of the bisphosphonates are very similar to those of pyrophosphate. Thus, most of them inhibit the formation and aggregation of calcium phosphate crystals (Francis, 1969; Francis et al., 1969; Fleisch et al., 1970) and slow down their dissolution (Fleisch et al., 1969; Russell et al., 1970). All these effects are related to the marked affinity of these compounds for the surface of solid phase calcium phosphate (Jung et al., 1973) where they act as a crystal poison on both growth and dissolution.

V. BIOLOGICAL EFFECTS

The bisphosphonates have two fundamental biological effects: inhibition of bone resorption and, at high doses, inhibition of calcification.

A. Inhibition of Bone Resorption

Bisphosphonates are very powerful inhibitors of bone resorption when tested in a variety of conditions both *in vitro* and *in vivo. In vitro,* bisphosphonates block bone resorption induced by various means in organ culture (Fleisch et al., 1969; Russell et al, 1970; Reynolds et al., 1972). It has been shown that, using the mouse calvaria system, there is a satisfactory correlation between the results obtained *in vitro* and those found *in vivo* (Green et al., 1994). An inhibition is also found when the effect of osteoclasts is investigated on various mineralized matrices *in vitro* (Flanagan and Chambers, 1989; Sato and Grasser, 1990). However, only certain of the models used give a relation of potency with that found *in vivo*. *In vivo*, bisphosphonates inhibit bone resorption both in intact animals and in animals in which resorption has been experimentally increased.

In normal growing rats, the bisphosphonates block the degradation of both bone and cartilage, thus impeding the remodeling of the metaphysis, which becomes club-shaped and radiologically denser than normal (Schenk et al., 1973). This effect is used as a model for studying the potency of new compounds (Schenk et al., 1986). The inhibition of endogenous bone resorption has also been documented by ^{45}Ca kinetic studies (Gasser et al., 1972), and by markers of bone resorption. The effect occurs between 24 and 48 hours and is therefore slower than that of calcitonin.

The decrease in resorption is accompanied by an increase in calcium balance (Gasser et al., 1972), and in the mineral content of bone. This is possible because of an increase in intestinal absorption of calcium, consequential to an elevation of $1{,}25(OH)_2$ vitamin D. This increase in balance is the reason why these compounds are administered to humans suffering from osteoporosis. The increase is, however, smaller than predicted, considering the dramatic decrease in bone resorption, because bone formation also decreases after a certain time, possibly because of the so-called "coupling" between formation and resorption. The main effect of bisphosphonates is therefore a reduction of bone turnover.

Bisphosphonates can also prevent an experimentally induced increase in bone resorption. They impair, among others, resorption induced by agents such as parathyroid hormone (PTH) (Fleisch et al., 1969; Russell et al., 1990), $1{,}25(OH)_2$ vitamin D, and retinoids. The effect on retinoid-induced hypercalcemia has been used to develop a powerful and rapid screening assay for new compounds (Trechsel et al., 1987). The bisphosphonates are also effective in preventing bone loss in a number of experimental osteoporosis models. These include: sciatic nerve section, which was the first model investigated (Mühlbauer et al., 1971), spinal cord section, hypokinesis, ovariectomy, heparin, lactation, low calcium diet, and corticosteroids. All of the bisphosphonates investigated, in alphabetical order, alendronate, clodronate, etidronate, ibandronate, incadronate, olpadronate, pamidronate, risedronate, tiludronate, and YH 529, have been effective.

It is now clear that, unless given in excess, bisphosphonates also improve biomechanical properties both in normal animals and in experimental models of osteoporosis. This is the case with alendronate, clodronate, etidronate (although with this drug the effect is more ambiguous since at higher doses it is obscured by an inhibition of mineralization), incadronate, neridronate, olpadronate, pamidronate, tiludronate, and YH 529. It is seen in various animals such as the rat, chick, and baboon.

Bisphosphonates partially or entirely correct the increase in bone resorption in experimental tumor bone disease (Martodam et al., 1983). In the model in which tumor cells are implanted subcutaneously, the effect is generally more pronounced on calciuria than on calcemia. This is explained by the fact that hypercalcemia is often due to the systemic production of PTH-related peptide, which increases both bone resorption and tubular reabsorption of calcium, while bisphosphonates act only on the former. Bone resorption due to actual tumor invasion is also slowed down (Jung et al., 1984). The bisphosphonates shown to be active were, among others, clodronate, etidronate, incadronate, pamidronate, and risedronate.

The bisphosphonates do not directly inhibit the multiplication of tumor cells and are, therefore, not active on the tumor itself but exert their action by inhibiting the osteolytic process. As a secondary consequence, multiplication of tumor cells may be decreased, possibly in part due to a decrease in local cytokines stimulating tumor cell replication, which are released when bone is resorbed.

B. Relative Activity of Various Bisphosphonates on Bone Resorption

The activity of bisphosphonates on bone resorption varies greatly from compound to compound (Shinoda et al., 1983). One of the aims of bisphosphonate

Potency to inhibit bone resorption					
~1 ×	~10 ×	~100 ×	>100 – <1000 ×	>1000 – <10 000 ×	>10 000 ×
Etidronate	Clodronate Tiludronate	Pamidronate Neridronate	Alendronate EB-1053 Incadronate Olpadronate	Ibandronate Risedronate	YH 529 Zoledronate

Figure 2. Potency of some bisphosphonates to inhibit bone resorption in the rat. From Fleisch, 1997.

research has been to develop compounds with a more powerful antiresorptive activity but without a higher inhibition of mineralization. This has proved to be possible, and today compounds have been developed which are 5–10,000 times more powerful inhibitors of bone resorption than etidronate. The gradation of potency evaluated in the rat corresponds quite well to that found in humans.

C. Structure-Activity Relationship in Bone Resorption

Up to now, no clear-cut relationship could be worked out. The length of the aliphatic carbon is important, and adding a hydroxyl group to the carbon atom at position 1 increases potency (Shinoda et al., 1983). Derivatives of an amino group at the end of the side-chain are very active. Again the length of the side-chain is relevant, the highest activity being found with a backbone of four carbons, as present in alendronate (Schenk et al., 1986). A primary amine is not necessary for this activity since dimethylation of the amino nitrogen of pamidronate, as seen in olpadronate, increases efficacy (Boonekamp et al., 1986). Activity is still further increased when other groups are added to the nitrogen, as seen in ibandronate which is extremely active (Mühlbauer et al., 1991). Cyclic geminal bisphosphonates are also very potent, especially those containing a nitrogen atom in the ring, such as risedronate. The most active compounds described so far, zoledronate (Green et al., 1994) and YH 529, belong to this class. This effect of nitrogen is very intriguing and has not yet been explained. A three-dimensional structural requirement appears to be involved. Indeed, stereoisomers of the same chemical structure have shown 10-fold differences in activity. This opens the possibility of a binding to some kind of receptor.

D. Mechanisms of Action in Bone Resorption

The effects of bisphosphonates on bone resorption can be considered at three levels: tissue, cellular, and molecular. At the tissue level, the action of all active bisphosphonates appears to be identical, namely a reduction in bone turnover. The latter is evidenced by a decrease in both bone resorption and bone formation, as assessed by biochemical markers. Since some bone is lost at each bone multicellular unit (BMU), because a greater amount of bone is resorbed than formed, the decrease in turnover per se will slow down total bone loss. In addition, the bisphosphonates also act at the individual BMU by decreasing the depth of the resorption site (Storm et al., 1993; Boyce et al., 1995). Both effects will lead to a decreased number of trabecular perforations, thus slowing down the decrease in bone strength and the occurrence of fractures.

Not only resorption but also formation is decreased, as evidenced by a reduction in the bone formation surface (Balena et al., 1993; Storm et al., 1993). This reduction in bone formation is thought to be secondary to the diminished resorption and to reflect reduced remodeling. Indeed, in the normal remodeling process, formation follows resorption, thus allowing bone mass to remain stable. This conclusion is supported by various experimental data. Thus, there is no evidence for reduced osteoblastic activity at individual bone formation sites, as judged by the amount of bone produced per unit time. On the contrary, the amount of bone formed at each individual basic structural unit (BSU), measured by the thickness of the newly formed bone, is, if anything, even increased (Balena et al., 1995; Boyce et al., 1995). This finding is also consistent with the observation that clodronate increases collagen synthesis by bone and cartilage cells in culture (Guenther et al., 1981) and that alendronate increases mineralized nodule formation in bone marrow cultures (Tsuchimoto et al., 1994).

The effect on formation might give one explanation for the finding that bisphosphonates produce a positive calcium balance in animals (Gasser et al., 1972), and increase the amount of bone in animals and in humans. There are, however, several other explanations for this gain. One is inherent to bone turnover. Thus, a decrease in bone resorption is not immediately followed by the diminution of formation, thus inducing a temporary increase in balance through a reduction in the so-called remodeling space. The second explanation is that, following the decrease in turnover, the remodeling of the newly formed bone will be delayed, giving it more time to finish mineralization which needs a long time for completion. Thirdly, if the decrease in resorption depth at individual remodeling sites is not matched by a decrease in formation in the individual BMU, which seems to be the case, the local bone balance in the BMU will be positive.

At the cellular level, there is general agreement that the final target of bisphosphonate action is the osteoclast. Three mechanisms appear to be involved: 1) inhibition of osteoclast recruitment; 2) inhibition of osteoclast activity; 3) shortening of the osteoclast life span. All these effects could be due to either a direct action on the osteoclast, or indirectly via action on cells which modulate the osteoclast. The initial theory that resorption is decreased because of a physicochemical decrease in mineral dissolution after adsorption of the compound on the crystal surface is no longer tenable.

Several bisphosphonates inhibit osteoclast differentiation in various culture systems of both cells and bones (Boonekamp et al., 1986; Hughes et al., 1989). Furthermore, bisphosphonates are powerful inhibitors of macrophage proliferation, cells which are of the same lineage as osteoclasts (Cecchini et al., 1987). It also seems that the potency rank of bisphosphonates, when assessed *in vitro,* correlates in some conditions with effects *in vivo* only when such systems are used which also detect osteoclast recruitment and not only osteoclast activity. However, several facts suggest that the inhibition of recruitment is not the primary, or at least not the only, mode of action of bisphosphonates *in vivo.* Indeed following bisphosphonate administration, the number of multinucleated osteoclasts on the bone surface often increases at first, but the cells look inactive (Schenk et al., 1973). It is only later, after chronic administration, that the osteoclast number decreases. The cause for the initial increase is unknown. One possibility is that it could reflect stimulation of osteoclast formation to compensate for the decrease in osteoclast activity.

The effect on osteoclast recruitment may be direct and/or indirect. Recent results speak for the latter. It is now generally accepted that cells of osteoblastic lineage control the recruitment and activity of osteoclasts under physiological and pathological conditions. This control is thought to be due to the production of an as yet unknown activity, generated by osteoblast-lineage cells stimulating bone resorption. Recently, we found that when osteoblastic cells of rats are exposed to low concentrations of potent bisphosphonates, they liberate a factor(s) which reduces osteoclastic bone resorption in culture (Sahni et al., 1993). The decrease in resorption was subsequently found to be due to an inhibitor of osteoclast recruitment or survival, with a molecular weight between 1 and 10 kDa, released by the osteoblasts (Vitté et al., 1996). Other groups have also found that the effect on osteoclasts can be mediated by osteoblasts (Nishikawa et al., 1996). Whether the action of the inhibitor also involves cell survival is not yet known, nor is it known which cells of the osteoblastic lineage are involved in this mechanism. The lining cells would be an interesting possibility.

Another possibility is an inhibition of osteoclast activity. This is supported by the fact that osteoclasts can show changes in morphology both *in vitro* and *in vivo.* These include changes in the cytoskeleton, especially actin, and the ruffled border (Sato et al., 1991; Muramaki et al., 1995). This mode of action is also supported by the facts that: (i) under certain conditions bisphosphonates can enter cells (Felix et al., 1984), especially those of the macrophage lineage; and (ii) bisphosphonate concentration can attain very high values under the osteoclasts, partly because they deposit specifically under these cells (Sato et al., 1991).

The third possibility is the shortening of the lifespan of the osteoclast. It has been proposed that this might be due to a toxic effect, but the results were obtained at very high concentrations. Recently it was reported that

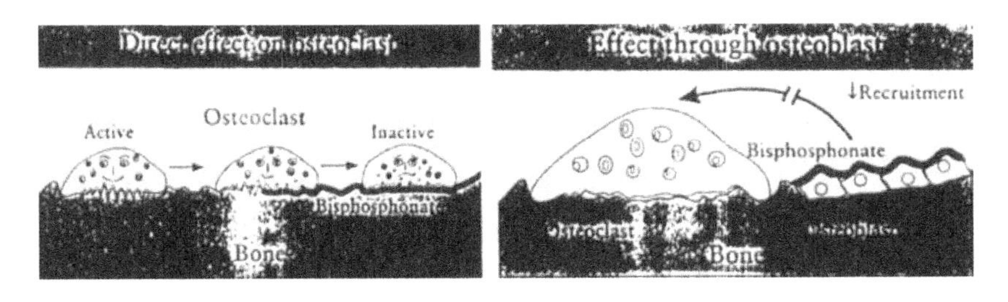

Figure 3. Modulation of osteoclasts by bisphosphonates. From Fleisch, 1997.

risedronate increases osteoclast apoptosis, a phenomenon which is different from toxic cell death, and this was proposed as a general mechanism for bisphosphonate action (Hughes et al., 1995). Whether this effect is direct or indirect through the effect on other cells or both is not known.

At the molecular level, events that lead to osteoclast inactivation or decreased formation, following direct or indirect exposure to a given bisphosphonate, have not yet been fully elucidated. The possibilities include either bisphosphonate action on a cell surface receptor, and/or bisphosphonate uptake by the cell, where it affects cellular metabolism. Until now no such receptor has been identified. In contrast, it is known that cells can take up bisphosphonates (Felix et al., 1984), possibly via fluid pinocytosis or adsorptive pinocytosis. The cells of the macrophage phagocytic system are particularly sensitive to bisphosphonates. This is shown by a decrease in proliferation *in vitro* (Cecchini et al., 1987) and an acute phase response in some patients, which is thought to be due to an increase in interleukin-6 and tumor necrosis factor α release (Schweitzer et al., 1995; Sauty et al., 1996).

It has been known for a long time that bisphosphonates decrease acid production of various cells (Fast et al., 1978). Recently bisphosphonates were shown to decrease the extrusion of acid by actual osteoclasts (Zimolo et al., 1995). Possibly part of this effect is due to the inhibition of the vacuolar type proton ATPase, which was found to be inhibited by tiludronate (Murakami et al., 1995). However, no correlation between the effect *in vitro* and *in vivo* was evidenced for these two effects. But the various bisphosphonates do not necessarily have a common mechanism of action.

In view of the homology between pyrophosphate and bisphosphonates, various enzymes involving pyrophosphate or ATP have been examined. Protein tyrosine phosphatases (PTPs), which control tyrosine phosphorylation involved in signal transduction pathways, have been found to be inhibited (Endo et al., 1996; Schmidt et al., 1996). However, again the potency of

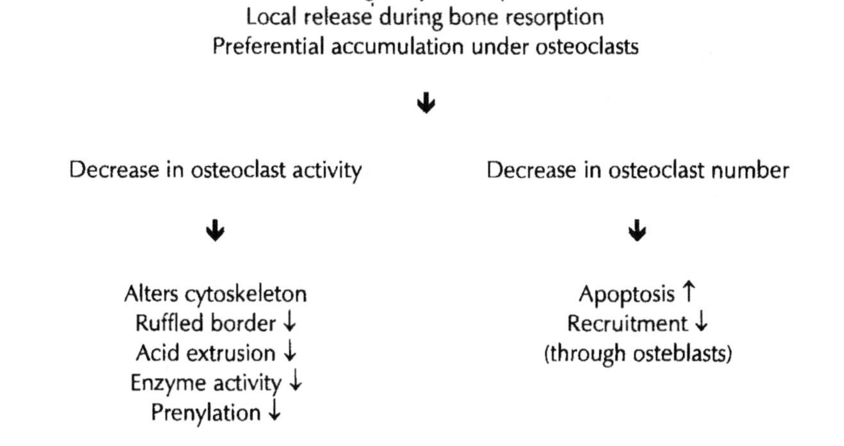

Figure 4. Cellular effects of bisphosphonates. From Fleisch, 1997.

various bisphosphonates in inhibiting the PTPs tested so far does not correspond to their pharmacological potency.

Recently an interesting new mechanism was proposed for at least some of the bisphosphonates. It has been shown that certain bisphosphonates, especially the nitrogen-containing compounds, inhibit the prenylation of proteins (Luckman et al., 1998). This is, however, not the case for etidronate and clodronate. This effect might be due to the inhibition of several enzymes, among them farnesylpyrophosphate synthase, which has indeed been found to be inhibited by bisphosphonates. This mechanism might be the cause of apoptosis.

Another interesting observation is that some cells can metabolize primary etidronate and clodronate, but not the bisphosphonates containing a nitrogen atom, to a non-hydrolyzable toxic ATP analogue containing a β, γ-methylene group (AppCp nucleotides) (Frith et al., 1997). Therefore, at least these compounds might act through this mechanism to induce apoptosis and necrotic cell death. Thus the aminobisphosphonates may have a mechanism of action different from that of the bisphosphonates containing no nitrogen.

E. Inhibition of Calcification

Like pyrophosphate, bisphosphonates also very efficiently inhibit calcification *in vivo.* Thus, among others, they prevent experimentally induced calcification of many soft tissues both when given parenterally and orally (Fleisch et al., 1970), as well as ectopic ossification. Topical administration leads to a decreased formation of dental calculus.

If administered in sufficient doses, bisphosphonates can also impair the mineralization of normal calcified tissues such as bone, cartilage, dentine, and enamel (King et al., 1971; Schenk et al., 1973). In contrast to bone resorption where the different compounds vary greatly in their activity, they do not do so in the case of inhibition of mineralization. For most species the effective daily dose is in the order of 5–20 mg of compound phosphorus per kg parenterally. There is a close relationship between the ability of an individual bisphosphonate to inhibit calcium phosphate *in vitro* and its effectiveness on calcification *in vivo* (Trechsel et al., 1987). Therefore, the mechanism is likely to be a physicochemical one. The inhibition of mineralization can lead to fractures and subsequently to impaired healing of the fractures. The inhibition is eventually reversed after discontinuation of the drug. The propensity to inhibit the calcification of normal bone has hampered the therapeutic use of bisphosphonates in ectopic calcification. This is, on the contrary, not the case for their use in bone resorption, since compounds have been developed which can inhibit this process at doses 1,000 times lower than those inhibiting mineralization.

VI. SUMMARY

The bisphosphonates are synthetic compounds characterized by a P-C-P bond. They are characterized by a strong affinity to calcium phosphates and hence to bone mineral. Many of the bisphosphonates inhibit bone resorption, the newest compounds being between 5 and 10,000 times more active in this respect than etidronate, the first bisphosphonate described. Bisphosphonates also prevent various types of experimental osteoporosis, as well as bone destruction induced by several tumor models.

It seems likely that part of the action on bone resorption is due to a direct effect on the osteoclasts which become inhibited after engulfing the compounds and cytoskeleton as well as series of intracellular effects, especially on the phosphorylative and prenylation of proteins. The cellular effect involves a decrease in acid production and changes in the ruffled border. Another part of the action appears to be mediated through the osteoblasts, which are induced to produce an inhibitor(s) of osteoclast recruitment. Large amounts of bisphosphonates can also inhibit mineralization through a physicochemical mechanism of inhibition of crystal growth.

REFERENCES

Adami, S., Bhalla, A.K., Dorizzi, R., Montesanti, F., Rosini, S., Salvagno, G., and Lo Cascio, V. (1987). The acute-phase response after bisphosphonate administration. Calcif. Tissue Int. 41, 326-331.

Balena, R., Toolan, B.C., Shea, M., Markatos, A., Myers, E.R., Lee, S.C., Opas, E.E., Seedor, J.G., Klein, H., Frankenfield, D., Quartuccio, H., Fioravanti, C., Clair, J., Brown, E., Hayes, W.C., and Rodan, G.A. (1993). The effects of two-year treatment with the aminobisphosphonate alendronate on bone metabolism, bone histomorphometry, and bone strength in ovariectomized nonhuman primates. J. Clin. Invest. 92, 2577-2586.

Blomen, L.J.M.J. (1995). History of bisphosphonates: Discovery and history of the nonmedical uses of bisphosphonates. In: *Bisphosponate on Bones.* (O.L.M.Bijvoet, O.L.M., Fleisch, H.A., Canfield, R.E., Russell, R.G.G., Eds.), pp. 11-124. Elsevier, Amsterdam.

Boonekamp, P.M., van der Wee-Pals, L.J.A., van Wijk–van Lennep, M.M.L., Thesing, C.W., and Bijvoet, O.L.M. (1986). Two modes of action of bisphosphonates on osteoclastic resorption of mineralized matrix. Bone Miner. 1, 27-39.

Boyce, R.W., Paddock, C.L., Gleason, J.R., Sletsema, W.K., and Eriksen, E.F. (1995). The effects of risedronate on canine cancellous bone remodeling: Three-dimensional kinetic reconstruction of the remodeling site. J. Bone Min. Res. 10 (2), 211-221.

Cecchini, M.G., Felix, R., Fleisch, H., and Cooper, P.H. (1987). Effects of bisphosphonates on proliferation and viability of mouse bone marrow-derived macrophages. J. Bone Min. Res. 2, 135-142.

Endo, N., Rutledge, S.J., Opas, R. Vogel, E.E., Rodan, G.A., and Schmidt, A. (1996). Human protein tyrosine phosphatase-σ alternative splicing and inhibition by bisphosphonates. J. Bone Min. Res. 11, 535-543.

Fast, D.K., Felix, R., Dowse, C., Neuman, W.F., and Fleisch H. (1978). The effects of diphosphonates on the growth and glycolysis of connective-tissue cells in culture. Biochem. J. 172, 97-107.

Flanagan, A.M. and Chambers, T.J. (1989). Dichloromethylenebisphosphonate (Cl_2MBP) inhibits bone resorption through injury to osteoclasts that resorb Cl_2MBP-coated bone. Bone Miner. 6, 33-43.

Felix, R., Guenther, H.L., and Fleisch H. (1984). The subcellular distribution of ^{14}C dichloromethylenebisphosphonate and ^{14}C 1-hydroxyethylidene-1,1-bisphosphonate in cultured calvaria cells. Calcified Tissue International 36, 108-113.

Fleisch, H. (1996). Bisphosphonates: Mechanisms of action and clinical use. In: *Principles of Bone Biology* (Bilezikian, J.P., Raisz, L.J., and Rodan, G.A., Eds.), pp. 1037-1052. Academic Press, San Diego.

Fleisch, H. (1997). *Bisphosphonates in Bone Disease. From the Laboratory to the Patient.* The Parthenon Publishing Group, New York.

Fleisch, H. (1997a). Bisphosphonates: Mechanisms of Action. Endocr. Rev. 19, 80-100.

Fleisch, H., Russell, R.G.G., Bisaz, S., Mühlbauer, R.C., and Williams, D.A. (1970). The inhibitory effect of phosphonates on the formation of calcium phosphate crystals in vitro and on aortic and kidney calcification in vivo. Eur. J. Clin. Invest. 1, 12-18.

Fleisch, H., Russell, R.G.G., and Francis, M.D. (1969). Diphosphonates inhibit hydroxyapatite dissolution in vitro and bone resorption in tissue culture and in vivo. Science 165, 1262-1264.

Francis, M.D. (1969). The inhibition of calcium hydroxyapatite crystal growth by polyphosphonates and polyphosphates. Calcif. Tissue Res. 3, 151-162.

Francis, M.D., Russell, R.G.G., and Fleisch H. (1969). Diphosphonates inhibit formation of calcium phosphate crystals in vitro and pathological calcification in vivo. Science 165, 1264-1266.

Frith, J.C., Mönkkönen, J. Blackburn, G.M., Russell, R.G.G., and Rogers, M.J. (1997). Clodronate and liposome-encapsulated clodronate are metabolized to a toxic ATP analog, adenosine 5'-(β,γ-dichloromethylene) triphosphate, by mammalian cells in vitro. J. Bone Miner. Res. 12, 1358-1367.

Gasser, A.B., Morgan, D.B., Fleisch, H.A., and Richelle, L.J. (1972). The influence of two diphosphonates on calcium metabolism in the rat. Clin. Sci. 43, 31-45.

Geddes, A.D., D'Souza, S.M., Ebetino, F.H., and Ibbotson, K.J. (1994). Bisphosphonates: structure-activity relationships and therapeutic implications. In: *Bone and Mineral Research.* (Heersche, J.N.M. and Kanis, J.A., Eds.), pp. 265-306. Elsevier, Amsterdam, London.

Green, J.R., Müller, K., and Jaeggi, K.A. (1994). Preclinical pharmacology of CGP 42'446, a new, potent, heterocyclic bisphosphonate compound. J. Bone Miner. Res. 9, 45-751.

Guenther, H.L., Guenther, H.E., and Fleisch, H. (1981). The effects of 1-hydroxyethane-1,1-diphosphonate and dichloromethanediphosphonate on collagen synthesis by rabbit articular chondrocytes and rat bone cells. Biochem. J. 196, 293-301.

Hughes, D.E., MacDonald, B.R., Russell, R.G.G., Gowen, M. (1989). Inhibition of osteoclast-like cell formation by bishposphonates in long-term cultures of human bone marrow. J. Clin. Invest. 83, 1930-1935.

Hughes, D.E., Wright, K.R., Uy, H.L., Sasaki, A., Yoneda, T., Roodman, G.D., Mundy, G.R., and Boyce, B.F. (1995). Bisphosphonates promote apoptosis in murine osteoclasts in vitro and in vivo. J. Bone Min. Res. 10, 1478-1487.

Jung, A., Bisaz, S., and Fleisch, H. (1973). The binding of pyrophosphate and two diphosphonates by hydroxyapatite crystals. Calc. Tissue Res. 11, 269-280.

Jung, A., Bornand, J., Mermillod, B., Edouard, C., and Meunier, P.J. (1984). Inhibition by diphosphonates of bone resorption induced by the Walker tumor of the rat. Cancer Res. 44, 3007-3011.

Kasting, G.B. and Francis, M.D. (1992). Retention of etidronate in human, dog, and rat. J. Bone Miner. Res. 7, 513-522.

King, W.R., Francis, M.D. and Michael, W.R. (1971). Effect of disodium ethane-1-hydroxy-1,1-diphosphonate on bone formation. Clin. Orthop. 78, 251-70.

Luckman, S.P., Hughes, D.E., Coxon, F.P., Russell, R.G.G., and Rogers, M.J. (1998). Nitrogen-containing bisphosphonates inhibit the mevalonate pathway and prevent post-translational prenylation of GTP-binding proteins, including Ras. J. Bone Miner. Res. 13, 581-589.

Martodam, R.R., Thornton, K.S., Sica, D.A., D'Souza, S.M., Flora, L., and Mundy, G.R. (1983). The effects of dichloromethylene diphosphonate on hypercalcemia and other parameters of the humoral hypercalcemia of malignancy in the rat Leydig cell tumor. Calcif. Tissue Int. 35, 512-519.

Menschutkin, N. (1865). Ueber die Einwirkung des Chlorazetyls auf phosphorige Säure. Ann. Chem. Pharm. 133, 317-320.

Mühlbauer, R.C., Bauss, F., Schenk, R., Janner, M., Bosies, E., Strein, K., and Fleisch, H. (1991). BM 21.0955, a potent, new bisphosphonate to inhibit bone resorption. J. Bone Miner. Res. 6, 1003-1011.

Mühlbauer, R.C., Russell, R.G.G., Williams, D.A., and Fleisch, H. (1971). The effects of diphosphonates, polyphosphates, and calcitonin on "immobilisation osteoporosis" in rats. Europ. J. Clin. Invest. 1, 336-344.

Murakami, H., Takahashi, N., Sasaki, T., Udagawa, N., Tanaka, S., Nakamura, I., Zhang, D., Barbier, A., and Suda, T. (1995). A possible mechanism of the specific action of bisphosphonates on osteoclasts: Tiludronate preferentially affects polarized osteoclasts having ruffled borders. Bone 17, 137-144.

Nishikawa, M., Akatsu, T., Katayama, Y., Yasutomo, Y., Kado, S., Kugai, N., Yamamoto, M., and Nagata, N. (1996). Bisphosphonates act on osteoblastic cells and inhibit osteoclast formation in mouse marrow cultures. Bone 18, 9-14.

Rodan, G.A. and Fleisch, H. (1996). Bisphosphonates: Mechanisms of action. J. Clin. Invest. 97, 2692-2696.

Russell, R.G.G., Mühlbauer, R.C., Bisaz, S., Williams, D.A., and Fleisch, H. (1970). The influence of pyrophosphate, condensed phosphates, phosphonates, and other phosphate compounds on the dissolution of hydroxyapatite in vitro and on bone resorption induced by parathyroid hormone in tissue culture and in thyroparathyroidectomised rats. Calcif. Tissue Res. 6, 183-196.

Reynolds, J.J., Minkin, C., Morgan, D.B., Spycher, D., and Fleisch, H. (1972). The effect of two diphosphonates on the resorption of mouse calvaria in vitro. Calcif. Tissue Res. 10, 302-313.

Sahni, M., Guenther, H.L., Fleisch, H., Collin, P., and Martin, T.J. (1993). Bisphosphonates act on rat bone resorption through the mediation of osteoblasts. J. Clin. Invest. 91, 2004-2011.

Sato, M. and Grasser, W. (1990). Effects of bisphosphonates on isolated rat osteoclasts as examined by reflected light microscopy. J. Bone Miner. Res. 5, 31-40.

Sato, M., Grasser, W., Endo, N., Akins, R., Simmons, H., Thompson, D.D., Golub, E., and Rodan, G.A. (1991). Bisphosponate action. Alendronate localization in rat bone and effects on osteoclast ultrastructure. J. Clin. Invest. 88, 2095-2105.

Sauty, A., Pecherstorfer, M., Zimmer-Roth, I., Fioroni, P., Juillerat, L., Markert, M., Ludwig, H., Leuenberger, P., Burckhardt, P., and Thiébaud, D. (1996). Interleukin-6 and tumor necrosis factor α levels after bisphosphonate treatment in vitro and in patients with malignancy. Bone 18, 133-139.

Schenk, R., Eggli, P., Fleisch, H., and Rosini, S. (1986). Quantitative morphometric evaluation of the inhibitory activity of new aminobisphosphonates on bone resorption in the rat. Calcif. Tissue Int. 38, 342-349.

Schenk, R., Merz, W.A., Mühlbauer, R., Russell, R.G.G., and Fleisch, H. (1973). Effect of ethane-1-hydroxy-1,1-diphosphonate (EHDP) and dichloromethylene diphosphonate (Cl_2MDP) on the calcification and resorption of cartilage and bone in the tibial epiphysis and metaphysis of rats. Calcif. Tissue Res. 11, 196-214.

Schmidt, A., Rutledge, S.J., Endo, N., Opas, E.E., Tanaka, H., Wesolowski, G., Leu, C.T., Huang, Z., Ramachandaran, C., Rodan, S.B., and Rodan, G.A. (1996). Protein-tyrosine phosphatase activity regulates osteoclast formation and function: Inhibition by Alendronate. 1996. Proc. Natl. Acad. Sci. USA 93, 3068-3073.

Schweitzer, D.H., Oostendorp-van de Ruit, M., van der Pluijm, G., Löwik, C.W.G.M., and Papapoulos, S.E. (1995). Interleukin-6 and the acute phase response during treatment of patients with Paget's disease with the nitrogen-containing bisphosphonate dimethylaminohydroxypropylidene bisphosphonate. J. Bone Miner. Res. 10, 956-962.

Shinoda, H., Adamek, G., Felix, R., Fleisch, H., Schenk, R., and Hagan, P. (1983). Structure-activity relationships of various bisphosphonates. Calcif. Tissue Int. 35, 87-99.

Storm, T., Steiniche, T., Thamsborg, G., Melsen, F. (1993). Changes in bone histomorphometry after long-term treatment with intermittent, cyclic etidronate for postmenopausal osteoporosis. J. Bone Miner. Res. 8, 199-208.

Trechsel, U., Schenk, R., Bonjour, J.-P., Russell, R.G.G., and Fleisch, H. (1977). Relation between bone mineralization, Ca absorption, and plasma Ca in phosphonate-treated rats. Am. J. Phyiol. 232, E298-E305.

Trechsel, U., Stutzer, A., and Fleisch, H. (1987). Hypercalcemia induced with an arotinoid in thyroparathyroidectomized rats. A new model to study bone resorption in vivo. J. Clin. Invest. 80, 1679-1686.

Tsuchimoto, M., Azuma, Y., Higuchi, O., Sugimoto, I., Hirata, N., Kiyoki, M., and Yamamoto, I. (1994). Alendronate modulates osteogenesis of human osteoblastic cells in vitro. Jpn. J. Pharmacol. 66, 25-33.

Vitté, C., Fleisch, H., and Guenther, H.L. (1996). Bisphosphonates induce osteoblasts to secrete an inhibitor of osteoclast-mediated resorption. Endocrinology 137, 2324-2333.

Zimolo, Z., Wesolowski, G., and Rodan, G.A. (1995). Acid extrusion is induced by osteoclast attachment to bone: Inhibition by alendronate and calcitonin. J. Clin. Invest. 96, 2277-2283.

NOVEL BONE-FORMING AGENTS

Ian R. Reid

I. INTRODUCTION

The development of a pharmaceutical which would produce sustained new bone formation has been the Holy Grail of osteoporosis research for a number of years. Increasing emphasis has been placed on this approach following the introduction of the potent bisphosphonates into clinical practice, which has carried with it the realization that it is now possible to very effectively inhibit bone resorption but that only a limited dividend, in terms of bone density, follows from such anti-osteoclastic therapies.

Advances in Organ Biology
Volume 5C, pages 851-867.

ISBN: 0-7623-0390-5

A highly effective method of increasing bone formation has been available for several decades, in the form of fluoride ion. Fluoride acts, possiblyas part of a fluoride-aluminum complex, to modulate a growth factor signaling pathway by enhancing tyrosine phosphorylation of several proteins in osteoblasts (Caverzasio et al., 1996). The resulting stimulation of osteoblast growth *in vitro* and its ability to produce sustained positive changes in bone density in clinical studies are well established. However, doubts have been cast upon its antifracture efficacy, though more data is accumulating which suggest that fluoride-induced increases in bone density are accompanied by increases in bone strength (Pak et al., 1995). However, with high doses of fluoride, there is interference with bone mineralization and bone strength is impaired. Thus, the therapeutic window for fluoride dose is narrow. Since bone fluoride appears to continue to accumulate with continued use, it is usual in most centers to limit therapy to periods of approximately four years. This limits the total increase in bone density that can be achieved with fluoride. Fluoride appears to exert its maximal effects on trabecular bone and its effects on non-vertebral fractures are, therefore, likely to be limited.

A second class of therapeutic agents sometimes regarded as promoting new bone formation are the anabolic steroids. Those used in osteoporosis therapy are testosterone analogues modified to reduce their virilizing effects. However, these modifications are only partially successful and the long-term clinical use of these agents continues to be severely limited by the development of acne, hirsutism, and voice changes. The extent to which they function as promoters of bone growth *in vivo* is uncertain, and some studies suggest their major action to be antiresorptive.

Of the novel agents, parathyroid hormone (PTH) and its analogues have received the greatest amount of study, with positive results emerging from several clinical trials. The use of growth hormone has met with mixed success and been accompanied by significant side effects. Insulin-like growth factor 1 (IGF-1) has also produced inconsistent results, but clinical studies are continuing. Considerable effort has been expended in studying the effects of other bone growth factors. This work will be reviewed. Their use in clinical practice is complicated by several considerations. Many of these growth factors are peptides and therefore both expensive to manufacture and not suitable for oral administration. Many of the major growth factors acting on bone (e.g., transforming growth factorβ, TGFβ) have actions on many other tissues. This makes the development of significant side effects likely with their long-term use to augment bone mass.

II. PARATHYROID HORMONE

PTH is an 84 amino acid peptide secreted by the parathyroid glands. Its principal function is to maintain extracellular fluid calcium concentration within the normal range, and the concentration of calcium in the pericellular fluid is the principal regulator of PTH secretion. It increases bone turnover, in particular bone resorption. Thus, patients with hypoparathyroidism have low rates of bone resorption, hypocalcemia, and increased bone mass. Conversely, those with hyperparathyroidism tend to have increased resorption, and there has been concern that they may have subnormal bone density and be at increased risk of fractures, though more recent evidence tends not to support this view (Grey et al., 1994). It is therefore somewhat paradoxical that PTH should now be looked upon as a potential therapy for osteoporosis. However, animal studies of injections of parathyroid extract as early as the 1930s suggested that this compound could produce increased bone mass (Selye, 1932; Shelling et al., 1933). These observations have led to limited clinical studies and an increasing effort to understand the cell biology of these differing effects.

PTH receptors are found on many cells. Those most important in calcium homeostasis are in the kidney (where PTH increases tubular reabsorption of calcium and increases 1-hydroxylation of 25-hydroxyvitamin D) and in cells of the osteoblast lineage. PTH receptors are not found on osteoclasts, and the hormone's effects on resorption are believed to be indirectly mediated through some form of osteoblast-osteoclast communication. The PTH receptor has seven membrane-spanning domains and is homologous with other G protein coupled receptors, such as that for calcitonin. The PTH receptor is also the target for amino-terminal fragments of parathyroid hormone related peptide. Coupling of PTH to its receptor results in activation of adenylyl cyclase and of phospholipase C. The latter results in production of diacylglycerol (which is involved in activation of protein kinase C) and inositol trisphosphate, which increases intracellular calcium concentration leading to the activation of calmodulin-dependent protein kinases. PTH thus has the potential to influence a large number of processes in target cells.

In many cells, activation of protein kinase C results in activation of the Na^+-H^+ antiporter with a resultant increase in intracellular pH. This is permissive to cell growth. However, cAMP-dependent protein kinase has the opposite effect on the activity of this ion exchanger, tending to reduce cytosolic pH. In the osteoblast-like cell line UMR 106, both these opposing effects operate following PTH treatment of the cells, but the cAMP effect dominates, pH falls and cell proliferation is reduced (Reid et al., 1988).

Impaired osteoblast proliferation has also been seen in other studies of UMR cells (Partridge et al., 1985) and in fetal rat osteoblasts (Jones and Boyde, 1976). However, PTH increases osteoblast proliferation in other models, such as isolated human osteoblasts (MacDonald et al., 1986) and in bone organ culture (Howard et al., 1981; DeBartolo et al., 1982). This inconsistency between different studies may be contributed to by different balances in the activity of the two second messenger systems, either between different models or according to the regimen of PTH administration. Cell proliferation following PTH exposure may also be influenced by other mediators since PTH increases production of IGF-1 and TGFβ by osteoblast-like cells (Centrella et al., 1988; Canalis et al., 1989) which are likely to have an autocrine effect on cell growth and matrix production. Some of the effects of PTH on bone may also be contributed to by its stimulation of c-*fos*, c-*jun*, and c-*myc*, early response genes that have been shown to be involved in the regulation of both cell differentiation and proliferation (Onyia et al., 1995).

A rapidly increasing number of animal studies more consistently demonstrate a positive effect of PTH on bone. Over the last 60 years, beneficial effects of PTH have been found in most animal models used, including the intact rat (Hock and Gera, 1992), the ovariectomized rat (Ibbotson et al., 1992; Sogaard et al., 1994; Li and Wronski, 1995; Li et al., 1995; Whitfield et al., 1995), the hypophysectomized rat (Schmidt et al., 1995), the immobilized rat (Yuan et al., 1995), the aged sheep (Delmas et al., 1995), and the dog (Podbesek et al., 1983). The anabolic effects are more marked when the peptide is administered intermittently rather than continuously (Tam et al., 1982; Podbesek et al., 1983; Hock and Gera, 1992): infusions of PTH often cause increased bone resorption and hypercalcemia in animal models.

In contrast, intermittent injections of the peptide increase a variety of indices of bone formation including osteoblast number, mineral apposition rate and mineralizing surface, and bone mass/volume increases at the spine, femur, and tibia. PTH has beneficial effects on cortical as well as trabecular bone (Wronski and Yen, 1994; Baumann and Wronski, 1995). Its effects on trabecular bone are manifest as an increase in trabecular thickness which results in decreased marrow space and an increase in indices of trabecular connectivity. However, in some circumstances its bone resorbing effects can be dominant even when administered intermittently. Cornish et al. (1995b) demonstrated marked bone loss following daily injections of PTH over the calvariae of adult mice.

Recently, some groups have examined the effects of combined therapy with PTH and an antiresorptive agent—either estrogen or a bisphosphonate.

Generally, these combinations have produced more positive effects on bone mass than either therapy alone (Shen et al., 1992, 1993; Mosekilde et al., 1994a) though an added benefit from the addition of the antiresorptive has not always been found (Mosekilde et al., 1994b; Sogaard et al., 1994; Wronski and Yen, 1994). Indeed, one study has been interpreted as demonstrating a diminution of the anabolic effect of PTH when tiludronate was coadministered (Delmas et al., 1995). However, this conclusion arises from consideration of turnover data alone: bone mass showed more beneficial trends with combined therapy.

Clinical studies of PTH in osteoporosis began in the 1970s (Reeve et al., 1980). These studies have used either PTH(1-34) or PTH(1-38), usually administered by daily injection. Most studies have been uncontrolled. They have generally shown substantial increases in spinal bone density or trabecular bone volume. However, at cortical sites such as the iliac or femoral cortices, there was a tendency for bone loss to occur (Hesp et al., 1981), suggesting that the gain in trabecular bone may have been occurring at the expense of cortical bone mass. This problem was addressed by combining PTH with calcitriol, the latter compound being added to increase intestinal calcium absorption. Slovik et al. (1986) demonstrated a doubling of vertebral trabecular bone density in osteoporotic men during 12 months of therapy with this combined regimen. Forearm bone density was stable throughout this period. The same group subsequently conducted a similar controlled trial in postmenopausal women in which trabecular bone density increased by one-third over a one- to two-year period but cortical density diminished by 6%. However, the increase in bone density did not appear to be progressive (Neer et al., 1990).

As in the animal studies, combined therapy with estrogen has also been assessed. Bradbeer et al. (1992) demonstrated increased trabecular bone volume in 10 women managed with such a combined regimen. Lindsay et al. (1993) have demonstrated 10% increases in integral spinal bone density after 18 months of therapy with PTH in women whose bone mass had previously stabilized on estrogen. There were also small increases in proximal forearm bone density in the PTH-treated women. Perhaps the most rigorous clinical study published in full is that of Finkelstein et al. (1994). Subjects who were estrogen deficient as a result of gonadotropin-releasing hormone therapy were randomly allocated to also receive PTH(1-34) daily. The addition of PTH led to preservation of spinal bone mass in contrast to a 3% loss in subjects not receiving PTH. A similar beneficial trend was seen at the proximal radius.

Thus, the clinical data remain few, but those available suggest that the significant therapeutic benefits seen with this compound in animal studies may well translate to the clinical arena. As a result, a number of drug companies are initiating trials with PTH or its analogues in the management of osteoporosis.

III. GROWTH HORMONE

Growth hormone is a single chain 191 amino acid peptide secreted from the anterior pituitary. It is well established as having a pivotal role in the regulation of longitudinal skeletal growth prior to the closure of the epiphyses at puberty. In the past, it has been regarded as having no significant role in post-pubertal life but this view has been increasingly challenged in recent years. Studies have demonstrated significant metabolic effects in growth hormone deficient adults given this hormone (Salomon et al., 1989). Treatment of such individuals produces striking changes in body composition with reductions in fat mass and increases in nitrogen balance and lean tissue mass. It has also become apparent that it has effects on bone and this has led to a number of studies exploring a possible role for growth hormone in both the etiology and therapy of osteoporosis. These suggestions have been supported by the observation that growth hormone secretion declines with advancing age (Pyka et al., 1992) and that states of growth hormone deficiency are associated with reduced bone density (Rosen et al., 1993; Holmes et al., 1994) whereas growth hormone excess in adults is associated with increased skeletal mass (Diamond et al., 1989).

Growth hormone secretion is episodic. Its principal regulators are growth hormone releasing hormone and somatostatin, but IGF-1, glucose, sex hormones, and amino acids also influence its secretion. Growth hormone pulses are most frequent early after the onset of sleep. Growth hormone acts on bone cells both directly and indirectly. Its direct action on osteoblast-like cells results in the local production of IGF-1 (Stracke et al., 1984) which, in turn, acts in an autocrine manner to stimulate osteoblast proliferation and protein synthesis (Canalis, 1980; Canalis et al., 1988; Hock et al., 1988; Chenu et al., 1990). Systemic growth hormone also regulates hepatic production and, thus, circulating levels of IGF-1. This provides another pathway by which bone cell function might be influenced.

The effect of growth hormone administration on biochemical indices of calcium metabolism has now been addressed in several studies. The most consistent finding is a marked increase in bone turnover involving markers of both bone formation (e.g., alkaline phosphatase and osteocalcin) and bone resorption (hydroxyproline and pyridinoline). The responses in markers of formation and resorption are comparable (Holloway et al., 1994) and responses are not different between subjects who are osteopenic and those with normal bone density (Kassem et al., 1994). The increase in turnover persists throughout at least 12 months of continuous therapy (Holloway et al., 1994). Growth hormone administration does not appear to influence circulating PTH concentrations. It causes transient increases in calcitriol which do not persist with long-term therapy. It increases renal tubular reabsorption of phosphate and circulating phosphate concentrations and appears to promote urinary calcium loss (Lieberman et al., 1994).

There is an increasing body of data describing changes in bone mass/density following growth hormone administration. Probably the first such study was that presented by Harris and Heaney (1969) in which dogs were treated with growth hormone for three months. They demonstrated marked changes in histomorphometric and kinetic indices of bone turnover and inferred that bone mass was substantially increased, though direct measurements of this parameter were not presented. Several years later Aloia et al. (1976) presented the first of series of studies of growth hormone administration in subjects with osteoporosis. In the first of these, eight subjects were treated over a period of six months. No changes in total body calcium, sodium, potassium, phosphorus, or chloride were demonstrable. There was an increase in urinary excretion of calcium and hydroxyproline and forearm bone mineral content declined significantly. As in many subsequent human studies, there were problems with side effects, particularly joint pain, carpal tunnel syndrome, hyperglycemia, and hypertension. Subsequently, a two-year, randomized study of women with postmenopausal osteoporosis was carried out. Subjects were treated with either calcitonin alone or calcitonin plus intermittent growth hormone. There was a significant increase in total body calcium in both groups, with no evidence of benefit from the addition of growth hormone. The group given growth hormone showed a significant loss of forearm bone mineral content which did not occur in those given calcitonin alone. Again, the use of growth hormone appeared to confer no benefit (Aloia et al., 1985). Subsequently, a further two-year study was reported in which 14 women were randomly allocated to receive cycles of growth hormone and calcitonin or intermittent calcitonin alone. In contrast to the two previous studies, a beneficial effect of growth hormone on total body calcium was observed with a change of almost 4% over the two-year period, in contrast to a small decline in the calcitonin only group (Aloia et al., 1987).

More recently, Rudman et al. (1990) have given growth hormone treatment to a group of healthy elderly men with low IGF-1 levels. They found the expected fall in fat mass and increase in lean body mass, but there was no change in bone density at the radius or proximal femur, though there was a 1.6% increase in the lumbar spine at six months. In a subsequent follow-up study of 45 of these patients, no changes in bone density were found with 12 months of treatment at any site, including the lumbar spine (Rudman et al., 1991). Holloway et al. (1994) have now reported a similar study in postmenopausal women. Side effects required reductions in doses of the peptide. Despite clear effects on bone turnover there was no increase in bone mineral density at the lumbar spine or hip in the treated group, though the placebo group showed declines of 2–3% at the trochanter and Ward's triangle. The authors concluded that growth hormone was unlikely to provide improvement in established osteoporosis and that its use in the elderly was complicated by fluid retention and carpal tunnel syndrome. A preliminary report of a further controlled trial in the elderly shows similar results and suggests that bone loss, if anything, might be accelerated by growth hormone treatment (MacLean et al., 1995).

In parallel to these studies, there have been a number of reports of the effects of growth hormone replacement in growth hormone deficient adults. These studies show very inconsistent results with some reporting improvements in bone density (O'Halloran et al., 1993; Vandeweghe et al., 1993), some no change (Beshyah et al., 1994, 1995), and others diminished density following treatment (Thoren et al., 1993; Holmes et al., 1995). In those studies demonstrating treatment effects, these have, as in the normal subjects, been small. The data collectively are consistent with a nil effect.

In contrast to the discouraging results with the use of growth hormone in either normal, osteoporotic, or growth hormone deficient humans, a number of investigators are now reporting positive results with its use in rats (Andreassen et al., 1995; Yeh et al., 1995). When balanced against the negative clinical evidence this suggests more that the rat is an inappropriate model for human osteoporosis rather than that a modification to the method of human administration is likely to alter the effectiveness of growth hormone in man. The continued growth of the adult rat skeleton may explain its different responsiveness to growth hormone from that of mature human bone.

IV. INSULINLIKE GROWTH FACTOR 1

The generally discouraging nature of the clinical studies with growth hormone has led to exploration of IGF-1 as a possible alternative therapy for osteoporosis. The role of IGF-1 in bone has been reviewed in detail (Rosen et al., 1994). As discussed above, IGF-1 acts directly on osteoblast-like cells to stimulate their proliferation and protein synthesis. IGF-1 is also a major component of bone matrix. It is a single-chain polypeptide (7.6 kDa) containing three disulfide bridges. It acts principally via the IGF-1 receptor but also binds to the IGF-2 receptor. Both receptors are found in bone cells. While most interest has concentrated on the potential effect of IGF-1 on osteoblast-like cells, there is also some evidence that it influences osteoclast biology. Bone marrow cultures treated with IGF-1 show increased numbers of multinucleated cells positive for tartrate-resistant acid phosphatase (Scheven and Hamilton, 1991). Mochizuki et al. (1992) have also shown IGF-1 to promote the differentiation of osteoclasts from precursor cells as well as demonstrating increased activity of mature osteoclasts following exposure to IGF-1. It is not clear, however, whether the latter finding represents a direct effect of IGF-1 on osteoclasts or whether it was mediated by contaminating osteoblasts.

Several studies have now demonstrated increased bone turnover in animals treated with IGF-1. Ibbotson et al. (1992) found that the dominant increase in turnover was of bone resorption in studies of an aged ovariectomized rat model. Bone mineral content tended to decrease in the IGF-1-treated animals in their study. In contrast, Bagi et al. (1995a,b) found that the dominant increase was in indices of formation and that IGF-1 treatment was associated with increased bone mass. When IGF-1 was co-administered with its binding protein (IGFBP3) the positive trend in bone mass was even more marked. Jerome et al. (1995) have documented trends towards increased bone formation in ovulating monkeys treated with IGF-1, though the trend in bone density in these animals was downward. Ammann (1993) has documented increases in bone density following six weeks treatment with IGF-1 in ovariectomized rats. Thus, the animal data is consistent in suggesting activation of turnover by IGF-1, but inconsistent with regard to whether resorption or formation dominates.

Similar results have now been demonstrated in limited human studies. Two case reports have shown stimulation of turnover in individuals with osteoporosis (Johansson et al., 1992; Rubin et al., 1994). Ebeling et al. (1993) administered various doses of IGF-1 over a period of 6 days to 18 normal postmenopausal women. There were dose-dependent increases in serum type 1 procollagen carboxyl-terminal propeptide concentration (an index of collagen synthesis) and of urinary deoxypyridinoline excretion, an indicator of bone resorption. Both these effects were seen at doses not associated with significant side effects, though the higher dose levels used were associated with tachycardia and edema in some patients. Ghiron et al. (1995) have now reported more detailed studies in a similar group of women. IGF-1 had no effect on sérum calcium but significantly increased urinary calcium excretion, though to a lesser extent than growth hormone. Treatment with IGF-1 at a dose of 30 μg/kg/day over a period of 28 days had no effect on indices of bone resorption, though it increased serum osteocalcin without changing serum alkaline phosphatase or type 1 procollagen carboxy-terminal propeptide. At a four-fold higher dose, there were significant increases in both formation and resorption markers. Data on the effects of IGF-1 on bone mass in humans are awaited.

V. OTHER POTENTIAL TREATMENTS

There is considerable research activity investigating the potential of other peptide growth factors. This is not as far advanced as that with growth hormone and IGF-1 as a result of concern with respect to side effects from these factors which act on many different cell types. TGFβ and the bone morphogenic proteins (BMPs) belong to the same peptide family. They are produced by bone cells, are found in bone matrix, and probably have a role in the paracrine regulation of bone cell function (Canalis et al., 1991). TGFβ has been shown to stimulate bone formation when injected locally over bone *in vivo* (Mackie and Trechsel, 1990; Marcelli et al., 1990); however, the development of bone nodules in osteoblast culture is inhibited by TGFβ, and the mRNAs for type 1 collagen, alkaline phosphatase, osteopontin, osteocalcin, and BMP2 are reduced by it (Harris et al., 1994). These results imply that TGFβ is most likely to act on osteoblast precursors to promote bone growth, since it appears to have the opposite effect in mature osteoblasts. Studies have now been reported of the systemic administration of TGFβ in rats in whom local osteopenia was produced by skeletal unloading (Machwate et al., 1995). In this context, TGFβ promoted bone formation, decreased bone resorption and prevented the development of osteopenia. In control rats its only effect was to reduce indices of resorption.

BMP2 has been shown to produce formation of both cartilage and bone when implanted subcutaneously with an inactive carrier (Wang et al., 1990). Subsequent studies with this protein have demonstrated that it can promote fracture healing when introduced into a fracture site in animals (Toriumi et al., 1991; Yasko et al., 1992). It has been shown that bone marrow cells exposed to BMP2 in the presence of a vascular supply will form bone within a silastic mould (Tomin et al., 1995). This opens up exciting possibilities for bone grafting. However, there seems to be no immediate prospect of this protein being used as a systemic therapy for osteoporosis. Other factors are being investigated in this role, however, including platelet derived growth factor (Lacey et al., 1995b) and megakaryocyte growth and differentiation factor (Lacey et al., 1995a) which have produced promising results in animals.

New factors with potentially beneficial activities on bone continue to be discovered. Amylin, a 37 amino acid peptide co-secreted with insulin from the pancreatic β cell, has been demonstrated to promote osteoblast proliferation *in vitro* and bone formation *in vivo* (Cornish et al., 1995a). It also blocks osteoclastic bone resorption, possibly via the calcitonin receptor. Its local administration over the calvariae of adult mice results in substantial increases in bone mass. The zeolites are another class of compounds that appear promising, based on laboratory studies. They are composed of $(SiO_4)^{4-}$ and $(AlO_4)^{5-}$ tetrahedra and have been shown in the past to increase egg shell thickness in chickens. The Mayo Clinic group has demonstrated that zeolite A increases the proliferation of human osteoblast-like cells and also augments alkaline phosphatase activity and osteocalcin release. mRNA levels for TGFβ are increased in the zeolite-treated cells (Keeting et al., 1992). More recently, the same group has demonstrated a substantial reduction in the resorbing activity of avian osteoclasts exposed to zeolite A (Schutze et al., 1995). Thus, like amylin, this compound has a very favorable profile for the treatment of osteoporosis.

VI. CONCLUSIONS

The intense research taking place to develop a clinically useful bone anabolic factor offers much promise in the therapy of osteoporosis and local bone defects. The recent, more positive, evidence relating to fluoride suggests that this agent will continue to be important in the coming years. The available clinical data with PTH is also exciting and it is possible that this agent will be introduced into clinical practice before the end of the decade. It will be some considerable time before the other agents reviewed are available for routine clinical prescription.

REFERENCES

Aloia, J.F., Zanzi, I., Ellis, K., Jowsey, J., Roginsky, M., Wallach, S., and Cohn, S.H. (1976). Effects of growth hormone in osteoporosis. J. Clin. Endocrinol. Metab., 992-999.

Aloia, J.F., Vaswani, A., Kapoor, A., Yeh, J.K., and Cohn, S.H. (1985). Treatment of osteoporosis with calcitonin, with and without growth hormone. Metabolism 34, 124-129.

Aloia, J.F., Vaswani, A., Meunier, P.J., Edouard, C.M., Arlot, M.E., Yeh, J.K., and Cohn, S.H. (1987). Coherence treatment of postmenopausal osteoporosis with growth hormone and calcitonin. Calcif. Tissue Int., 253-259.

Ammann, P., Rizzoli, R., Muller, K., Slosman, D., and Bonjour, J.P. (1993). IGF-I and pamidronate increase bone mineral density in ovariectomized adult rats. Am. J. Physiol. 265, E770-E776.

Andreassen, T.T., Melsen, F., and Oxlund, H. (1995). Effect of growth hormone on vertebral body strength, bone mass and formation in old rats. J. Bone Miner. Res. 10, S230.

Bagi, C.M., Deleon, E., Brommage, R., Adams, S., Rosen, D., and Sommer, A. (1995a). Systemic administration of rhIGF-I or rhIGF-I/IGFBP-3 increases cortical bone and lean body mass in ovariectomized rats. Bone 16, S263-S269.

Bagi, C.M., Deleon, E., Brommage, R., Rosen, D., and Sommer, A. (1995b). Treatment of ovariectomized rats with the complex of rhIGF-I/IGFBP-3 increases cortical and cancellous bone mass and improves structure in the femoral neck. Calcif. Tissue Int. 57, 40-46.

Baumann, B.D. and Wronski, T.J. (1995). Response of cortical bone to antiresorptive agents and parathyroid hormone in aged ovariectomized rats. Bone 16, 247-253.

Beshyah, S.A., Thomas, E., Kyd, P., Sharp, P., Fairney, A., and Johnston, D.G. (1994). The effect of growth hormone replacement therapy in hypopituitary adults on calcium and bone metabolism. Clin. Endocrinol. 40, 383-391.

Beshyah, S.A., Kyd, P., Thomas, E., Fairney, A., and Johnston, D.G. (1995). The effects of prolonged growth hormone replacement on bone metabolism and bone mineral density in hypopituitary adults. Clin. Endocrinol. 42, 249-254.

Bradbeer, J.N., Arlot, M.E., Meunier, P.J., and Reeve, J. (1992). Treatment of osteoporosis with parathyroid peptide (hPTH 1-34) and oestrogen—increase in volumetric density of iliac cancellous bone may depend on reduced trabecular spacing as well as increased thickness of packets of newly formed bone. Clin. Endocrinol. 37, 282-289.

Canalis, E. (1980). Effect of insulinlike growth factor I on DNA and protein synthesis in cultured rat calvaria. J. Clin. Invest. 66, 709-719.

Canalis, E., McCarthy, T., and Centrella, M. (1988). Isolation and characterization of insulinlike growth factor 1 (somatomedin-C) from cultures of fetal rat calvariae. Endocrinology 122, 22-27.

Canalis, E., Centrella, M., Burch, W., and McCarthy, T.L. (1989). Insulinlike growth factor I mediates selective anabolic effects of parathyroid hormone in bone cultures. J. Clin. Invest. 83, 60.

Canalis, E., McCarthy, T.L., and Centrella, M. (1991). Growth factors and cytokines in bone cell metabolism. Ann. Rev. Med. 42, 17-24.

Caverzasio, J., Imai, T., Ammann, P., Burgener, D., and Bonjour, J.-P. (1996). Aluminium potentiates the effect of fluoride on tyrosine phosphorylation and osteoblast replication in vitro and bone mass in vivo. J. Bone Miner. Res. 11, 46-55.

Centrella, M., McCarthy, T.L., and Canalis, E. (1988). Parathyroid hormone modulates, transforming growth factor β activity and binding in osteoblast-enriched cell cultures from fetal rat parietal bone. Proc. Nat. Acad. Sci. USA 85, 5889-5893.

Chenu, C., Valentin-Opran, A., Chavassieux, P., Saez, S., Meunier, P.J., and Delmas, P.D. (1990). Insulinlike growth factor I hormonal regulation by growth hormone and

1,25(OH)2D3 and activity on human osteoblastlike cells in short-term cultures. Bone 11, 81-86.

Cornish, J., Callon, K.E., Cooper, G.J.S., and Reid, I.R. (1995a). Amylin stimulates osteoblast proliferation and increases mineralized bone volume in adult mice. Biochem. Biophys. Res. Commun. 207, 133-139.

Cornish, J., Callon, K.E., and Reid, I.R. (1995b). An in vivo model for the rapid assessment of the local effects of parathyroid hormone on bone histomorphometry. Bone 17, S249-S254.

DeBartolo, T.F.L., Pegg, L.E., Shasserre, C., and Hahn, T.J. (1982). Comparison of parathyroid hormone and calcium ionophore A23187 effects on bone resorption and nucleic acid synthesis in cultured fetal rat bone. Calcif. Tissue Int. 34, 495-500.

Delmas, P.D., Vergnaud, P., Arlot, M.E., Pastoureau, P., Meunier, P.J., and Nilssen, M.H. (1995). The anabolic effect of human PTH (1-34) on bone formation is blunted when bone resorption is inhibited by the bisphosphonate tiludronate—is activated resorption a prerequisite for the in vivo effect of PTH on formation in a remodeling system? Bone 16, 603-610.

Diamond, T., Nery, L., and Posen, S. (1989). Spinal and peripheral bone mineral densities in acromegaly; the effects of excess growth hormone and hypogonadism. Ann. Intern. Med. 111, 567-573.

Ebeling, P.R., Jones, J.D., O'Fallon, W.M., Janes, C.H., and Riggs, B.L. (1993). Short-term effects of recombinant human insulinlike growth factor I on bone turnover in normal women. J. Clin. Endocrinol. Metab. 77, 1384-1387.

Finkelstein, J.S., Klibanski, A., Schaefer, E.H., Hornstein, M.D., Schiff, I., and Neer, R.M. (1994). Parathyroid hormone for the prevention of bone loss induced by estrogen deficiency. N. Eng. J. Med. 331, 1618-1623.

Ghiron, L.J., Thompson, J.L., Holloway, L., Hintz, R.L., Butterfield, G.E., Hoffman, A.R., and Marcus, R. (1995). Effects of recombinant insulinlike growth factor-I and growth hormone on bone turnover in elderly women. J. Bone Miner. Res. 10, 1844-1852.

Grey, A.B., Evans, M.C., Stapleton, J.P., and Reid, I.R. (1994). Body weight and bone mineral density in postmenopausal women with primary hyperparathyroidism. Ann. Intern. Med. 121, 745-749.

Harris, S.E., Bonewald, L.F., Harris, M.A., Sabatini, M., Dallas, S., Feng, J.Q., Ghosh-Choudhury, N., Wozney, J., and Mundy, G.R. (1994). Effects of transforming growth factor β on bone nodule formation and expression of bone morphogenetic protein 2, osteocalcin, osteopontin, alkaline phosphatase, and type-I collagen mRNA in long-term cultures of fetal rat calvarial osteoblasts. J. Bone Miner. Res. 9, 855-863.

Harris, W.H. and Heaney, R.P. (1969). Effect of growth hormone on skeletal mass in adult dogs. Nature 223, 403.

Hesp, R., Hulme, P., Williams, D., and Reeve, J. (1981). The relationship between changes in femoral bone density and calcium balance in patients with involutional osteoporosis treated with human parathyroid hormone fragment (hpth 1-34). Metabol. Bone Dis. Rel. Res. 2, 231-234.

Hock, J.M. and Gera, I. (1992). Effects of continuous and intermittent administration and inhibition of resorption on the anabolic response of bone to parathyroid hormone. J. Bone Miner. Res. 7, 65-72.

Hock, J.M., Centrella, M., and Calalis, E. (1988). Insulinlike growth factor I has independent effects on bone matrix formation and cell replication. Endocrinology 122, 254-260.

Holloway, L., Butterfield, G., Hintz, R.L., Gesundheit, N., and Marcus, R. (1994). Effects of recombinant human growth hormone on metabolic indices, body composition, and bone turnover in healthy elderly women. J. Clin. Endocrinol. Metab. 79, 470-479.

Holmes, S.J., Economou, G., Whitehouse, R.W., Adams, J.E., and Shalet, S.M. (1994). Reduced bone mineral density in patients with adult onset growth hormone deficiency. J. Clin. Endocrinol. Metab. 78, 669-674.

Holmes, S.J., Whitehouse, R.W., Swindell, R., Economou, G., Adams, J.E., and Shalet, S.M. (1995). Effect of growth hormone replacement on bone mass in adults with adult onset growth hormone deficiency. Clin. Endocrinol. 42, 627-633.

Howard, G.A., Bottenmiller, B.L., and Baylink, D.J. (1981). Parathyroid hormone stimulates bone formation and resorption in organ culture: Evidence for a coupling mechanism. Proc. Natl. Acad. Sci. USA. 78, 3204-3208.

Ibbotson, K.J., Orcutt, C.M., D'Souza, S.M., Paddock, C.L., Arthur, J.A., Jankowsky, M.L., and Boyce, R.W. (1992). Contrasting effects of parathyroid hormone and insulinlike growth factor I in an aged ovariectomized rat model of postmenopausal osteoporosis. J. Bone Miner. Res. 7, 425-432.

Jerome, C.P., Sass, D.A., Bennett, A., Bowman, A.R., LeRoth D, and Epstein, S. (1995). Effects of short-term treatment with growth hormone and insulinlike growth factor-1 on vertebral bone in aged monkeys. J. Bone Miner. Res. 10, S254.

Johansson, A.G., Lindh, E., and Ljunghall, S. (1992). Insulinlike growth factor I stimulates bone turnover in osteoporosis (letter). Lancet 339, 1619.

Jones, S.J. and Boyde, A. (1976). Experimental study of changes in osteoblastic shape-induced by calcitonin and parathyroid extract. Cell Tissue Res. 169, 449-465.

Kassem, M., Brixen, K., Blum, W.F., Mosekilde, L., and Eriksen, E.F. (1994). Normal osteoclastic and osteoblastic responses to exogenous growth hormone in patients with postmenopausal spinal osteoporosis. J. Bone Miner. Res. 9, 1365-1370.

Keeting, P.E., Oursler, M.J., Wiegand, K.E., Bonde, S.K., Spelsberg, T.C., and Riggs, B.L. (1992). Zeolite A increases proliferation, differentiation, and transforming growth factor β production in normal adult human osteoblastlike cells in vitro. J. Bone Miner. Res. 7, 1281-1289.

Lacey, D., Benson, W., Hill, D., McNiece, I., and Yan, X.O. (1995a). High levels of systemically delivered megakaryocyte growth and differentiation factor (MGDF) induce bone formation in rodents. J. Bone Miner. Res. 10, S438.

Lacey, D., Kenney, W., Chu, J., Benson, W., Hill, D., Ruegg, P., and Huffer, W.E. (1995b). Platelet-derived growth factor (pdgf) stimulation of in vivo bone formation is initiated by site-specific stromal and cellular interactions. J. Bone Miner. Res. 10, S441.

Li, M. and Wronski, T.J. (1995). Response of femoral neck to estrogen depletion and parathyroid hormone in aged rats. Bone 16, 551-557.

Li, M., Mosekilde, L., Sogaard, C.H., Thomsen, J.S., and Wronski, T.J. (1995). Parathyroid hormone monotherapy and cotherapy with antiresorptive agents restore vertebral bone mass and strength in aged ovariectomized rats. Bone 16, 629-635.

Lieberman, S.A., Holloway, L., Marcus, R., and Hoffman, A.R. (1994). Interactions of growth hormone and parathyroid hormone in renal phosphate, calcium, and calcitriol metabolism and bone remodeling in postmenopausal women. J. Bone Miner. Res. 9, 1723-1728.

Lindsay, R., Cosman, F., Nieves, J., Dempster, D.W., and Shen, V. (1993). A controlled clinical trial of the effects of 1-34hPTH in estrogen treated osteoporotic women. J. Bone Miner. Res. 8, S130.

MacDonald, B.R., Gallagher, J.A., and Russell, R.G.G. (1986). Parathyroid hormone stimulates the proliferation of cells derived from human bone. Endocrinology 118, 2445-2449.

Machwate, M., Zerath, E., Holy, X., Hott, M., Godet, D., Lomri, A., and Marie, P.J. (1995). Systemic administration of transforming growth factor β-2, prevents the impaired bone formation and osteopenia induced by unloading in rats. J. Bone Miner. Res. 10, S176.

Mackie, E.J. and Trechsel, U. (1990). Stimulation of bone formation in vivo by transforming growth factor-β—remodeling of woven bone and lack of inhibition by indomethacin. Bone 11, 295-300.

MacLean D, Kieri, D.P., and Rosen, C.J. (1995). Low-dose growth hormone for frail elders stimulates bone turnover in a dose-dependent manner. J. Bone Miner. Res. 10, S458.

Marcelli, C., Yates, A.J., and Mundy, G.R. (1990). In vivo effects of human recombinant transforming growth factor-β on bone turnover in normal mice. J. Bone Miner. Res. 5, 1087-1096.

Mochizuki, H., Hakeda, Y., Wakatsuki, N., Usui, N., Akashi, S., Sato, T., Tanaka, K., and Kumegawa, M. (1992). IGF-I supports formation and activation of osteoclasts. Endocrinology 131, 1075-1080.

Mosekilde, L., Danielsen, C.C., and Gasser, J. (1994a). The effect on vertebral bone mass and strength of long-term treatment with antiresorptive agents (estrogen and calcitonin), human parathyroid hormone-(1-38), and combination therapy, assessed in aged ovariectomized rats. Endocrinology 134, 2126-2134.

Mosekilde, L., Sogaard, C.H., McOsker, J.E., and Wronski, T.J. (1994b). PTH has a more pronounced effect on vertebral bone mass and biomechanical competence than antiresorptive agents (estrogen and bisphosphonate)—assessed in sexually mature, ovariectomized rats. Bone 15, 401-408.

Neer, R., Slovik, D., Daly, M., Lo, C., Potts, J., and Nussbaum, S. (1990). In: *Osteoporosis.* (Christiansen, C. and Overgaard, K., Eds.), pp. 1314-1317, Osteopress, Copenhagen.

O'Halloran, D.J., Tsatsoulis, A., Whitehouse, R.W., Holmes, S.J., Adams, J.E., and Shalet, S.M. (1993). Increased bone density after recombinant human growth hormone (GH) therapy in adults with isolated GH deficiency. J. Clin. Endocrinol. Metab. 76, 1344-1348.

Onyia, J.E., Bidwell, J., Herring, J., Hulman, J., and Hock, J.M. (1995). In vivo, human parathyroid hormone fragment (hPTH 1-34) transiently stimulates immediate early-response gene expression, but not proliferation, in trabecular bone cells of young rats. Bone 17, 479-484.

Pak, C.Y.C., Sakhaee, K., Adamshuet, B., Piziak, V., Peterson, R.D., and Poindexter, J.R. (1995). Treatment of postmenopausal osteoporosis with slow-release sodium fluoride. Final report of a randomized controlled trial. Ann. Intern. Med. 123, 401-408.

Partridge, N.C., Opie, A.L., Opie, R.T., and Martin, T.J. (1985). Inhibitory effects of parathyroid hormone on growth of osteogenic sarcome cells. Calcif. Tissue Int. 37, 519-525.

Podbesek, R.D., Edouard, C., Meunier, P.J., Parsons, J.A., Reeve, J., Stevenson, R.W., and Zanelli, J.M. (1983). Effects of two treatment regimes with synthetic human parathyroid hormone fragment on bone formation and the tissue balance of trabecular bone in greyhounds. Endocrinology 112, 1000-1006.

Pyka, G., Wiswell, R.A., and Marcus, R. (1992). Age-dependent effect of resistance exercise on growth hormone secretion in people. J. Clin. Endocrinol. Metab. 75, 404-407.

Reeve, J., Meunier, P.J., and Parsons, J.A. (1980). The anabolic effect of human parathyroid hormone fragment (hpth 1-34) therapy on trabecular bone in involutional osteoporosis. BMJ 280, 1340-1344.

Reid, I.R., Civitelli, R., Avioli, L.V., and Hruska, K.A. (1988). Parathyroid hormone depresses cytosolic pH and DNA synthesis in osteoblastlike cells. Am. J. Physiol. 255, E9-E15.

Rosen, C.J., Donahue, L.R., and Hunter, S.J. (1994). Insulinlike growth factors and bone: The osteoporosis connection. Proc. Soc. Exp. Biol. Med. 206, 83-102.

Rosen, T., Hansson, T., Granhed, H., Szucs, J., and Bengtsson, B.A. (1993). Reduced bone mineral content in adult patients with growth hormone deficiency. Acta. Endocrinol. 129, 201-206.

Rubin, C.D., Reed, B., Sakhaee, K., and Pak, C.Y. (1994). Treating a patient with the Werner syndrome and osteoporosis using recombinant human insulinlike growth factor. Ann. Intern. Med. 121, 665-668.

Rudman, D., Feller, G., Nagraj, H.S., Gergens, G.A., Lalitha, P.Y., Goldberg, A.F., Schlenker, R.A., Cohn, L., Rudman, I.W., and Mattson, D.E. (1990). Effects of human growth hormone in men over 60 years old. N. Eng. J. Med. 323, 52-54.

Rudman, D., Feller, A.G., Cohn, L., Shetty, K.R., Rudman, I.W., and Draper, M.W. (1991). Effects of human growth hormone on body composition in elderly men. Horm. Res. 36 (Suppl. 1), 73-81.

Salomon, F., Cuneo, R.C., Hesp, R., and Sonksen, P.H. (1989). The effects of treatment with human recombinant growth hormone on body composition and metabolism in adults with growth hormone deficiency. N. Engl. J. Med. 321, 1797-1803.

Scheven, B.A.A. and Hamilton, N.J. (1991). Stimulation of macrophage growth and multinucleated cell formation in rat bone marrow cultures by IGF-I. Biochem. Biophys. Res. Comm. 174, 647-653.

Schmidt, I.U., Dobnig, H., and Turner, R.T. (1995). Intermittent parathyroid hormone treatment increases osteoblast number, steady-state messenger ribonucleic acid levels for osteocalcin, and bone formation in tibial metaphysis of hypophysectomized female rats. Endocrinology 136, 5127-5134.

Schutze, N., Oursler, M.J., Nolan, J., Riggs, B.L., and Spelsberg, T.C. (1995). Zeolite A inhibits osteoclast-mediated bone resorption in vitro. J. Cell. Biochem. 58, 39-46.

Selye, H. (1932). On the stimulation of new bone formation with parathyroid extract and irradiated ergosterol. Endocrinology 16, 547-558.

Shelling, D.H., Asher, D.E., and Jackson, D.A. (1933). Calcium and phosphorus studies. VII. The effects of variations in dosage of parathormone and calcium and phosphorus in the diet on the concentrations of calcium and phosphorus in the serum and on the histology and chemical composition of the bones of rats. Bull. Johns Hopkins Hosp. 53, 348-389.

Shen, V., Dempster, D.W., Mellish, R.W., Birchman, R., Horbert, W., and Lindsay, R. (1992). Effects of combined and separate intermittent administration of low-dose human parathyroid hormone fragment (1-34) and 17 β-estradiol on bone histomorphometry in ovariectomized rats with established osteopenia. Calcif. Tissue Int. 50, 214-220.

Shen, V., Dempster, D.W., Birchman, R., Xu, R., and Lindsay, R. (1993). Loss of cancellous bone mass and connectivity in ovariectomized rats can be restored by combined treatment with parathyroid hormone and estradiol. J. Clin. Invest. 91, 2479-2487.

Slovik, D.M., Rosenthal, D.I., Doppelt, S.H., Potts, J.T., Daly, M.A., Campbell, J.A., and Neer, R.M. (1986). Restoration of spinal bone in osteoporotic men by treatment with human PTH and 1,25 dihydroxyvitamin D. J. Bone Miner. Res. 1, 377-381.

Sogaard, C.H., Wronski, T.J., McOsker, J.E., and Mosekilde, L. (1994). The positive effect of parathyroid hormone on femoral neck bone strength in ovariectomized rats is more pronounced than that of estrogen or bisphosphonates. Endocrinology 134, 650-657.

Stracke, H., Schultz, A., Moeller, D., Rossol, S., and Schatz, H. (1984). Effect of growth hormone on osteoblasts and demonstration of somatomedin C/IGF-1 in bone organ culture. Acta Endocrinol 107, 16-24.

Tam, C., Heersche, J.N.M., Murray, T.M., and Parsons, J.A. (1982). Parathyroid hormone stimulates the bone apposition rate independently of its resorptive action: Differential effects of continuous and intermittent administration. Endocrinology 110, 506-512.

Thoren, M., Soop, M., Degerblad, M., and Saaf, M. (1993). Preliminary study of the effects of growth hormone substitution therapy on bone mineral density and serum osteocalcin levels in adults with growth hormone deficiency. Acta Endocrinol. 128, 41-43.

Tomin, E., Panossian, V., Heberbrand, D., Schumacker, L., Jones, N., Lane, J.M., and Wozney, J.M. (1995). The effects of rhbmp-2 and bone marrow on the formation of a molded vascularized bone graft in vivo. J. Bone Miner. Res. 10, S436.

Toriumi, D.M., Kotler, H.S., Luxenberg, D.P., Holtrop, M.E., and Wang, E.A. (1991). Mandibular reconstruction with a recombinant bone-inducing factor. Functional, histologic, and biomechanical evaluation. Arch. Otolaryngol. Head Neck Surg. 117, 1101-1112.

Vandeweghe, M., Taelman, P., and Kaufman, J.M. (1993). Short-term and long-term effects of growth hormone treatment on bone turnover and bone mineral content in adult growth hormoneBdeficient males. Clin. Endocrinol. 39, 409-415.

Wang, E.A., Rosen, V., D'Alessandro, J.S., Pjauduy, M., Cordes, P., Harada, T., Israel, D.I., Hewick, R.M., Kerns, K.M., LaPan, P., Luxenberg, D.P., McQuiad, D., and Moutsatsos, I.K. (1990). Recombinant bone morphogenetic protein induces bone formation. Proc. Nat. Acad. Sci. USA 87, 2220-2224.

Whitfield, J.F., Morley, P., Ross, V., Isaacs, R.J., and Rixon, R.H. (1995). Restoration of severely depleted femoral trabecular bone in ovariectomized rats by parathyroid hormone-(1-34). Calcif. Tissue Int. 56, 227-231.

Wronski, T.J. and Yen, C.F. (1994). Anabolic effects of parathyroid hormone on cortical bone in ovariectomized rats. Bone 15, 51-58.

Yasko, A.W., Lane, J.M., Fellinger, E.J., Rosen, V., Wozney, J.M., and Wang, E.A. (1992). The healing of segmental bone defects, induced by recombinant human bone morphogenetic protein (rhbmp-2). A radiographic, histological, and biomechanical study in rats. J. Bone Joint Surg. 74A, 659-670.

Yeh, J.K., Chen, M.M., and Aloia, J.F. (1995). Effects of growth hormone and estradiol administration on bone turnover in pituitary and ovarian hormone deficient rats. J. Bone Miner. Res. 10, S435.

Yuan, Z.Z., Jee, W.S.S., Ma, Y.F., Wei, W., and Ijiri, K. (1995). Parathyroid hormone therapy accelerates recovery from immobilization-induced osteopenia. Bone 17, S219-S223.

TRANSGENIC MODELS FOR BONE DISEASE

Pietro De Togni

Advances in Organ Biology
Volume 5C, pages 869-890.

ISBN: 0-7623-0390-5

I. INTRODUCTION

Genetically manipulated mice have been used extensively for the study of the function of a wide variety of eukaryotic genes in an intact organism. The genetically-manipulated mouse also provides a means of creating models of human disease, thus increasing our understanding of normal physiology as well as pathophysiology of several important human diseases. Two sorts of genetic manipulations are normally carried out. One of these, the classical "transgenic," involves the overexpression of the gene of interest in a mouse. In essence, this approach provides for a useful model to study the function and regulation of a particular gene in a tissue-specific manner. In this technique, recombinant DNA molecules are introduced by microinjection into pronuclei of fertilized mouse eggs. The transgene, in most instances, is randomly integrated into the genome, and is then transmitted in the germline. The other more recent approach, popularly termed the "knockout mouse" or gene targeting, is based on the inactivation of a specific gene through its targeted disruption. This model provides information on whether a gene product is required for the specific function that has been previously attributed to the gene, as well as to identify its novel functions. It is a particularly powerful tool in evaluating the function of a gene product during embryonic development.

Both techniques have more recently provided vital insights into skeletal development, as well as in understanding the pathophysiology of important bone and mineral disorders. Depending upon the time point in embryonic or fetal development at which a specific gene is expressed, its effects on the resulting phenotype can be varied (Copp, 1995). For example, inactivation of the E-cadherin gene, which encodes a calcium-dependent cell adhesion protein, is associated with early death of the embryos due to failure to form trophectodermal epithelium at the blastocyst stage (Larue et al., 1994). This defect appears to be cell-autonomous and not to result from the disruption of housekeeping functions. Similarly, the disruption of the transcription factor *evx1* gene, results in peri-implantational death in utero due to failure to develop extraembryonic tissues (Spyropoulos and Cappecchi, 1994). Other mutants died during early organogenesis due to failure to establish a normal hematopoiesis in the yolk sac, such as the *c-myc* (Davis et al., 1993), *GATA-2* (Tsai et al., 1994), and *Rbtn2* (Warren et al., 1994) knockout mice, or leakage of blood from the yolk sac into the exocoelomic cavity, such as *α5-integrin* (Yang et al., 1993) and *fibronectin* (George et al., 1993). Other causes of death during early fetal development are the failure to develop cardiovascular circulation and liver hematopoiesis resulting in severe anemia in the early- to mid-fetal period. An example of these mutants include the *RXRa* (Sucov et al., 1994) and the *gp130* (Yoshida et al., 1996) knockout mutants which show hypoplastic ventricular myocardium. Disruption of the β-*catenin, BMP4,* or the *BMP2* type I receptor genes lead to early embryonic lethality due to defects in mesoderm formation (Hagel et al., 1995).

Inactivation of other genes, such as the retinoic acid receptor and the *Hox* gene family, can produce multiple morphogenetic defects including major axial, limb, and craniofacial abnormalities that are particularly obvious in double knockouts (Lohnes et al., 1995; Mendelhson et al., 1994; Davis et al., 1995). With other genes, such as the interleukin-6 gene, the bone phenotype is very much more subtle, and only becomes evident following a particular intervention, in this case, ovariectomy (Kopf et al., 1994; Fattori et al., 1994; Poli et al., 1994).

II. GENERAL APPLICATIONS OF GENETICALLY MANIPULATED MOUSE MODELS

A. Transgenic Mice

Transgenic mice can be used to understand a variety of biological processes. These are too numerous to emunerate in this chapter; nevertheless, I will highlight certain important concepts, particularly in relation to their use in defining the development and function of skeletal tissues.

Several methods have been described to produce transgenic mice (Jaenisch, 1988). The most established method involves the isolation of fertilized ova from pregnant females and microinjection of their pronuclei with several hundred copies of DNA (Figure 1). The manipulated eggs are then transferred into the oviduct of pseudopregnant females. The latter are generated by breeding female mice with vasectomized sterile males.

The foreign DNA will normally integrate randomly at a single site per diploid genome, although multiple insertion sites have been described. If the integration occurs in the ova before cell division, the foreign gene will be expressed with the same copy number in all the cells including the germ line. However, if the integration occurs at a later embryonic stage after a few divisions, a different number of copies will be expressed in different cells of the mouse including its germ line. The mouse is then referred to as a mosaic for that particular gene. The rate of integration varies in different laboratories, but usually is between 10 and 30% in all surviving embryos, particularly when linearized DNA is used.

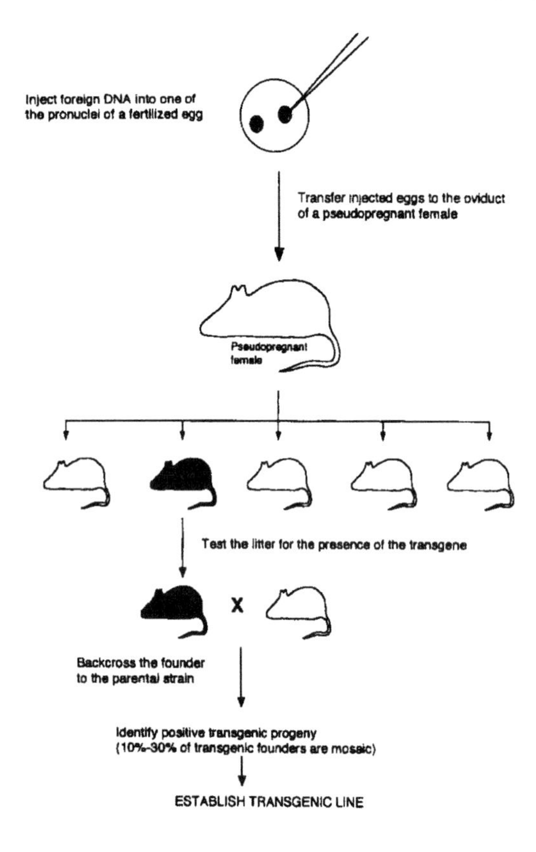

Figure 1. General procedure for producing transgenic mice.

The expression of an exogenous gene following its integration into the genome is affected by several factors (Gordon, 1989). The use of a genomic clone instead of cDNA, removal of prokaryotic DNA from the plasmid, and inclusion of introns in the construction (minigenes) all increase the rate of gene expression. Other factors affecting the rate of expression are the integration site, the strain of mice from which the fertilized eggs were derived, and the degree of DNA methylation.

A major application of the transgenic mouse technology is to study the regulation of gene expression by *cis* acting elements and to identify both hormonal-response elements and elements that confer tissue specificity to the promoter of any given gene. Hybrid genes, which contain the putative promoter and a receptor gene, are constructed to direct expression of marker proteins to a selected cell type. Expression of the reporter gene can be scored by enzymatic assays, for example those using chloramphenecol acetyltransferase (CAT), β-galactosidase, or luciferase. Quantitative data can be obtained by Northern blot analysis, primer extension, and RNAase protection assay. To study the function of the osteocalcin gene promoter, Kesterson et al. made transgenic mice using the hybrid gene made up of the 3.9 kb human osteocalcin gene promoter and CAT (Kesterson et al., 1993). The study suggested the human osteocalcin gene promoter was sufficient to direct the expression of a reporter gene to the osteoblast in a 1, 25-$(OH)_2$-vitamin D_3-sensitive manner.

Another application of the transgenic technology is to investigate the effect of overexpression of a given gene in the context of a whole organism. Either a gene can be expressed widely by its fusion to a promoter such as metallothionein, or else can be directed in a tissue-specific (or cell-line-specific) manner by using a tissue specific promoter, such as the thymus-specific *lck* (McGuire et al., 1992) or the β-cell specific insulin promoter (Picarella et al., 1992).

A further major application, particularly relevant to bone, is the creation of mouse models of human disease, including dominant-negative mutant mice. This has been used to create transgenic mouse models for the genetic bone disorder, osteogenesis imperfecta, as well as for the cartilage disorder, chondrodysplasia. More recently, *Ets-2*, a proto-oncogene and transcription factor that has been found to be highly expressed in newly forming cartilage, has been overexpressed in mice using the metallothionein promoter. The resulting phenotype was found to display the skeletal abnormalities of Down syndrome (trisomy 21). These are discussed below.

B. Knockout Mice

This technology is based on the use of targeted mutagenesis in embryonic stem (ES) cells (Capecchi, 1994) (Figure 2). These cells are derived from the inner cell mass of normal blastocysts and can be manipulated *in vitro* without losing their ability to differentiate into every cell type including the germ line when introduced back into blastocysts. To produce a knockout mouse, replacement vectors containing the desired mutation are prepared using a portion of the cloned gene (Figure 3). Usually, the coding region of the gene of interest is disrupted by the insertion of a drug resistance gene, such as the neomycin-resistance (*neo^r^*) cassette, which interrupts the complete transcription of the gene and also serves as a positive selection marker. The replacement vector is then linearized and introduced into ES cells by electroporation with the purpose to introduce that specific mutation into the gene of interest by homologous recombination. Because gene targeting is a

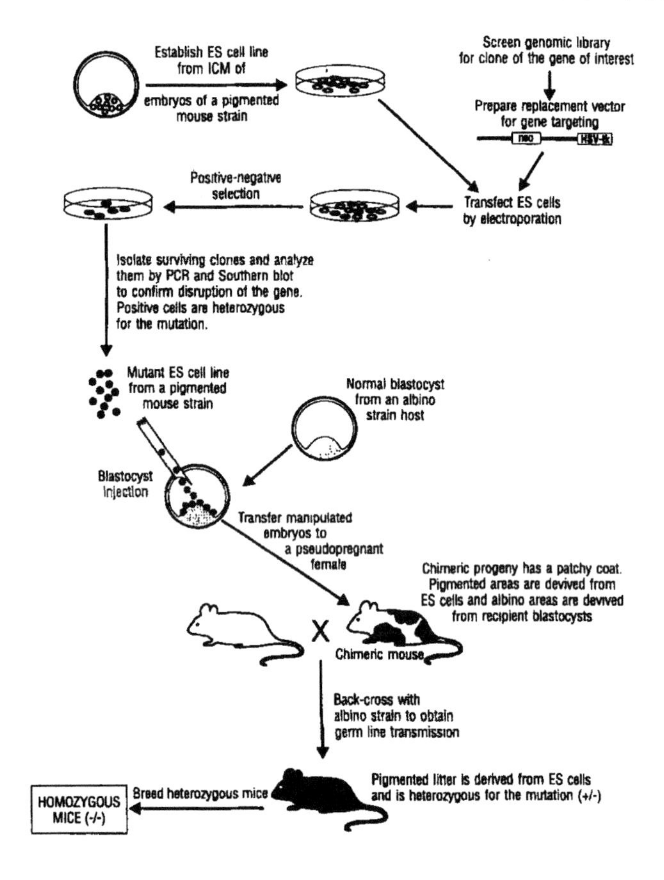

Figure 2. Schematic representation of the generation of knockout mice by gene targeting in ES cells.

rare event and the majority of the linearized DNA integrates randomly into the genome, negative selection genes have been used to increase the rate of detection of homologous recombinant clones. One of these techniques, the positive-negative selection, was first proposed by Capecchi (Mansour et al., 1988). It involves the addition of a herpes simplex virus thymidine kinase gene (*HSV-TK*) at the end of the linearized vector. Crossover on either sides of the *neo*^{r} gene will result in loss of the *HSV-TK* gene if homologous recombination occurs. On the contrary, with random integration the whole linearized vector including the *HSV-TK* gene will be integrated in the genome. Presence of the *HSV-TK* gene can be selected against, because any cells expressing the gene will be killed by gancyclovir, which is metabolized to a toxic metabolite by the viral thymidine kinase.

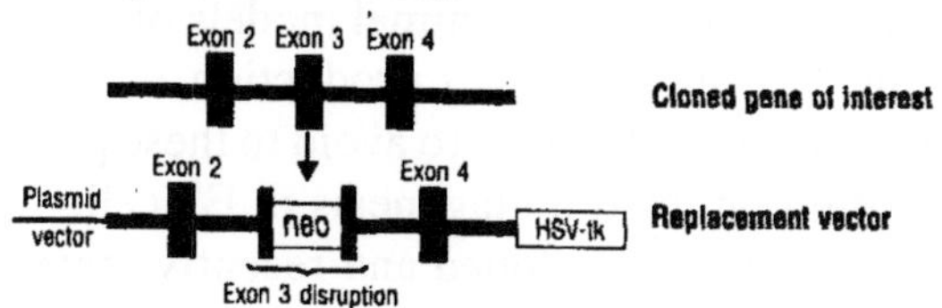

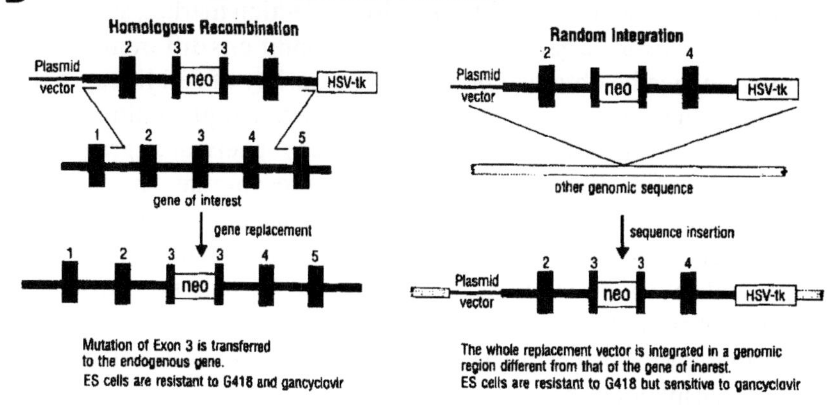

Figure 3. Schematic representation of the strategy of gene targeting in ES cells. (**A**) Preparation of the targeting vector from the cloned gene of interest. The vector contains a neo cassette inserted in one of the exons and a *HSV-tk* gene at the end of the linearized vector. (**B**) Positive negative selection in ES cells. Comparison of homologous recombination and random integration.

To maximize the frequency of homologous recombination, the two regions of homology flanking the positive selection marker in the targeting vector should contain at least 6 kb or more of homologous DNA and each of them should be greater than 1 kb (Deng and Capecchi, 1992). Also the degree of homology between the construct and the gene of interest is essential to obtain optimal recombination (Riele et al., 1992). Therefore, the homologous DNA is usually prepared from DNA isogenic with the target gene in ES cells.

Once a clone is selected, ES cells with a mutation in one of the alleles are then injected into blastocysts to produce chimeric mice (Figure 2). If the germ cells carry the mutated gene (germ line transmission), the animal can be bred and transmit the mutation to its offspring. The offspring will be heterozygous with respect to the knockout gene. Breeding these animals to homozygosity will result in both alleles being mutant.

Gene targeting is an invaluable tool for analyzing the function of gene products *in vivo* and to provide animal models of human disease. However, germ line transmission and production of homozygous mutant mice is expensive and laborious. To avoid to these problems, alternative methods based on targeted mutagenesis in ES cells without germ line transmission have been developed and recently reviewed by Nagy.and Rossant (1996) . One of these methods includes the generation of homozygous mutant ES cells by direct targeting of the second allele using a targeting vector containing a different drug selection marker, such as the hygromycin resistance gene. Under various conditions differentiation of ES cells can then be obtained *in vitro* with a variety of cell lineages that recapitulate the embryonal development. In fact, in liquid culture, wild-type ES cells form embryoid bodies, which are spherical aggregates of differentiated cells . Because embryoid bodies recapitulate the early embryonic development, these models have been used to study the role of different genes in hematopoiesis, vasculogenesis, neuronal cell differentiation, and myogenesis (Weiss and Orin, 1996). It is not difficult to imagine that similar approaches in the future might be applied as a form of gene therapy to correct inherited mutations or to generate cells and tissues for transplantation.

III. MAJOR SKELETAL PHENOTYPES

Numerous skeletal phenotypes have been described in the literature as a result of either overexpressing genes of interest or inactivating genes by targeted disruption (Fassler et al., 1995; Shastry, 1995). The description of each phenotype is beyond the scope of this chapter. Table 1 enumerates some of the models with skeletal abnormalities and the reader is referred to the appropriate references for further details. Here, I shall describe several key phenotypes that have proven critical to the understanding of skeletal physiology and pathology.

IV. GENE DELETIONS CAUSING OSTEOPETROSIS

The *c-fos* oncogene is the cellular homologue of oncogenes carried by two mouse sarcoma viruses (MSV), FBJ-MSV and FBR-MSV, which induce osteosarcomas in newborn mice (Muller, 1986; Verma, 1986). The *c-fos* protein is a major component of the AP-1 transcription factor complex,

Table 1. Mouse Mutants with Alternation of Skeletal Development and Function

Gene	*Skeletal Phenotype*
A. Gene expression (Transgenic models)	
α-amilase promoter/SV40 T antigen	Osteosarcoma
Protamine 1 promoter/SV40 T antigen	temporal osteosarcoma
metallothionein promoter/*cfos*	osteosarcoma, condrosarcoma
metallothionein promoter/*Ets2*	Down syndrome–like skeletal abnormalities
Hox-4.2	homeotic transformation of the occipital bones
Hox-2.3	craniofacial abnormalities
mutant α1(II) collagen gene	chondrodysplasia
mutant α1(I) collagen gene	osteogenesis imperfecta
osteocalcin promoter/TGF-beta 2	osteoporosis
SR α promoter/G-CSF	osteoporosis
type II collagen promoter/PTHrP	chondrodysplasia, delayed endochondral ossification
B. Gene Inactivation	
1. Natural mutants	
CSF-1	*op/op* mouse, osteopetrosis
Pax-1	*undulated* mouse, deformities of axial skeleton
fro/fro mouse	osteogenesis imperfects
pa/we mouse	deficient bone growth
Bmp-5	*short ear* mouse, defect of skeletal patterning
basic-helix-loop-helix-leucine	*microphthalmia* (mi) mouse, Osteopetrosis
2. Insertional mutagenesis	
gene on chromosome 2	defect in pattern of limb formation
unknown	fused digits
3. Homologous recombination (Knock-out mice)	
a. Genes involved in embryonal morphogenesis (prior to bone differentiation):	
retinoic acid receptor genes	malformations of head, vertebrae and limbs
retinoic acid inducible transcription factor AP-2	malformations of head and trunk
Hox genes	malformations of head, vertebrae and limbs
Engrailed-1	disruption of the patterning of the forelimb paws, sternum and ribs
b. Genes involved in skeletal morphogenesis:	
c-src	osteopetrosis
c-src and *hck* double mutant	severe osteopetrosis
c-fos	osteopetrosis
GM-CSF	none described
IL-6	protection from bone loss caused by estrogen depletion
FGFR-3	bone dysplasia with enhanced endochondrial bone growth
PTHrP	osteochondrodysplasia
PTH/PTHrP receptor	osteochondrodysplasia
Tartrate-resistant Acid Phosphatase	deformities of limbs and axial skeleton, defect of endochondral bone formation late-onset osteopetrosis
Msx1	cleft palate, abnormalities of craniofacial and tooth development
TGF-β3	cleft palate
int-2	abnormalities of tail and inner ear
Krox-20	reduced length and thickness of newly formed bone due to defect of endochondral ossification
Osteocalcin	higher bone mass due to increase bone formation

which includes members of the *jun* family . It is expressed constitutively in extra-embryonic tissues during early development, while its expression in the embryo is restricted to later development during osteogenesis (Muller et al., 1983; Dony and Gruss, 1987). Notably, high levels are present in the chondrogenic layer which precedes the formation of bone in the fetus (De-Togni et al., 1988). Post-natally, *c-fos* is associated with the induction of myelomonocytic differentiation and macrophage proliferation (Muller et al., 1985).

To study the function of *c-fos* in normal development, cell growth control, and the induction of differentiation, genetically manipulated mice have been generated. Two groups have produced mice lacking a functional *c-fos* gene (Johnson et al., 1992; Wang et al., 1992). Homozygous mutants show reduced placental and fetal weight and only 40% of these survived this being consistent with their mortality in the perinatal period. Those that survived were found to develop severe osteopetrosis. The bone disorder was characterized by long bone shortening, marrow space ossification and absent tooth eruption. The long bone cavity was most often completely obliterated by bone and cartilage trabeculae. The growth plates were highly disorganized and showed a reduction in the zone of proliferating chondrocytes.

This osteopetrotic phenotype results from lack of bone remodeling and failure of *c-fos* deficient mice to remove bone. Grigoriadis et al. (1994) reported a lack of multinucleate tartrate-resistant acid phosphatase-positive cells. They believe that the absence of osteoclasts was due to an intrinsic defect in the osteoclast lineage, and did not come from altered osteoblast or stromal cell environment. They found that both wild-type and homozygous embryonic limbs transplanted under the kidney capsule of an adult host develop normally with adequate osteoclast formation. In contrast, when similarly transplanted into mutant mice, they develop abnormally into osteopetrotic bones lacking osteoclasts. It has thus been concluded that *c-fos*-deficient osteoblasts and stromal cells are fully capable of supporting osteoclast differentiation. Moreover, bone marrow transplantation from a wild-type to an irradiated homozygous mutant can rescue osteopetrosis, confirming that the osteoclast defect lies within the hematopoietic cell, not in the stromal cell compartment.

In addition to reduced osteoclast numbers, the long bones of *c-fos* deficient mice have an increased number of terminally differentiated macrophages which retain their ability to phagocytize latex beads in culture (Grigoriadis et al., 1994). Although a similar osteopetrotic phenotype has been described for the *op/op* mouse, which has a naturally occurring point mutation in the macrophage colony stimulating factor (M-CSF/CSF-1) gene (46, 47), the latter has a deficiency of both macrophages and osteoclasts (Naito et al., 1991). It appears, therefore, that *c-fos*-deficient mice have a defect in trafficking, so that fewer osteoclasts and more macrophages result from the progenitor cells.

The osteopetrosis of *c-fos* deficiency is distinct in certain ways from that observed in *c-src* knockout mice (Soriano et al., 1991). Notably, *c-src* is a member of a large family of related kinases, including *fyn, yes, hck, lck, fgr, lyn,* and *blk* (Eastman et al., 1990). The major phenotypic difference between *c-fos* and *c-src* deficiency lies in the presence of functionless osteoclasts in the latter versus the absence of osteoclasts in the former. The precise nature of the defect is presently uncharacterized, but it is known from other studies that *src* kinases are involved in osteoclast activation. The lack of *src* may therefore predispose an osteoclast to quiescence; hence the observed phenotype. Figure 4 illustrates the various points at which a

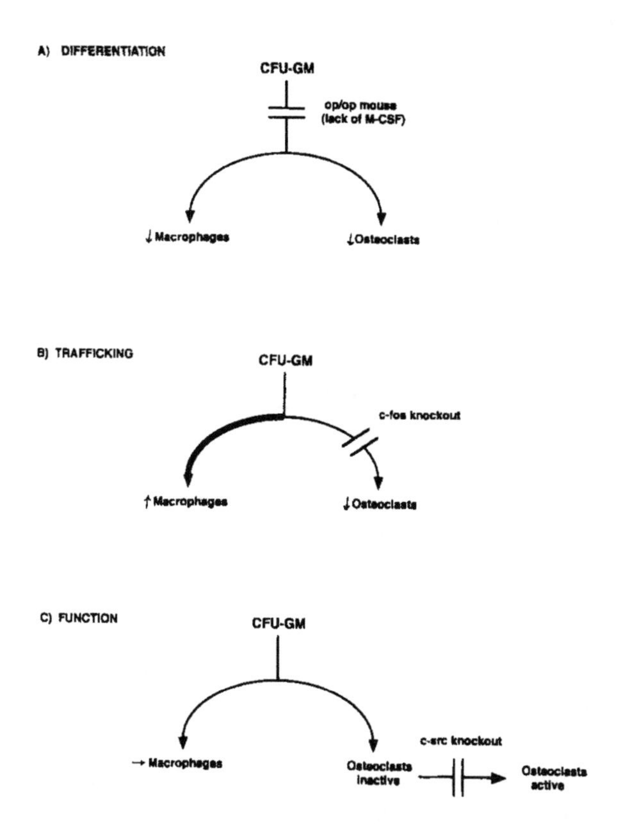

Figure 4. Different levels at which gene disruption can affect osteoclastogenesis. (**A**) *op/op* mouse; (**B**) *c-fos* knockout; (**C**) *c-src* knockout.

known mutation can disrupt osteoclastogenesis and impair osteoclast function.

Another intriguing observation is the lack of central nervous system involvement in *c-src* deficient mice despite the fact that *c-src* is abundantly expressed in the nervous tissue of wild-type mice. This may be due to redundancy, in that the function of *c-src* is taken over by a related kinase. On the other hand, bone tissue may exhibit greater *src*-specificity, which we believe is consistent with the essential function of the enzyme in osteoclast activation. This remains hypothetical.

In contrast to the osteopetrosis that results from *c-fos* deficiency, overexpression of the gene in transgenic animals is associated with the production of chondrogenic tumors with areas of osteogenesis. The earliest transgenic mice were made using the heavy metal-inducible, human metallothionein promoter, and the complete *c-fos* gene (Ruther et al., 1987). Exogenous *c-fos* expression in these mice was undetectable because of rapid RNA degradation. However, when a posttranscriptional regulatory element located in the 3' non-coding region of the gene was removed and substituted with a viral long terminal repeat (LTR), the mice showed high levels of protein expression. Typically, a few weeks after birth, the *c-fos* transgenic mice developed swellings at the end of the long bones that were associated with bone marrow fibrosis and new bone formation. In 15% of the mice, these lesions were followed several months later by tumors resembling those induced by the FBJ-MSV and FBR-MSV viruses (Ruther et al., 1989).

V. GENE MANIPULATIONS AND OSTEOCHONDRODYSPLASTIC MODELS

Several genetically manipulated models show evidence of osseous and chondral dysplasia either in isolation or in combination. Three mice models are worthy of mention: the parathyroid hormone-related protein (PTHrP) gene knockout, the fibroblast growth factor receptor III knockout, and the *Ets*-2 transgenic mouse.

Homozygous PTHrP knockout mutants (Karapalis et al., 1994), whether spontaneously delivered or delivered by cesarean section, mostly died at birth possibly from asphyxia. Analysis of their skeletons showed evidence of a multiplicity of skeletal deformities, including a domed skull, short snout and mandible, protruding tongue, and short limbs, all of which were consistent with a form of osteochondrodysplasia. A striking feature was an excessive degree of mineralization, often resulting in premature ossification and bone union, particularly in the rib cage. Bones that develop by intramembranous ossification were not affected. These macroscopic features resulted from diminished chondrocyte proliferation, early chondrocytic differentiation, and active perichondral osteoblastic bone formation. The chondrocytic components of growth plates of long bones were also shortened, as were the bony trabeculae of the primary spongiosa. Multinucleated chondroclasts and osteoclasts were morphologically normal.

The authors conclude from their studies that PTHrP plays a vital role in the interaction between the developing cartilaginous skeleton and the perichondrium during endochondral ossification. Nevertheless, an extrapolation to human chondrodysplasias is unclear, in that neither does the phenotype resemble classical achondroplasia, nor have the known inherited dysplastic syndromes been mapped to chromosome 12p12.1-p11.2 which is the locus of the human PTHrP gene. However, an analogy exists in that the PTHrP mutant phenotype somewhat resembles Albright's hereditary osteodystrophy, a human disease that is known to result from defective signaling through G_{sa}, a G protein that is involved in PTH/PTHrP signal transduction. Poor downstream signaling in Albright's osteodystrophy may therefore amount to being a phenotypic equivalent of a deficient ligand.

In contrast to the PTHrP knockout mouse, a mouse mutant overexpressing the *Ets-2* gene shows a somewhat different phenotype (Sumarsono et al., 1996). *Ets-2* is a protooncogene and a transcription factor with a widespread expression pattern. The metallothionein promoter was used to overexpress the *Ets-2* cDNA. The mice showed abnormal cranial development, hypoplastic facies, and kyphotic spines. In contradistinction to the PTHrP knockout mouse that showed only a defect in endochondral ossification, in the *Ets-2* transgenic, both endochondral and intramembranous ossification were defective, particularly in the skull. The skeletal phenotype resembled that of Down syndrome (trisomy 21) in humans and trisomy 16 in mice. Notable is the observation that the gene encoding Ets-2 is located in chromosome 21 in humans and chromosome 16 in mice.

Intriguing are the more recent observations by Deng et al (1996) that demonstrate a negative regulation of bone growth by fibroblast growth factor receptor 3 (FGFR-3). Contrary to what has been expected from human genetic studies wherein point mutations of the FGFR-3 receptor result in achondroplastic syndromes, deletion of the FGFR-3 gene in mice produced the opposite phenotype. FGFR-3-deficient mice showed evidence of increased chondrocyte proliferation, consistent with enhanced bone growth.

It appeared that the FGFR-3 gene was not essential for embryonic or postnatal survival. Worthy of note were much longer and distorted tails, kyphotic spines with multiple bends, and late femoral bowing. Increased osteogenesis, together with an increased number of osteoblasts, osteoclasts, and chondrocytes, appeared to underlay the abnormal bone growth. This model identifies FGFR-3 as a key negative regulator of bone growth. On the basis of more recent results using the osteocalcin gene knockout model, a similar function has been attributed to osteocalcin that has until now been recognized as a positive bone forming agent (in press).

VI. NEGATIVE-DOMINANT MUTANTS AND COLLAGEN DISORDERS

By expressing a mutated type I procollagen gene (COL1A1) with a single amino acid substitution, Gly with Cys at position 859, Stacey et al. (1988) developed a mouse model mimicking osteogenesis imperfecta. None of these mice survived after birth, and they were characterized by markedly underdeveloped skeleton with short wavy ribs, short and broad long bones, and poor bone mineralization. Expression of as little as 10% of the mutant gene resulted in a reduced amount of type I collagen and consequently in a dominant lethal perinatal phenotype. The defect arose from poor folding of the triple helix of type I collagen and the consequent more rapid degradation of the abnormal molecule. This phenotype is also similar to that of the naturally occurring mutant *fro/fro* mouse, in which cartilage formation appears normal but osteoid is significantly demineralized.

Similarly transgenic mice expressing mutant type II procollagen (COL2A1) genes containing either in-frame deletions of some of the exons (Vandenberg et al., 1991; Metsaranta et al., 1992) or a point mutation in the coding region (Garofalo et al., 1991) were affected with a severe form of short-limbed dwarfism resembling human chondrodysplasia. Histologically, there was reduction of the cartilagenous extracellular matrix and disorganization of the growth plate with marked reduction of the proliferating chondrocytes and loss of the typical columnar arrangement. These phenotypes were attributed to the presence of abnormal α chains which reduced the stability of the triple helix and increased intracellular degradation of the mutant procollagen molecules. Impaired cartilage production resulted in retarded bone growth and mineralization. Electronmicroscopic analysis showed a marked reduction in the number of type II fibrils in the cartilage (Garafalo et al., 1991).

VII. LIMITATIONS

A. Transgenic Technology

There are some major limitations to the use of both the transgenic and knockout technology. In the case of transgenic mice, the site of integration of the recombinant gene and the presence or absence of a complete promoter can affect its expression, so that the latter may vary with respect to its endogenous counterpart. Furthermore, as insertion of the recombinant DNA into the host genome is a random process, in 15% of cases it can be associated with disruption of an endogenous gene and therefore may result in a loss of function mutation. This can result in a visible phenotype such as major limb deformities in the legless mutant when the transgenic mice are bred to homozygosity. This approach has, however, been used advantageously to identify potentially interesting mutations.

In addition, occasionally, the structural component of the fusion gene can modify the activity of the promoter to which it is fused, so as to result in the ectopic expression of the fusion gene. Finally, as in the case of the *c-fos* gene, the presence of mRNA-destabilizing sequences in the 3' nontranslated region of a gene have been found to negatively affect the expression of the transgene. However, when this region was exchanged with a retroviral LTR, a high level of exogenous *c-fos* expression was detected and the mice developed characteristic bone lesions. Thus, the presence of those sequences that confer transcript instability can reduce the level of transgene expression.

B. Knockout Technology

Different problems arise with knockout mice. One of the major advantages of this technology is to help identification of those biological processes for which an individual member of a family of related molecules is essential. Nevertheless, this is often limited by the fact that genes exhibit redundancy, and the phenotype may not always be obvious. An example stems from our own work on the lymphotoxin-α gene (De Togni et al., 1994). Lymphotoxin-α, previously termed tumor necrosis factor β, is a member of a family of closely related molecules. That we observed only a lack of lymph nodes in the lymphotoxin-α knockout model may indeed reflect the replacement of lymphotoxin-α other functions with that of a closely related member, such as tumor necrosis factor α. Such redundancy, which may extend to cytokines as well as to their receptors, will prevent the development of phenotype that would be otherwise expected by disrupting the gene of a given member of the cytokine family.

A further limitation of the knockout technology relates to genes that are essential for embryonic morphogenesis, but at the same time are involved in postnatal development. Their disruption often causes lethality in the embryonic stage, and it is not possible to assess their function in adult life. An example for this is the gene for the ubiquitous and essential nuclear enzyme, polymerase β. Such an approach is however useful in studying the role of the given gene at various stages of embryonic development.

Another problem of the knockout technology relates to the nonidentity of mice and man, such that a mutation of the same gene cannot lead to an identical phenotype in two different species. Although the knockout mice can offer valuable models for human genetic diseases, the different genetic background can influence the phenotypic consequences of a gene mutation.

VIII. FUTURE PERSPECTIVES

As discussed above, one of the main limitations of the original knockout technique is that it is not possible to study the function of a gene that is essential for normal embryonic development other than in embryonic life due to intrauterine mortality or lack of survival postnatally. This problem has recently been overcome by the introduction of a new strategy based on the *Cre-loxP* recombination system of the bacteriophage P1 which allows gene inactivation in a tissue-specific manner (tissue specific knockout) and in an inducible way (inducible knockout). Specifically, this technique has been used for the first time by Rajewsky and colleagues to inactivate the DNA polymerase β gene only in the T cells (Gu et al., 1994; Kuhn et al., 1995). This mouse represents the first model in which a gene is knocked out only in certain tissues, after the early critical period in which it is necessary for embryonal development.

This procedure is based on introducing into the target gene two 34 nucleotide DNA sequences called *loxP* sites by homologous recombination in ES cells. These two *loxP* sites are then recognized by an enzyme called Cre recombinase which slices the intervening portion of the gene, leaving behind only one *loxP* site and a deleted gene. Mice that have *loxP* sites in their gene are perfectly normal, because they are able to produce a normal protein. However, if they are mated with transgenic mice in which the Cre recombinase gene was expressed under the control of a T cell specific promoter, the T cells of the offspring could not make the DNA polymerase, because an essential portion of the DNA polymerase β gene had been deleted. Similarly, by using an interferon-responsive promoter to control the expression of the Cre recombinase, the same authors were able to induce inactivation of the DNA polymerase β gene in different tissue after treating adult mice with interferon.

Recently the problem of functional redundancy for members of a same family of genes has been successfully overcome by the creation of double knockout mice. For instance, in the case of the retinoic acid receptors, the disruption of one of the genes alone is associated with the lack of any detectable phenotype. However, compound null mutations lead to lethality *in utero* or shortly after birth and to numerous developmental abnormalities (Lohnes et al., 1994, 1995; Mendelsohn et al., 1994). Generation of knockout mice with point mutations will also provide additional models to study genetic diseases that are associated only with subtle mutations. As these techniques become more available, they will open a broad range of possibilities for genetic manipulations and the understanding of complex processes such as bone development and function.

REFERENCES

Balling, R., Deutsch, U., and Gruss, P. (1988). Undulated, a mutation affecting the development of the mouse skeleton, has a point mutation in the paired box of Pax 1. Cell 55, 531-535.

Behringer, R.R., Peschon, J.J., Messing, A. et al. (1988). Heart and bone tumors in transgenic mice. Proc. Natl. Acad. Sci. USA 85 (8), 2648-2652.

Capecchi, M.R. (1994). Targeted gene replacement. Sci. Amer. 270 (3), 52-59.

Condie, B.G., and Capecchi, M.R. (1994). Mice with targeted disruption of the paralogous genes hoxa-3 and hoxd-3 reveal synergistic interactions. Nature 370, 304-307.

Copp, A.J. (1995). Death before birth: clues from gene knockout and mutations. Trends Genetics. 11 (3), 87-93.

Davis, A.P., and Capecchi, M.R. (1994). Axial homeosis and appendicular skeleton defects in mice with targeted disruption of hoxd-11 Development. 120 (8), 2187-2198.

Davis, A.C., Wims, M., Spotts, G.D., Hann, S.R., and Bradley, A. (1993). A null c-myc mutation causes lethality before 10.5 days of gestation in homozygous and reduced fertility in heterozygous females. Gene Deve. 7, 671-682.

Davis, A.P., Witte, D.P., Hsieh-Li, H.M., Potter, S.S., and Capecchi, M.R. (1995). Absence of radius and ulns in mice lacing hoxa-11 and hoxd-11. Nature 375, 791-795.

Deng, C., and Capecchi, R. (1992). Reexamination of gene targeting frequency as a function of the extent of homology between the targeting vector and the target locus. Mol. Cell. Biol. 12, 3365-3371.

Deng, C., Wynshaw-Boris, A., Zhou, F., Kuo, A., and Leder, P. (1996). Fibroblast growth factor receptor 3 is a negative regulator of bone growth. Cell 84, 911-921.

De Togni, P., Goellner, J., Ruddle, N. et al. (1994). Abnormal development of peripheral lymphoid organs in lymphotoxin deficient mice. Science 264, 703-707.

De Togni, P., Niman, H., Raymond, V., Sawchenko, P., and Verma, I.M. (1988). Detection of fos protein during osteogenesis by monoclonal antibodies. Mol. Cell. Biol. 8, 2251-2256.

Dony, C., and Gruss, P. (1987). Nature. 328, 711-714.

Eiseman, E., and Bolen, J.B. (1990). Src-related tyrosine kinases as signaling components in hematopoietic cells. Cancer cells 2, 303-310.

Fassler, R., Martin, K., Forsberg, E., Litzenburger, T., and Iglesias, A. (1995). Knockout mice: how to make them and why. The immunological approach. Intl. Arch. Allergy Immunol. 106 (4), 323-334.

Fattori, E., Cappelletti, M., Costa, P. et al. (1994). Defective inflammatory response in interleukin-6-deficient mice. J. Exp. Med. 190, 1243-1250.

Garofalo, S., Vuorio, E., Metsaranta, M. et al. (1991). Reduced amount of cartilage collagen fibrils and growth plate anomalies in trangenic mice harboring a glycine-to-cysteine mutation in the mouse type II alpha1-chain gene. Proc. Natl. Acad. Sci. 88, 9648-9651.

George, E.L., Georges-Labouesse, E.N., Patel-King, R.S., Rayburn, H., and Hynes, R.O. (1993). Defects in mesoderm, neural tube and vascular development in mouse embryos lacking fibronectin. Development 119, 1079-1091.

Green, M.C. (1989). Catalog of mutant genes and polymorphic loci. In: *Genetic Variants and Strains of Laboratory Mouse.* (Lyon, M.S., Searle, A.G., eds. 2nd. ed.) Oxford University Press, Oxford.

Gordon, J.W. (1989). Transgenic animals. Intl. Rev. Cyt. 115, 171-229.

Graff, R.J., Simmons, D., Meyer, J., Martin-Morgan, D., and Kurtz, M. (1986). Abnormal bone production associated with mutant mouse genes pa and we. J. Heredity. 77 (2), 109-113.

Grigoriadis, A.E., Wang, Z., Cecchini, M.G. et al. (1994). c-fos: A key regulator of osteoclast-macrophage lineage determination and bone remodeling. Science 266, 443-448.

Gu, H., Marth, J.D., Orban, P.C., Mossmann, H., and Rajewsky, K. (1994). Deletion of a DNA polymerase beta gene segment in T cells using cell type-specific gene targeting. Science. 265, 103-106.

Hayman, A.R., Foster, D., Colledge, W.H., Evans, M.J., and Cox, T.M. (1995). Tartrate-resistant acid phosphatase is essential for normal bone development in the mouse. J. Bone Min. Res. 10, A77.

Haegel, H., Larue, L., Ohsugi, M., Fedorov, L., Herrenknecht, K., and Kemler, R. (1995). Lack of β-catenin affects mouse development at gastrulation. Development 121. 3529-3537.

Jaenisch, R. (1988). Transgenic animals. Nature. 240, 1468-1473.

Johnson, R.S., Spiegelman, B.M., and Papaioannou, V. (1992). Pleiotropic effects of null mutation in the c-fos proto-oncogene. Cell 77, 577-586.

Karaplis, A.C., Luz, A., Glowacki, J. et al., (1994). Lethal skeletal dysplasia from targeted disruption of the parathyroid hormone-related paptide gene. Genes Develop. 8, 277-289.

Kesterson, R.A., Stanley, L., DeMayo, F., Finegold, M., and Pike, J.W. (1993). The human osteocalcin promoter directs bone-specific vitamin D–regulatable gene expression in transgenic mice. Mol. Endocr. 7, 462-467.

Kingsley, D.M., Bland, A.E., Grubber, J.M. et al. (1992). The mouse short ear skeletal morphogenesis locus is associated with defects in a bone morphogenetic member of the TGF-β superfamily. Cell 71, 399-410.

Kopf, M., Baumann, H., Freer, G. et al. (1994). Impaired immune and acute phase responses in interleukin-6-deficient mice. Nature 368, 339-342.

Kostic, D., and Capecchi, M.R. (1994). Targeted disruption of the murine Hoxa-4 and Hoxa-6 genes results in homeotic transformation of components of the vertebral column. Mech. Develop. 46, 231-247.

Knowles, B.B., McCarrick, J., Fox, N.. Solter, D., and Damjanov, I. (1990). Osteosarcomas in transgenic mice expressing an α-amylase-SV40 T-antigen hybrid gene. Am. J. Path. 137 (2), 259-262.

Kuhn, R., Schwenk, F., Aguet, M., and Rajeqsky, K. (1995). Inducible gene targeting in mice. Science 269, 1427-1429.

Larue, L., Ohsugi, M., Hirchenhain, J., and Kempler, R. (1994). E-cadherin null mutant embryos fail to form a trophectoderm epithelium. Proc. Natl. Acad. Sci. USA 91, 8263-8267.

Le-Mouellic, H., Lallemand, Y., and Brulet, P. (1992). Homeosis in the mouse induced null mutation in the Hox-3.1 gene. Cell 69, 251-264.

Levi, G., Topilko, P., Schneider-Maunoury, S. et al. (1996). Defective bone formation in Krox-20 mutant mice. Development 122 (1), 113-120.

Lohnes, D., Kastner, P., Dierich, A., Mark, M., LeMeur, M., land Chambon, P. (1993). Function of retinoic acid receptor γ in the mouse. Cell 73, 643-658.

Lohnes, D., Mark, M., Mendelsohn, C. et al. (1995). Developmental roles of the retinoic acid receptors. J. Ster. Bioch. Mol. 53, (1-6), 475-486.

Lufkin, T., Dierich, A., LeMeur, M., Mark, M., and Chambon, P. (1991). Disruption of the *Hox-1.6* homeobox gene results in defects in a region corresponding to its rostral domain of expression. Cell 66, 1105-1119.

Lufkin, T., Mark, M., Hart, C.P., Dolle, P., LeMeur, M., and Chambon, P. (1992). Homeotic transformation of the occipital bones of the skull by ectopic expression of a homeobox gene. Nature 359, 835-841.

Lohnes, D., Mark, M., Mendelsohn, C. et al. (1994). Function of the retinoic acid receptors (RARs) during development (I). Craniofacial and skeletal abnormalities in RAR double mutants. Development. 120 (10, 2723-2748.

Mansour, S.L., Thomas, K.R., and Capecchi, M.R. (1988). Disruption of the proto-oncogene int-2 in mouse embryo-derived stem cells: a general strategy for targeting mutations to non-selectable genes. Nature 336, 348-352.

Mark, M., Lohnes, D., Mendelsohn, C. et al. (1995). Roles of retinoic acid receptors and of Hox genes in the patterning of the teeth of the jaw skeleton. Int. J. Dev. Biology 39 (1), 111-121.

McGuire, E., Rintoul, C.E., Sclar, G.M., and Korsmeyer, S.J. (1992). Thymic overexpression of Ttg-1 in transgenic mice results in T-cell acute lymphoblastic leukemia/lymphomo. Mol. Cell. Biol. 12, 4186-4196.

McLain, K., Schreiner, C., Yager, K.L., Stock, J.L., and Potter, S.S. (1992). Ectopic expression of Hox-2.3 induced craniofacial and skeletal malformations in transgenic mice. Mech. Develop. 39, 3-16.

Mendelsohn, C., Lohnes, D., Decimo, D. et al. (1994). Function of the retinoic acid receptors (RARs) during development (II). Multiple abnormalities at various stages of organogenesis in RAR double mutants. Development 120, 2749-2771.

Metsarranta, M., Garofalo, S., Decker, G., Rintala, M., de Crombrugghe, B., and Vuorio, E. (1992). Chondrodysplasia in transgenic mice harboring a 15-amino acid deletion in the triplet helical domain of pro-α (II) collagen chain. J. Cell Biol. 118, 203-212.

Mishina, Y., Suzyi, A., Ueno, N., and Behringer, R.R. (1995). Bnpr encodes a type I bone morphogenetic protein receptor that is essential for gastrulation during mouse embryogenesis. Genes Devel. 9, 00-00.

Muller, R. (1986). Cellular and viral fos genes: Structure regulation of expression and biological properties of their encoded products. Biochim. Biophys. Acta. 823, 207-225.

Muller, R. (1994). *The fos and jun families of transcription factors.* (Angel, P., and Herrlich, P., Eds.) CRC, Boca Raton, FL.

Muller, R., Curran, T., Muller, D., and Guilbert, L. (1985). Induction of c-fos during myelomonocytic differentiation and macrophage proliferation. Nature 314, 546-548.

Muller, R., Verma, I.M., and Adamson, E.D.. (1983). Expression of c-onc genes: c-fos transcripts accumulate to high levels during development of mouse placenta, yolk sac and amnion. EMBO J. 4, 1775-1781.

Nagy, A., and Rossant, J. (1996). Targeted mutagenesis: analysis of phenootype without germ line transmission. J. Clin. Inve. 97, 1360-1365.

Naito, M., Hayashi, S., Yoshida, H., Nishikawa, S., Shultz, L.D., and Takahashi, K. (1991). Abnormal differentiation of tissue macrophage populations in "osteopetrosis" (op) mice defective in the production of macrophage colony-stimulating factor. Am. J. Path. 139, 657-667.

Picarella, D.E., Kratz, A., Li, C., Ruddle, N.H., and Flavell, R.A. (1992). Insulitis in transgenic mice expressing tumor necrosis factor β (lymphotoxin) in the pancreas. Proc. Natl. Acad. Sci. USA 89, 10036-10040.

Poli, V., Balena, R., Fattori, E. et al. (1994). Interleukin-6-deficient mice are protected from bone loss caused by estrogen depletion. EMBO J. 13, 1189-1196.

Proetzel, G., Pawlowski, S.A., Wiles, M.V. Et al. (19??). Transforming growth factor-beta3 is required for secondary palate fusion.

Riele, H.T., Maandog, E.R., and Berns, A. (1992). Highly efficient gene targeting in embryonic stem cells through homologous recombination with isogenic DNA constructs. Proc. Natl. Acad. Sci. USA 89, 5128-5132.

Rijli, F.M., Mark, M., Lakkaraju, S., Dierich, A., Dolle, P., and Chambon, P. (1993). A homeotic transformation is generated in the rostral branchial region of the head by disruption of Hoxa-2, which acts as a selector gene. Cell 75, 1333-1249.

Robertson, E.J. (Ed.) (1987). Teratocarcinomas and Embryonic Stem Cells. A Practical Approach. IRL Press, Oxford.

Ruther, U., Garber, C., Komitowski, D., Muller, R., and Wagner, E.F. (1987). Deregulated c-fos expression interfere with normal bone development in transgenic mice. Nature. 325, 412-416.

Ruther, U., Komitowski, D., Shubert, F.R., and Wagner, E.F. (1989). c-fos expression induces bone tumors in transgenic mice. Oncogene. 4, 861-865.

Satokata, I., and Maas, R. (1994). Msx1 dificient mice exhibit cleft palate and abnormalities of craniofacial and tooth development (see comments). Nature Genetics. 6, 348-356.

Shastry, B.S. (1995). Genetic knockouts in mice: an update. Experientia. 51, 1028-1039.

Sillence, D.O., Ritchie, H.E., Dibbayawan, T., Eteson, D., and Brown, K. (1993). Fragilitas ossium (fro-fro) in the mouse: A model for a recessively inherited type of osteogenesis imperfecta. Am. J. Med. Genet. 45, 276-283.

Soriano, P., Montgomery, C., Geska, R., and Bradley, A. (1991). Targeted disruption of the c-src proto-oncogene leads to osteopetrosis in mice. Cell 64, 693-702.

Spyropoulos, D.D., and Capecchi, M.R. (1994). Targeted disruption of the even-skipped gene, *evx1*, causes early postimplantation lethality of the mouse conceptus. Genes Devel. 8, 1949-1961.

Stacey, A., Bateman, J., Choi, T., Mascara, T., Cole, W., and Janish, R. (1988). Perinatal lethal osteogenesis imperfecta in transgenic mice bearing an engineered mutant pro-α1(I) collagene gene. Nature 332, 131-136.

Stanley, E., Lieschke, G.J., Grail, D. et al. (1994). Granulocyte-macrophage colony-stimulating factor-deficient mice show no major perturbation of hematopoiesis but develop a characteristic pulmonary pathology. Proc. Natl. Acad. Sci. USA. 91, 5591-?.

Sucov, H.M., Dyson, E., Gumeringer, C.L., Price, J., Chien, K.R., and Evans, R.M. (1994). RXR α mutant mice establish a genetic basis for vitamin A signaling in heart morphogenesis. Genes Devel. 8, 1007-1018.

Sumarsono, S.H., Wilson, T.J., Tymms, M. et al. (1996). Down syndrome-like skeletal abnormalities in Ets2 transgenic mice. Nature 379, 534-537.

Tsai, F.Y., Eller, G., Kuo, F.C. et al. (1994). An early haematopoietic defect in mice lacking the transcription factor GATA-2. Nature 371, 221-226.

Vandenberg, P., Kkhillan, J.S., Prockop, D.J., Helminen, H., Kontusaari, S., and Ala-koko, L. (1991). Expression of a partially deleted gene of human type II procollagen (COL2A1) in transgenic mice produces a chondrodysplasia. Proc. Natl. Acad. Sci. 88, 7640-7644.

Verma, I.M. (1986). Trends in GEnetics. 2, 93-96.

Wallin, J., Wilting, J., Koseki, H., Fritsch, R., Christ, B., and Balling, R. (1994). The role of Pax-1 in axial skeleton development. Development 120, 1109-1121.

Wang, Z., Ovitt, C., Grigoriadis, A.E., Mohle-Ssteinlein, U., Ruther, U., and Wagner, E.F. (1992). Bone and haematopoietic defects in mice lacking c-fos. Nature 360, 741-745.

Warren, A.J., Colledge, W.H., Carlton, M.B., Evans, M.J., Smith, A.J., and Rabbitts, T.H. (1994). The oncogenic cysteine-rich LIM domain protein rbtn2 is essential for erythroid development. Cell 45-57.

Weiss, M.J., and Orin, S.H. (1996). In vitro differentiation of murine embryonic stem cells. J. Clin. Inv. 97, 591-595.

Wiktor-Jedrzejczak, W., Bartocci, A., Ferrante, A.W.J. et al. (1990). Total absence of colony-stimulating factor 1 in macrophage-deficient osteopetrotic (op/op) mouse. Proc. Natl. Acad. Sci. 87, 4828-4832.

Winnier, G., Blessing, M., Llabosky, P.A., and Hogan, B.L.M. (1995). Bone morphogenetic protein-4 is required for mesodermal formation and patterning in the mouse. Genes Devel. 9, 2105-2116.

Woychik, R.P., Stewart, T.A., Davis, L.G., D'Eustacchio, P., and Leder, P. (1985). An inherited deformity created by insertional mutagenesis in a transgenic mouse. Nature 318, 36-40.

Wurst, W., Auerbach, A.B., and Joyner, A.L. (1994). Multiple Developmental defects in Engrailed-1 mutant mice: an early mid-hindbrain deletion and patterning defects in forelimbs and sternum. Development 120, 2065-2075.

Yang, J.T., Rayborn, H., and Hynes, R.O. (1993). Development 119, 1093-1105.

Yoshida, H., Hayashi, S., Kunisada, T. et al. (1990). The murine mutation osteopetrosis is in the coding region of the macrophage colony stimulating factor gene. Nature 345, 442-444.

Yoshida, K., Taga, T., Saito, M. et al. (1996). Targeted disruption of gp130, a common signal transducer for the interleukin 6 family of cytokines, leads to myocardial and hematological disorders. Proc. Natl. Acad. Sci. USA 93, 407-?.

INDEX

Printed and bound by CPI Group (UK) Ltd, Croydon, CR0 4YY
08/05/2025
01865012-0002

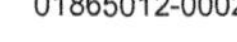